Advanced Construction Mathematics

Advanced Construction Mathematics covers the range of topics that a student must learn in order to achieve success in Level 3 and 4 mathematics for the Pearson BTEC National and BTEC HNC/HND in Construction, Building Services, and Civil Engineering.

Packed with easy to follow examples, its 18 chapters cover algebra (equations, transposition and evaluation of formulae), differentiation, integration, statistics and numerous other core concepts and their application in the construction/civil engineering field. The book explains technical processes before applying mathematical techniques to solve practical problems which gradually build in complexity. Each chapter contains self-test exercises and answers and numerous illustrations to simplify the essential maths required at Levels 3 and 4. The book is also a useful recap or primer for students on BSc or non-cognate MSc Construction and Civil Engineering degrees.

Surinder S. Virdi is a lecturer at South and City College Birmingham, UK. He has taught construction-related subjects for over 30 years and has written and co-authored two books, *Construction Science and Materials*, 2nd edition, 2017, and *Construction Mathematics*, 2nd edition, 2014.

Advanced Construction Mathematics

Surinder S. Virdi

LONDON AND NEW YORK

First published 2019
by Routledge
2 Park Square, Milton Park, Abingdon, Oxon OX14 4RN

and by Routledge
52 Vanderbilt Avenue, New York, NY 10017

Routledge is an imprint of the Taylor & Francis Group, an informa business

British Library Cataloguing-in-Publication Data
A catalogue record for this book is available from the British Library

Library of Congress Cataloging-in-Publication Data
Names: Virdi, Surinder Singh, author.
Title: Advanced construction mathematics / Surinder S. Virdi.
Description: Abingdon, Oxon : Routledge, 2019. | Includes bibliographical references and index. |
Identifiers: LCCN 2018051849 (print) | LCCN 2018055294 (ebook) | ISBN 9780429683602 (ePub) | ISBN 9780429683596 (Mobipocket) | ISBN 9780429683619 (Adobe PDF) | ISBN 9780367002107 (hardback) | ISBN 9780367002138 (pbk.) | ISBN 9780429400742 (ebook)
Subjects: LCSH: Building—Mathematics. | Civil engineering—Mathematics.
Classification: LCC TH437 (ebook) | LCC TH437 .V568 2019 (print) | DDC 624.01/51—dc23 .
LC record available at https://lccn.loc.gov/2018051849

ISBN: 978-0-367-00210-7 (hbk)
ISBN: 978-0-367-00213-8 (pbk)
ISBN: 978-0-429-40074-2 (ebk)

Typeset in Helvetica
by Swales & Willis Ltd, Exeter, Devon, UK

Visit the eResources: www.routledge.com/9780367002138

To
the memory of my parents, Mrs Rattan Kaur Virdi
and Mr Amar Singh Virdi

Contents

Units, symbols and prefixes xiii
Preface xv

1 Introduction to some basic techniques **1**
1.1 Introduction 1
1.2 Order of operations 1
1.3 Rounding 2
1.3.1 To the nearest whole number 2
1.3.2 Truncation 2
1.3.3 Significant figures 2
1.4 Standard form 3
1.5 Estimation 4
1.6 Error 4
1.7 Indices 5
1.7.1 Laws of indices 5
Exercise 1.1 7
Answers – Exercise 1.1 7

2 Algebra 1 **9**
2.1 Introduction 9
2.2 Multiplication and division 9
2.3 Brackets 11
2.4 Factorisation 12
2.5 Simple equations 12
2.6 Dimensional analysis 14
2.6.1 Introduction 14
2.6.2 Principle of homogeneity 15
2.6.3 Limitations of dimensional analysis 15
2.7 Arithmetic progression (AP) 17
2.8 Geometric progression (GP) 19
2.8.1 The sum of a series in GP 19
Exercise 2.1 20
Exercise 2.2 22
Answers – Exercise 2.1 23
Answers – Exercise 2.2 23

3 Algebra 2 **25**
3.1 Introduction 25
3.2 Transposition of formulae 25
3.2.1 Transposition involving addition/subtraction 26
3.2.2 Transposition involving multiplication/division 26

3.2.3 Transposition involving addition/subtraction and multiplication/division 27
3.2.4 Transposition involving squares/square roots 27
3.3 Evaluation of formulae 30
3.4 Binomial theorem 31
3.4.1 Introduction 31
3.4.2 Pascal's triangle 31
3.4.3 Binomial series for fractional or negative indices 33
3.4.4 Approximations 34
3.4.5 Practical problems 35
Exercise 3.1 36
Exercise 3.2 37
Answers – Exercise 3.1 38
Answers – Exercise 3.2 39

4 Simultaneous and quadratic equations **41**
4.1 Simultaneous equations 41
4.1.1 The elimination method 41
4.1.2 The substitution method 42
4.1.3 Application of simultaneous equations 44
4.2 Quadratic equations 45
4.2.1 Factorising 46
4.2.2 The quadratic formula 48
4.2.3 Completing the square 48
4.2.4 Application of quadratic equations 49
Exercise 4.1 50
Exercise 4.2 51
Answers – Exercise 4.1 52
Answers – Exercise 4.2 52

5 Graphical solutions **53**
5.1 Introduction 53
5.2 Linear equations 55
5.3 Linear simultaneous equations 56
5.4 The law of straight line 58
5.5 Quadratic equations 60
5.6 Cubic equations 61
5.7 Curve fitting 62
Exercise 5.1 65
Answers – Exercise 5.1 66

6 Geometry, areas and volumes **67**
6.1 Geometry 67
6.1.1 Angles 67
6.1.2 Triangles 68
6.1.3 Similar triangles 69
6.1.4 Pythagoras' theorem 71
6.1.5 The circle 71

6.2 Area 75
6.2.1 Area of regular shapes 76
6.2.2 Area of quadrilaterals 77
6.2.3 Area of circle 78
6.3 Area of irregular shapes 80
6.3.1 Mid-ordinate rule 80
6.3.2 Trapezoidal rule 81
6.3.3 Simpson's rule 81
6.4 Volume 83
6.4.1 Volume of a sphere 83
6.5 Volume of irregular objects 86
6.5.1 Mass-haul diagram 87
6.6 The theorem of Pappus 88
Exercise 6.1 89
Answers – Exercise 6.1 96

7 Trigonometry 1 **97**
7.1 Introduction 97
7.2 The trigonometrical ratios 98
7.3 Angles of elevation and depression 99
7.4 Roofs 101
7.5 The sine rule and the cosine rule 103
7.5.1 The sine rule 103
7.5.2 The cosine rule 104
7.5.3 Sine rule: the ambiguous case 107
7.6 Frames 108
7.7 Area of triangles 110
Exercise 7.1 111
Exercise 7.2 113
Answers – Exercise 7.1 116
Answers – Exercise 7.2 116

8 Trigonometry 2 **117**
8.1 Introduction 117
8.2 Trigonometric identities 117
8.3 Trigonometric ratio of compound angles 119
8.4 Double angle formulae 121
8.5 Trigonometric equations 121
8.6 Trigonometric graphs 124
8.7 Addition of sine waves 127
Exercise 8.1 129
Answers – Exercise 8.1 130

9 Logarithmic, exponential and hyperbolic functions **133**
9.1 Logarithmic function 133
9.1.1 Antilogarithm 134
9.1.2 Laws of logarithms 135

9.2 Exponential function 137
9.2.1 Compound interest on savings 138
9.2.2 Newton's law of cooling 139
9.2.3 Thermal movement of building components 140
9.2.4 Laws of growth and decay 140
9.2.5 Decay of sound energy 141
9.3 Hyperbolic function 142
Exercise 9.1 144
Exercise 9.2 145
Exercise 9.3 146
Answers – Exercise 9.1 146
Answers – Exercise 9.2 147
Answers – Exercise 9.3 147

10 Differentiation **149**
10.1 Introduction 149
10.2 Differentiation from first principles 150
10.2.1 Differentiation of $y = x^n$ 152
10.3 Trigonometric functions 154
10.4 Differentiation of function of a function 155
10.5 The chain rule by recognition 157
10.6 Differentiation of exponential and logarithm functions 157
10.7 Differentiation of a product 159
10.8 Differentiation of a quotient 160
10.9 Numerical values of differential coefficients 161
Exercise 10.1 162
Exercise 10.2 162
Exercise 10.3 163
Exercise 10.4 163
Answers – Exercise 10.1 164
Answers – Exercise 10.2 164
Answers – Exercise 10.3 165
Answers – Exercise 10.4 165

11 Applications of differentiation **167**
11.1 Application in structural mechanics 167
11.2 Second derivatives 169
11.3 Velocity and acceleration 170
11.4 Maximum and minimum 171
Exercise 11.1 175
Answers – Exercise 11.1 176

12 Integration **177**
12.1 Introduction 177
12.2 Indefinite integrals 178
12.2.1 Integration of a sum 179
12.3 Definite integrals 180

12.4 Integration by substitution 181
12.5 Change of limits 182
12.6 Integration by parts 183
Exercise 12.1 184
Answers – Exercise 12.1 186

13 Applications of integration 187
13.1 Introduction 187
13.2 Area under a curve 187
13.3 Area enclosed between a curve and a straight line 189
13.4 Volumes of revolution 190
13.5 Earth pressure on retaining walls 192
13.6 Permeability of soils 193
13.7 Bending moment and shear force in beams 195
Exercise 13.1 196
Answers – Exercise 13.1 197

14 Properties of sections 199
14.1 Centroids of simple shapes 199
14.2 Centroids of simple/complex shapes by integration 202
14.3 Second moment of area 206
14.4 Radius of gyration 207
14.5 The parallel axis theorem 209
Exercise 14.1 213
Answers – Exercise 14.1 216

15 Matrices and determinants 217
15.1 Introduction 217
15.2 Square matrix 217
15.3 Diagonal matrix 218
15.4 Unit matrix 218
15.5 Addition and subtraction 218
15.6 Multiplication/division by a scalar 218
15.7 Transpose 220
15.8 Determinants 220
15.9 Cofactors 220
15.10 Inverse of a square matrix 221
15.11 Properties of determinants 222
15.12 Application of matrices 223
Exercise 15.1 225
Answers – Exercise 15.1 226

16 Vectors 229
16.1 Introduction 229
16.2 Addition of vectors 230
16.3 Subtraction of vectors 232
16.4 Unit vectors 233

16.5 Resolution of vectors 234
16.6 Addition and subtraction of two vectors in Cartesian form 236
Exercise 16.1 237
Answers – Exercise 16.1 239

17 Statistics **241**
17.1 Introduction 241
17.2 Types of data 241
17.2.1 Discrete data 241
17.2.2 Continuous data 242
17.2.3 Raw data 242
17.2.4 Grouped data 242
17.3 Averages 243
17.3.1 Comparison of mean, mode and median 243
17.4 Statistical diagrams 245
17.5 Frequency distributions 247
17.6 Measures of dispersion 251
17.6.1 Standard deviation 252
17.7 Distribution curves 256
17.8 Probability 258
17.8.1 Mutually exclusive events (the OR rule) 258
17.8.2 Independent events (the AND rule) 259
17.8.3 Tree diagrams 260
17.9 Binomial distribution 260
17.10 Poisson distribution 261
17.11 Normal distribution 262
17.12 Normal distribution test 266
Exercise 17.1 268
Answers – Exercise 17.1 269

18 Computer techniques **271**
18.1 Introduction 271
Exercise 18.1 282
Answers – Exercise 18.1 283

End of unit assignment 285
Appendix 1 289
Appendix 2 299
Appendix 3 301
Index 399

Units, symbols and prefixes

Units

Quantity	Name of unit	Symbol	Quantity	Name of unit	Symbol
Length	metre	m	Acceleration		m/s^2
Mass	kilogram	kg	Velocity		m/s
Time	second	s	Celsius temperature	degrees celsius	°C
Thermodynamic Temperature	kelvin	K	Thermal conductivity		W/mK
Plane angle	radian degrees	rad°	Moment of a force		kNm
Frequency	hertz	Hz	Potential difference		V
Force	newton	N	Sound reduction index	decibels	dB
Pressure or stress	pascal stress	Pa N/m^2	Wavelength	metre	m
Power	watt	W			

Greek alphabet

Upper case	Lower case	Name	Upper case	Lower case	Name
A	α	Alpha	N	ν	Nu
B	β	Beta	Ξ	ξ	Xi
Γ	γ	Gamma	O	ο	Omicron
Δ	δ	Delta	Π	π	Pi
E	ε	Epsilon	P	ρ	Rho
Z	ζ	Zeta	Σ	σ	Sigma

(continued)

(continued)

Upper case	Lower case	Name	Upper case	Lower case	Name
H	η	Eta	T	τ	Tau
Θ	θ	Theta	Y	υ	Upsilon
I	ι	Iota	Φ	φ	Phi
K	κ	Kappa	X	χ	Chi
Λ	λ	Lambda	Ψ	ψ	Psi
M	μ	Mu	Ω	ω	Omega

Prefixes

Name	Symbol	Factor
Tera	T	10^{12}
Giga	G	10^{9}
Mega	M	10^{6}
Kilo	K	10^{3}
Hecto	H	10^{2}
Deca	Da	10
Deci	D	10^{-1}
Centi	C	10^{-2}
Milli	M	10^{-3}
Micro	M	10^{-6}
Nano	N	10^{-9}
Pico	P	10^{-12}

Abbreviations/symbols

Description	Symbol
Approximately	Approx.
Change in length, temperature etc.	Δ
For example	e.g.
That is	i.e.
Greater than	>
Degrees, minutes, seconds	° ' "
Less than	<
Proportional to	∝
Angle	∠
Triangle	Δ
Therefore	∴

Preface

Mathematics is a mandatory unit on several Edexcel courses in civil engineering and other construction-related disciplines. This book is intended to provide the essential mathematics required by Level 3 and Level 4 programmes in these disciplines. Students pursuing studies at Level 5 may also find some of the topics useful. The book may also be useful for students on the access to H.E. and Foundation degree courses.

The contents of the book have been divided into 18 chapters starting from the very basic and progressing to trigonometry, calculus and other relevant topics. After explaining the concept of each topic, worked examples and exercises have been included to explain the application of mathematics in construction. There are end of chapter exercises so that the readers can check their learning by producing solutions. Answers and solutions of all exercises have also been included to provide help, should the readers require it. To check the readers' learning further, an assignment has also been included at the end of the book. Liberal use of diagrams has been made to supplement the text and make concepts/problems more understandable.

Many thanks are due to my family, students and colleagues for the interest they have shown in this project. A big thank you to Ed Needle (Editor) and Patrick Hetherington (Editorial Assistant) of Taylor & Francis Group for their support in the publication of this book.

Surinder S. Virdi

CHAPTER 1

Introduction to some basic techniques

Topics covered in this chapter:

- Order of operations, rounding off numbers: rounding off numbers to the nearest whole number, truncation, significant figures, decimal places
- Estimation of answers, absolute and relative errors
- Indices

1.1 Introduction

In this chapter some of the basic techniques are explained. Most of these do not involve any complexity, but it is very important to solve a number of problems and to show the answer in an acceptable format.

1.2 Order of operations

In mathematics and other analytical subjects, the calculations may involve one or more of:

multiplication, division, addition, subtraction and brackets.

There is a definite order of operations in algebra that needs to be followed to get accurate solutions. For example, in the following calculation, the answers could be 51, 13 or 19:

Evaluate: $20 - 4 + 1 \times 3$

The correct answer is 19, which results from following the right procedure. This procedure, known as the order of precedence of operations, involves dealing with brackets (B) first and then of (O), division (D), multiplication (M), addition (A), subtraction (S), in that order. This is usually remembered as BODMAS.

In the above example, $20 - 4 + (1 \times 3) = 20 - 4 + 3 = 20 - 1 = 19 \quad (-4 + 3 = -1)$

Example 1.1

Solve: a) $16 + 5 - 2 \times 1.5$

b) $20 - 2 + (2 \times 3)$

c) $25 + 5 - 3(1.5 \times 4) - 3 \times 2$

Solution:

a) $16 + 5 - 2 \times 1.5 = 16 + 5 - 3 = 21 - 3 = \mathbf{18}$

b) $20 - 2 + (2 \times 3) = 20 - 2 + 6 = 20 + 4 = \mathbf{24}$ $\quad (-2 + 6 = +4)$

c) $25 + 5 - 3(1.5 \times 4) - 3 \times 2 = 25 + 5 - 3(6) - 6 = 25 + 5 - 18 - 6$
$= 30 - 18 - 6 = \mathbf{6}$

1.3 Rounding

1.3.1 To the nearest whole number

The convention is to round 0.5 and above to the next highest whole number, and 0.499 (recurring) to the next lowest whole number. For example, 8.5 will become 9.0 when rounded to the nearest whole number. Similarly, 8.499 will become 8.0 when rounded to the nearest whole number.

1.3.2 Truncation

Truncation involves the omission of the unwanted digits at the end of a number. For example, 25.3458 truncated to four figures becomes 25.34.

1.3.3 Significant figures

3462, 346200 and 0.03462 all have four significant figures, not counting zeros at the beginning or end of the number. To write these numbers to 3 significant figures, the last figure is discarded. If the last figure is 5 or greater, the next figure is increased by 1. If the last figure is less than 5 then the next figure remains unchanged.

Zeros in the middle of a number are counted as significant figures, for example 3402, 340200 and 0.03402 have 4 significant figures.

Example 1.2

The area of a building plot is 648.4613 m^2. Show this value:

a) to the nearest whole number b) truncated to 4 figures c) to 4 significant figures (s.f.) d) to 1 decimal place (d.p.)

Solution:

a) **648** (nearest whole number)

b) **648.4** (truncated to 4 figures)

c) **648.5** (4 s.f.)

d) **648.5** (1 d.p.)

Example 1.3

Write: a) 260.362 to 5, 4, 3, 2 and 1 significant figures (s.f.)

b) 0.06734 to 3, 2 and 1 significant figures

Solution:

a) 260.362 = **260.36** correct to 5 s.f.

= **260.4** correct to 4 s.f.

= **260** correct to 3 s.f.

= **260** correct to 2 s.f.

= **300** correct to 1 s.f.

b) 0.06734 = **0.0673** correct to 3 s.f.

= **0.067** correct to 2 s.f.

= **0.07** correct to 1 s.f.

1.4 Standard form

The standard form, which is used in scientific and technical calculations, involves splitting a small or a large number into two parts: the first part is a number that is greater than 1 but less than 10, and the second part is 10^n, where n could be a negative or a positive integer. The two parts when multiplied are equal to the original number.

Example 1.4

Write the following numbers in the standard form:

a) 235 b) 3565 c) 58390 d) 0.25 e) 0.000356

Solution:

a) 235 should be split into two parts so that when multiplied they should result into 235.

The first part is 2.35 (this should be greater than 1 but less than 10)

The second part is 100 or 10^2, as we must multiply 2.35 by 100 (or 10^2) to get 235.

Therefore, the standard form of 235 is: $\mathbf{2.35 \times 10^2}$.

b) $3565 = 3.565 \times 1000 = \mathbf{3.565 \times 10^3}$

c) $58390 = 5.8930 \times 10000 = \mathbf{5.839 \times 10^4}$

d) $0.25 = \dfrac{2.5}{10} = \dfrac{2.5}{10^1} = \mathbf{2.5 \times 10^{-1}}$

e) $0.000356 = \dfrac{3.56}{10000} = \dfrac{3.56}{10^4} = \mathbf{3.56 \times 10^{-4}}$

1.5 Estimation

The process of finding an approximate answer of a calculation is known as **estimation.** The process involves replacing the actual numbers by those numbers which are convenient for manual calculations.

Example 1.5

Estimate the results of the following calculations and compare them to their accurate answers:

a) $25 + 89 + 110 - 190$

b) $\dfrac{145 \times 220}{80 \times 115}$

Solution:

a) The numbers in this question may be rounded to:

$30 + 100 + 100 - 200 = \mathbf{30}$

(The accurate answer is 34)

b) This question may be written as:

$$\frac{150 \times 200}{100 \times 100} = \frac{150 \times 2}{100} = \frac{300}{100} = \mathbf{3}$$

(The accurate answer is 3.47 (2 d.p.))

1.6 Error

'Error' in mathematics is the difference between what is acceptable as a true value and what is taken for an approximate or estimated value. The actual difference between an approximation or estimate and the true value is known as the absolute error. The absolute error expressed as a percentage of the estimated value is known as the relative error. Other types of errors, such as the random and systematic errors, are relevant in statistical and scientific investigations.

Example 1.6

Calculate the absolute and relative errors for the calculation in example 1.5(b).

Solution:

Estimated answer = 3.0 accurate answer = 3.47

a) Absolute error = $3.47 - 3.0 = \mathbf{0.47}$

b) Relative error $= \dfrac{\text{absolute error}}{\text{estimated value}} \times 100$

$$= \frac{0.47}{3.0} \times 100 = \mathbf{15.67\%}$$

1.7 Indices

The power of a number is called the index, and indices is the plural of index. If we have a number 5^3 (read as 5 raised to the power 3), it means that 5 is multiplied by itself 3 times; here 5 is called the base and 3 the index. The laws of indices may be used to solve problems where numbers raised to power are given.

1.7.1 Laws of indices

a) Multiplication

The indices are added when a number raised to a power, is multiplied by the same number raised to a power. If, number 'a' is raised to powers 'm' and 'n', the law of multiplication is written as:

$$a^m \times a^n = a^{m+n}$$

Number 'a' is called the base and it is important that in any one calculation, the base numbers are the same.

Example 1.7

Simplify: $b^3 \times b^4$

Solution:

$$b^3 \times b^4 = b^{3+4} = \mathbf{b^7}$$

b) Division

If the base numbers are the same, the indices are subtracted:

$$\frac{a^m}{a^n} = a^m \div a^n = a^{m-n}$$

Example 1.8

Solve $\frac{4^6}{4^5}$

Solution:

As the base numbers are the same, i.e. 4, the law of indices can be used to get the solution:

$$\frac{4^6}{4^5} = 4^{6-5} = 4^1 = \mathbf{4}$$

Example 1.9

Simplify $\frac{b^2 \times b^7}{b^2 \times b^4}$

Solution:

Use the multiplication and the division laws:

$$\frac{b^2 \times b^7}{b^2 \times b^4} = \frac{b^9}{b^6} = b^{9-6} = b^3$$

c) Power of a power

If a number raised to a power is raised to another power, the indices are multiplied:

$(a^m)^n = a^{mn}$

Example 1.10

Simplify: $(2m^2)^4$

Solution:

$(2m^2)^4 = 2^4 \times m^{2 \times 4} = \mathbf{16\ m^8}$

d) Negative powers

We have seen earlier that when the law of division is used the solution of $\frac{a^m}{a^n}$ is a^{m-n}. We can get the same solution by writing the question as $a^m \times a^{-n}$ and applying the law of multiplication:

$a^m \times a^{-n} = a^{m+(-n)} = a^{m-n}$

In other words, when a number is moved from numerator to denominator or vice versa, the sign of the power changes.

4^{-n} is a typical example of a number with a negative power; this can also be written as $\frac{1}{4^n}$.

Example 1.11

Solve $\frac{5^4}{5^2 \times 5^4}$

Solution:

$$\frac{5^4}{5^2 \times 5^4} = \frac{5^4}{5^6}$$

Apply the division rule, $\frac{5^4}{5^6} = 5^{4-6}$

$$= 5^{-2} = \frac{1}{5^2} = \mathbf{0.04}$$

e) Zero index

If number is raised to the power zero, the answer is 1. For example $a^0 = 1$, $285^0 = 1$, and $1065^0 = 1$.

Exercise 1.1

1. Solve: a) $6 + 5 \times 4 - 20 + 2 \times 3$
   b) $5 - 6 \times 1.5 - 3 \times 2.5 + 5 \times 6 \times 1.5$
2. The volume of an excavation is 7525.6484 m^3. Show this value:
   a) to the nearest whole number b) truncated to 6 figures c) to 6 significant figures (s.f.) d) to 2 decimal place (d.p.)
3. Write: a) 256.7238, correct to 5, 4, 3, 2, and 1 significant figures
   b) 285294, correct to 5, 4, 3, 2 and 1 significant figures
   c) 0.00847, correct to 2 and 1 significant figures.
4. Write the following numbers in standard form:
   a) 0.0256
   b) 25.8
   c) 475605
5. Estimate your answers for the following questions:
   a) $620 - 380 - 120$
   b) $\frac{45 \times 20}{50}$
6. Estimate the answer to the following question and hence calculate the absolute and relative errors:
   $\frac{39 \times 89 \times 143}{43 \times 108}$
7. Simplify/solve
   a) $4 \times 4^2 \times 3^4 \times 3^2$
   b) $\frac{x^6 \times x^2 \times x^2}{x^3 \times x \times x^{-2}}$
   c) $(2^2)^3 + (2x^2)^2$
   d) $\frac{m^6 \times m^4 \times m^2}{m^3 \times m^5 \times m^4}$

Answers – Exercise 1.1

1. a) 12
   b) 33.5
2. a) 7526
   b) 7525.64

c) 7525.65

d) 7525.65

3. a) 256.72; 256.7; 257; 260; 300

   b) 285290; 285300; 285000; 290000; 300000

   c) 0.0085; 0.009

4. a) 2.56×10^{-2}

   b) 2.58×10^{1}

   c) 4.75605×10^{5}

5. a) 100

   b) 20

6. Estimate = 100; Absolute error = 6.88; Relative error = 6.88%

7. a) 46656

   b) x^8

   c) $64 + 4x^4$

   d) 1

CHAPTER

Algebra 1

Topics covered in this chapter:

- Multiplication and division of algebraic expressions
- Solution of linear equations
- Factorisation
- Dimensional analysis
- Arithmetic progression and geometric progression

2.1 Introduction

In algebra, symbols are used to represent the unknown quantities that we need to determine. Equations are formed from the given information using symbols and the solution of these equations will give us the value of the unknown quantities. For example, in the equation 3w + 2d, w and d represent the price of a window and a door respectively. The solution of this equation, simultaneously with another, will give the price of a window and a door.

2.2 Multiplication and division

Multiplication and division of symbols is not much different from the multiplication and division of numbers, as long as the processes are applied to the same symbols. Different symbols have to be treated in a different way. For example:

i) $3 \times 3 = 3^2$; similarly, $x \times x = x^2$ (refer to indices – chapter 1)

ii) $x \times x \times y = (x \times x) \times y = x^2y$

iii) $\frac{x^2}{x} = x$; $\frac{x^2}{y \times y} = \frac{x^2}{y^2}$

Example 2.1

Simplify: a) $5\,x^2y \times 6\,xy^3$

b) $30\,x^2y^3z^2 \div 6\,xyz^2$

Solution:

a) The question can be divided into 3 parts: x terms, y terms and the numbers. These will be multiplied separately first and then combined to give the answer.

$x^2 \times x = x \times x \times x = x^3$ (1) (laws of indices can also be used)

$y \times y^3 = y \times y \times y \times y = y^4$ (2)

$5 \times 6 = 30$ (3)

Combining 1, 2 and 3

$5\,x^2y \times 6\,xy^3 = \mathbf{30\,x^3y^4}$

b) $30\,x^2y^3z^2 \div 6\,xyz^2$ may be written as $\dfrac{30x^2y^3z^2}{6xyz^2}$

The numbers and other terms will be simplified individually first, but combined later to give the answer:

$$\frac{30}{6} = 5 \qquad (1)$$

$$\frac{x^2}{x} = \frac{x \times x}{x} = x \qquad (2)$$

$$\frac{y^3}{y} = \frac{y \times y \times y}{y} = y \times y = y^2 \qquad (3)$$

$$\frac{z^2}{z^2} = \frac{z \times z}{z \times z} = 1 \qquad (4)$$

Combining 1, 2, 3 and 4:

$$\frac{30x^2y^3z^2}{6xyz^2} = 5xy^2 \times 1 = 5xy^2$$

Example 2.2

Multiply $2x + 3$ by $x - 1$

Solution:

Write the two expressions as shown:

$$\begin{array}{r} 2x + 3 \\ \times \quad x - 1 \\ \hline \end{array}$$

Multiplication in algebra proceeds from left to right. $2x + 3$ is multiplied by x first and then by -1. After multiplication the terms are added or subtracted, as appropriate:

$$
\begin{array}{r}
2x+3 \\
\times \quad x-1 \\
\hline
2x^2+3x \\
-2x-3 \\
\hline
2x^2+x-3
\end{array}
$$

Example 2.3

Multiply (3x + 3y + 4) by (x – 3)

Solution:

$$
\begin{array}{rl}
3x+3y+4 & \\
\times \quad x-3 & \\
\hline
3x^2+3xy+4x & \quad (\text{every item of } 3x+3y+4 \text{ is multiplied by } x) \\
-9x-9y-12 & \quad (\text{every item of } 3x+3y+4 \text{ is multiplied by } -3) \\
\hline
\mathbf{3x^2+3xy-5x-9y-12} &
\end{array}
$$

2.3 Brackets

Brackets are frequently used to enclose an expression. The solution of the expression within brackets takes precedence over the rest of the question. For example the solution of, $4 \times (6 \div 2)$ is derived at by solving $6 \div 2$ first. The answer is multiplied by 4 to get the final answer. Brackets may also be used in writing factors, for example, $3x + 9y$ may be written as $3(x + 3y)$.

The expression $3(5x + 2y)$ shows that the common factor is 3. The original expression was $15x + 6y$, and since 3 is a common factor the expression can be written as:

$3 \times 5x + 3 \times 2y$, or $3(5x + 2y)$

Sometimes the algebraic expression may be of the form $6 + (2a - 3b)$ or $6 - (2a - 3b)$. In this case there is no common factor, therefore the simplification involves the removal of the brackets, with due consideration to the + or – sign before the brackets.

Therefore, $6 + (2a - 3b) = 6 + 2a - 3b$

$6 - (2a - 3b) = 6 - 2a + 3b$

Example 2.4

Simplify: i) $4(2a - 0.5b)$

ii) $4 + (3a - 6b)$

iii) $4 - (3a - 6b - 2c)$

Solution:

Simplification of these expressions will involve the removal of brackets.

i) There is no sign, plus or minus, between 4 and the brackets, which means that 4 is a common factor. The expression within the brackets is multiplied by 4, for simplification.

$4(2a - 0.5b) = 4 \times 2a - 4 \times 0.5b = \mathbf{8a - 2b}$

ii) The plus sign before the brackets means that the expression within the brackets will remain unchanged:

$4 + (3a - 6b) = \mathbf{4 + 3a - 6b}$

iii) Due to the minus sign before the brackets, the signs of all terms within the brackets will change:

$4 - (3a - 6b - 2c) = \mathbf{4 - 3a + 6b + 2c}$

2.4 Factorisation

If two or more terms in an algebraic expression have a common factor, then the factors may be used to rewrite the expression. The common factors could be numbers, symbols or a combination of the two. The process is the reverse of removing brackets, as shown in example 2.5.

Example 2.5

Factorise: a) $4xy + 6yz$

b) $8x^2 - 6x$

Solution:

a) $4xy + 6yz$: The two terms of this expression have two common factors, 2 and y or 2y, as both terms are divisible by the common factors.

Dividing 4xy by 2y gives 2x.

Dividing 6yz by 2y gives 3z.

The terms 2x and 3z are written within the brackets, therefore:

$4xy + 6yz = \mathbf{2y(2x + 3z)}$ (the factors of the expression are 2, y and $(2x + 3z)$).

We can check the answer by multiplying out the brackets; we should get the original expression back.

b) $8x^2 - 6x$: In this question 2x is the common factor.

Dividing $8x^2$ by 2x gives 4x.

Dividing 6x by 2x gives 3.

So, $8x^2 - 6x = \mathbf{2x(4x - 3)}$

2.5 Simple equations

An **equation** is a mathematical statement that shows the equality of two expressions, which are separated by the 'equal to' sign. Depending on the type of the equation, there may be one or two unknowns denoted by letters such as x, y etc. The determination of the unknowns from an equation is known as 'solving the equation'.

If there is only 1 letter (1 unknown) in an equation, and its power is 1, the equation is known as a simple equation.

Example 2.6

Solve the following equations:

i) $x + 4 = 10$

ii) $3x - 6 = 12$

iii) $\frac{5x}{4} = 2.5$

iv) $2x - 7 - 3x = 3 - 5x$

Solution:

i) $x + 4 = 10$

In this question, the solution involves the determination of x (the unknown). This equation, like other equations, has two sides: the left side (LHS) and the right side (RHS), separated by the = sign. For determining the value of x, we must move + 4 to the right side, and it becomes −4 in that process.

$x + 4 = 10$

$x = 10 - 4 = 6$

Therefore x = 6, or the value of x is 6.

ii) $3x - 6 = 12$

Using the same method as used in the previous question:

$3x - 6 = 12$

$3x = 12 + 6 = 18$

$x = \frac{18}{3} = 6$ **(3x is 18, therefore, x must be equal to 6, because 3 × 6 is 18)**

iii) $\frac{5x}{4} = 2.5$

Multiply both sides of the equation by 4 to simplify:

$\frac{5x}{4} \times 4 = 2.5 \times 4$

$5x = 10$, or $\mathbf{x = 2}$

iv) $2x - 7 - 3x = 3 - 5x$

Transpose −7 to the right side and −5x to the left side to have the unknown terms on one side and the numbers on the other side.

$2x - 3x + 5x = 3 + 7$

$7x - 3x = 10$

$4x = 10$, or $\mathbf{x = \frac{10}{4} = 2.5}$

Example 2.7

Solve: a) $3(2x + 4) - 2(x - 3) = 4(2x + 4)$

b) $\frac{x}{2}+\frac{x+3}{4}=\frac{2x+1}{3}$

Solution:

a) Multiply 2x + 4 by 3, x – 3 by 2 and 2x + 4 by 4, to simplify:

$6x + 12 - 2x + 6 = 8x + 16$

Transpose the unknown terms to the LHS and the numbers to the RHS

$6x - 2x - 8x = 16 - 12 - 6$

$-4x = -2$, $\mathbf{x} = \frac{-2}{-4} = \mathbf{0.5}$

b) Multiply all terms of the equation by 12 (the common factor) to simplify the fractions

$\frac{12x}{2}+\frac{12(x+3)}{4}=\frac{12(2x+1)}{3}$

$6x + 3(x + 3) = 4(2x + 1)$

$6x + 3x + 9 = 8x + 4$

$9x - 8x = 4 - 9$

$\mathbf{x = -5}$

2.6 Dimensional analysis

2.6.1 Introduction

In science and engineering, a number of methods can be used to solve problems. Dimensional analysis is an effective method that can be used not only to establish relationship among physical quantities using their dimensions, but also to check the dimensional correctness of a given equation.

The dimension of a physical quantity is a combination of the basic physical dimensions:

[Mass]	M
[Length]	L
[Time]	T
[Temperature]	K
[Current]	A

Square brackets are used to denote the dimension of a physical quantity and it is worth noting that:

- The dimensions of a physical quantity are independent of the units used. For example:

  [Area] = length × width

  $= L \times L = L^2$ (the dimension of length/width is L)

  The units could be mm^2, cm^2, m^2, $feet^2$ etc. However, their power is always 2.

- We can write the unit of a physical quantity (speed, force etc.) if we know its dimensions. For example, if a physical quantity has the dimension $[MLT^{-1}]$ then its unit will be $kg\ m\ s^{-1}$.

2.6.2 Principle of homogeneity

For any given equation, the principle of homogeneity is used to check its correctness and consistency. According to this principle, an equation is dimensionally correct if the dimensions of each component on either side of the equation are the same. Consider the formula, $v = u + at$.

Here, v = final velocity; u = initial velocity; a = acceleration; t = time

Let us take the units of velocity as metres per second. The dimension of metre (length or distance) is L and the dimension of time is T. Therefore, the dimensions of velocity are:

$$\text{Velocity} = \frac{\text{metres}}{\text{second}} = \frac{L}{T} = LT^{-1}$$

Dimensions of 'v' = LT^{-1}

Dimensions of 'u' = LT^{-1}

Dimensions of 'a' = LT^{-2} ($a = m/sec^2 = m\ sec^{-2} = LT^{-2}$)

Dimensions of 't' = T

Replacing the symbols in the formula by their dimensions:

$$LT^{-1} = LT^{-1} + (LT^{-2}) \times T$$

$$= LT^{-1} + LT^{-1} = 2(LT^{-1})$$

As the dimensions of the RHS are the same as those of the LHS, the formula is dimensionally correct. 2 is a pure number having no dimension, therefore, like other numbers it is ignored in dimensional analysis.

2.6.3 Limitations of dimensional analysis

Limitations of dimensional analysis are:

i) The value of dimensionless constants cannot be determined by this method.

ii) It cannot be applied to an equation involving more than 3 physical quantities.

iii) The method cannot be applied to equations involving trigonometric and exponential functions.

iv) The method can only check whether a physical relation is dimensionally correct or not. It does not say whether the relation is absolutely correct or not.

Example 2.8

Find the dimensions of the following quantities in terms of M, L and T:

a) Force b) Kinetic energy c) Pressure

Solution:

a) Force = mass × acceleration

[Force] = M × m/sec^2

$$= M \times \frac{L}{T^2} = MLT^{-2}$$

b) Kinetic energy $= \frac{1}{2}mv^2$ (m = mass and v = velocity)

$[\text{Kinetic energy}] = \frac{1}{2}mv^2 = M(LT^{-1})^2$ (dimensions of velocity = LT^{-1})

$= \mathbf{ML^2T^{-2}}$

c) $\text{Pressure} = \frac{\text{Force}}{\text{Area}}$

$[\text{Pressure}] = \frac{MLT^{-2}}{L^2} = \mathbf{ML^{-1}T^{-2}}$

Example 2.9

Hooke's law states that the force F in a spring, extended by a length x, is given by F = – kx. Calculate the dimensions of the spring constant k.

Solution:

According to Hooke's law, $F = -kx$

After transposition, $k = -\frac{F}{x}$

The dimensions of force are MLT^{-2}, and the dimension of 'x' is L

Therefore spring constant k has the dimensions:

$[k] = \frac{MLT^{-2}}{L} = \mathbf{MT^{-2}}$

Example 2.10

The velocity (v) of sound through a medium may be assumed to depend on the density (ρ) and elasticity (E) of the medium. Use the method of dimensions to deduce the formula for the velocity of sound.

Solution:

As 'v' is proportional to 'ρ' and 'E'; this may be written as:

$v \propto \rho E$

As we don't know the powers of 'ρ' and 'E', the above expression is written as:

$v = \rho^a E^b$

$\rho = \frac{\text{Mass}}{\text{Volume}}$, therefore $[\rho] = ML^{-3}$

$E = \frac{\text{Stress}}{\text{Strain}} = \frac{\text{Force}}{\text{Area}}$, therefore $[E] = \frac{MLT^{-2}}{L^2} = ML^{-1}T^{-2}$

The formula for velocity in terms of its components' dimensions becomes:

$LT^{-1} = (ML^{-3})^a \times (ML^{-1}T^{-2})^b$

$= M^aL^{-3a} \times M^bL^{-1b}T^{-2b}$

$= M^{a+b}L^{-3a-1b}T^{-2b}$

The next step is to match the indices of M, L and T on the two sides of the equation. There is no 'M' on the left hand side, therefore, LT^{-1} may be considered to be M^0LT^{-1}. M^0 does not change anything as $M^0 = 1$.

(M) $0 = a + b;$ $a = -b$

(L) $1 = -3a - 1b;$ $1 = -3a + a,\ a = -\frac{1}{2}$

(T) $-1 = -2b;$ $b = \frac{1}{2}$

Replace 'a' and 'b' by their actual values in the equation, $v = \rho^a E^b$:

$$v = \rho^{-1/2} E^{1/2} = \frac{E^{1/2}}{\rho^{1/2}}$$

This can also be written as: $v = \sqrt{\frac{E}{\rho}}$

2.7 Arithmetic progression (AP)

A set of numbers which follow a pattern or rule is called a series or progression. Each of the numbers forming this set is known as a term. An arithmetic progression (AP) is a sequence of numbers in which the difference between two consecutive terms is constant; the difference is known as the common difference. Consider the sequence:

3, 6, 9, 12, 15, ...

The first term (denoted by a) = 3, and the common difference (d) = 6 – 3 = 3

The second term is $a + d = a + (2 - 1)d$

The third term is $a + 2d = a + (3 - 1)d$

The fourth term is $a + 3d = a + (4 - 1)d$

Therefore any term of the sequence can be given as:

$T_n = a + (n - 1)d$, where n the number of the term (second, third, fourth etc.)

The sum of n terms of an arithmetic progression (S_n) is given by:

$$S_n = \frac{n}{2}(2a + (n-1)d)$$

Example 2.11

Find the eleventh term and the sum of the series: 1, 6, 11, 16, 21, ... to 11 terms.

Solution:

Let T_{11} be the eleventh term, therefore $T_{11} = a + (n - 1)d$

n = 11, a = 1 (first term of the series),

d (the common difference) = 6 – 1 = 5

$T_{11} = 1 + (11 - 1)5$

$= 1 + (10)5 = 51$

The sum of first 11 terms is given by, $S_{11}=\frac{n}{2}(2a+(n-1)d)$

$$=\frac{11}{2}(2\times1+(11-1)5)$$

$$= 5.5\,(2 + 50) = \mathbf{286}$$

Example 2.12

The third term of an arithmetic progression is 34 and the eighth term is 19. Find the common difference and the first term.

Solution:

Let T_3 be the 3rd term and T_8 be the eighth term

$T_3 = 34 = a + (3 - 1)d$, therefore $a + 2d = 34$

$T_8 = 19 = a + (8 - 1)d$, therefore $a + 7d = 19$

Solving the above equations simultaneously gives, $\mathbf{d = -3}$

Put $d = -3$ in either of the above equations:

$a + 2(-3) = 34$, hence a (first term) $= 34 + 6 = \mathbf{40}$

Example 2.13

Atlantic Engineering Co. started the production of double-glazed windows last year and produced 3000 windows. The company plans to increase the production by 50 windows each year. Determine:

a) the production in the tenth year
b) the total production from start to when it has just increased the production by 20% over the initial production

Solution:

a) The first term, $a = 3000$ windows, $d = \mathbf{50\ windows}$

Production in the tenth year, $T_{10} = 3000 + (10 - 1)50$

$$= 3000 + 450 = 3450 \text{ windows}$$

b) Increased production $= 3000 + 0.2 \times 3000 = 3600$

If n is the year in which the specified increase occurs, then $3000 + (n - 1)\,50 = 3600$

$50n - 50 = 600,\ n = \frac{650}{50} = 13$ years

Total production in 13 years, $S_{13} = \frac{n}{2}(2a+(n-1)d)$

$$=\frac{13}{2}(2\times3000+(13-1)50)$$

$$= 6.5(6000 + 600) = 6.5(6600) = \mathbf{42900\ windows}$$

2.8 Geometric progression (GP)

The terms of a series are said to be in geometric progression if the ratio of each term to its preceding term is a constant value. This ratio is known as the common ratio. For example:

- In the series 3, 9, 27, 81, ... the common ratio is $\frac{9}{3} = 3$
- In the series 2, –4, 8, –16, 32, ... the common ratio is $\frac{-4}{2} = -2$

Let the first term of the series be a and the common ratio be r.

The second term is ar (the index of $r = 2 - 1 = 1$)

The third term is ar^2 (the index of $r = 3 - 1 = 2$)

The tenth term is ar^9

The nth term is ar^{n-1}

Example 2.14

Find the eighth term of the series 2, 6, 18, 54, ...

Solution:

The first term, $a = 2$

Common ratio, $r = \frac{6}{2} = 3$

The eighth term $= ar^{8-1} = ar^7$

$= 2 \times 3^7 = \mathbf{4374}$

Example 2.15

The second term of a GP is 3 and the fifth term is 81. Find the tenth term.

Solution:

The second term $= ar^{2-1} = ar = 3$ (1)

The fifth term $= ar^{5-1} = ar^4 = 81$ (2)

Divide equation (2) by (1), $\frac{ar^4}{ar} = \frac{81}{3}$

$r^3 = 27$, hence $r = 3$

From equation (1), $ar = 3$, therefore, $a = \frac{3}{r} = \frac{3}{3} = 1$

The tenth term $= ar^9 = 1 \times 3^9 = \mathbf{19683}$

2.8.1 The sum of a series in GP

Consider a series in geometric progression consisting of n terms

The first term is a

The second term is ar

The last term is ar^{n-1}

The sum (S_n) is given by: $S_n = a + ar + ar^2 + \ldots\ldots\ldots\ldots + ar^{n-1}$ (1)

Multiply the above equation by r

$rS_n = ar + ar^2 + ar^3 + \ldots\ldots\ldots\ldots + ar^n$ (2)

Subtract equation (2) from (1)

$$S_n - rS_n = a - ar^n$$

$$S_n(1 - r) = a(1 - r^n)$$

$$S_n = \frac{a(1-r^n)}{(1-r)}$$

Example 2.16

Find the sum of 7 terms of the series 2, 6, 18, 54, ...

Solution:

The first term, $a = 2$

Common ratio, $r = \frac{6}{2} = 3$

Sum of 7 terms, $S_7 = \frac{a(1-r^7)}{(1-r)}$

$$= \frac{2(1-3^7)}{(1-3)} = \frac{2(1-2187)}{-2} = \mathbf{2186}$$

Exercise 2.1

1. Multiply: i) $1.5x\,y^2$ by $6x\,y\,z$
   ii) $2a^2\,b^2\,c$ by $4a\,b^2\,c^3$
2. Divide: i) $4x^6\,y^4\,z^3$ by $2x^4\,y^2\,z^2$
   ii) $2x^8\,y^3\,z^3$ by $x^4\,y^3\,z^{-2}$
3. Simplify: i) $3 + (2a + 5b - 20)$
   ii) $6 - (3a - 2b - 10)$
   iii) $2(2a + 2.5b)$
   iv) $1.5(2a - 3b - 5)$
4. Factorise: a) $5xz - 15yz$
   b) $4xy^2 + 8x^2y$
   c) $x^3 + 3x^2 + x$

5. Solve the following equations:

   i) $3x - 3 = x + 7$

   ii) $5x + 2 = 12$

   iii) $\frac{2x}{3} = 3.2 - x$

6. Solve the following equations:

   i) $2(x + 4) + 4(x + 1) = 6(2x + \frac{1}{2})$

   ii) $\frac{6 - A}{2} + \frac{A - 5}{3} = \frac{5}{4}$

   iii) $\frac{3Y}{10} = \frac{Y}{3} - \frac{5}{6}$

7. For a particular building job, cement, sand, gravel and water are mixed to produce 1100 kg of concrete. Use the following information to determine the quantity (in kg) of each ingredient:

   Quantity of sand = 1.5 × quantity of cement

   Quantity of gravel = 200 kg more than the quantity of sand

   Quantity of water = 50% of the quantity of cement

8. The width of a rectangle is 4 cm less than its length. If the perimeter of the rectangle is 32 cm, find its length and width.

9. The length to width ratio of a rectangle is 1.5. If the perimeter of the rectangle is 40 cm, calculate its length and width.

10. Find the dimensions of the following quantities in terms of M, L and T:

    a) Potential energy b) Work c) Power

11. Use the method of dimensions to check whether the following formulae are dimensionally correct; u and v are velocities, a is acceleration, s is distance and t is time:

    a) $v^2 = u^2 + 2as$

    b) $v = u + at^2$

    c) $s = ut + \frac{1}{2}at^2$

12. A metal bar, L metres long, expands by ΔL when its temperature increases by ΔT. Find the dimensions of the co-efficient of thermal expansion, α, if the change in length is given by:

    $\Delta L = \alpha L \Delta T$

13. Find the dimensional formula of thermal conductivity, λ, if the flow of heat energy, Q, is given by:

    $Q = \frac{\lambda A(\theta_2 - \theta_1)t}{d}$, where A is the surface area, $(\theta_2 - \theta_1)$ is the temperature difference, t is time and d is the thickness of the material through which the heat energy flows.

14. The centripetal force (F) acting on an object, moving uniformly in a circle, depends on the mass (m) of the object, its velocity (v) and radius of the circle (r). Derive the dimensional formula for force F.

Exercise 2.2

1. Find the twelfth term and the sum of the first 12 terms of the series:

   1, 3.5, 6, 8.5, ...

2. The fifth term of an AP is 28 and the twelfth term is 63. Find the common difference and the first term.

3. The noise produced by a machine at a construction site ranges from 52 dB to 90 dB, at six different speeds. If the noise forms an arithmetic progression, calculate the other noise levels.

4. Alpha engineering company started the production of uPVC doors last year and produced 4000 doors by the end of the year. If the company intends to increase the production by 50 doors each year, calculate:

   a) the production in the eighth year

   b) the total production for the first 20 years

   c) the time (years) it will take for the production to increase by 25%.

5. Delta engineering company make precast concrete floor units and increase their production each year in an arithmetic progression. The total production from start to the end of the tenth year was 137500 units and after 18 years the total was 427500 units. Calculate:

   a) the production after the first year

   b) the yearly increase in production

   c) the total production for the first 40 years

6. Find the seventh term of the series 2.2, 4.4, 8.8, 16.16, ...

7. Find the sum of 8 terms of the series 1, 4, 16, 64, ...

8. The third term of a GP is 22.5 and the sixth term is 607.5. Find the eleventh term.

9. In a standard penetration test (SPT) on sandy clay the sampler penetrated by 18.4 mm into the soil due to the second blow of the hammer. If the sampler penetration decreases in geometric progression, and the sampler penetrated by 2.47 mm due to the eleventh blow of the hammer, calculate:

   a) the penetration of the sampler due to the seventh blow of the hammer

   b) total penetration of the sampler into the soil after 15 blows.

10. The annual rent of a building in the first year was £6000.00 and increases every year by 10%. Calculate:

    a) the rent in the tenth year

    b) total of all rents up to year 10.

Answers – Exercise 2.1

1. i) $9x^2 y^3 z$
   ii) $8a^3 b^4 c^4$
2. i) $2x^2 y^2 z$
   ii) $2x^4 z^5$
3. i) $2a + 5b - 17$
   ii) $16 - 3a + 2b$
   iii) $4a + 5b$
   iv) $3a - 4.5b - 7.5$
4. a) $5z(x - 3y)$
   b) $4xy(y + 2x)$
   c) $x(x^2 + 3x + 1)$
5. i) $x = 5$
   ii) $x = 2$
   iii) $x = 1.92$
6. i) $x = 1.5$
   ii) $A = 0.5$
   iii) $Y = 25$
7. Cement = 200 kg, Sand = 300 kg, Gravel = 500 kg, Water = 100 kg
8. Length = 10 cm, Width = 6 cm
9. Length = 12 cm, Width = 8 cm
10. a) ML^2T^{-2} b) ML^2T^{-2} c) ML^2T^{-3}
11. a) Correct b) Incorrect c) Correct
12. $[T^{-1}]$
13. $[MLT^{-3}K^{-1}]$
14. $F = \frac{kmv^2}{r}$

Answers – Exercise 2.2

1. $T_{12} = 28.5$, $S_{12} = 177$
2. $a = 8$, $d = 5$

3. 59.6, 67.2, 74.8, 82.4 dB
4. a) 4350 doors b) 89500 doors c) 21 years
5. a) 2500 units b) 2500 units c) 2050000 units
6. 140.8
7. 21845
8. 147622.5
9. a) 6.03 mm b) 110.95 mm
10. a) £14147.69 b) £95624.40

CHAPTER 3

Algebra 2

Topics covered in this chapter:

- Transposition of simple and complex formulae involving one or more of: addition, subtraction, multiplication, division, squares/square roots, cubes/cube roots and logarithms
- Evaluation of simple and complex formulae
- Binomial theorem and its application

3.1 Introduction

The formula for determining the volume of a sphere (V) is given by:

$V = \frac{4}{3}\pi r^3$, where r is the radius of the sphere.

In this formula V is known as the 'subject of the formula'. If we want to determine the volume of a sphere, that can be done by putting in the values of π and r^3 in this formula. If the volume of a sphere is given, and the question is to calculate the radius of the sphere, we need to rearrange the formula so that the radius becomes the 'subject of the formula'. This process of rearranging the symbols of formulae is known as the transposition of formulae.

3.2 Transposition of formulae

The process of transposition may involve one or more of:

a) Addition and/or subtraction of the components of a formula.
b) Multiplication and/or division of the components of a formula.
c) Square, square root, cube root etc.

3.2.1 Transposition involving addition/subtraction

A formula is an equation, with the equal to sign (=) creating the left side (LHS) and the right side (RHS). Consider the formula:

$x = y - w - z$

If we want to make 'w' the subject of the formula, the process will involve rearrangement of the symbols so that 'w' is on the left side, on its own, positive and on the numerator side. Applying these rules to $x = y - w - z$:

$x + w = y - z$

or, $w = y - z - x$

Formulae involving addition and/or subtraction require the symbols to change sides horizontally. There is no multiplication or division involved.

3.2.2 Transposition involving multiplication/division

Consider the formula to determine the circumference of a circle (c):

$c = 2\pi r$, where r is the radius of the circle

If we want to make 'r' the subject of the formula, then 2π must be shifted to the other side to leave 'r' on its own. This can be achieved by dividing both sides of the formula by 2π:

$$\frac{c}{2\pi} = \frac{2\pi r}{2\pi}$$

$$\frac{c}{2\pi} = r$$

$$\text{or } r = \frac{c}{2\pi}$$

Example 3.1

Transpose $v^2 - u^2 = 2$ as to make 's' the subject.

Solution:

$v^2 - u^2 = 2$ as may be written as $2\text{ as} = v^2 - u^2$

Divide both sides by 2a to leave 's' on its own:

$$\frac{2as}{2a} = \frac{v^2 - u^2}{2a}$$

$$\text{or, } s = \frac{v^2 - u^2}{2a}$$

Example 3.2

Transpose $v = ut + \frac{1}{2}at^2$, to make 'a' the subject.

Solution:

The question may be written as: $ut + \frac{1}{2}at^2 = v$

Transpose 'ut' to the RHS, $\frac{1}{2}at^2 = v - ut$

Multiply both side by 2 to simplify the fraction, $2 \times \frac{1}{2}at^2 = 2 \times (v - ut)$

$at^2 = 2(v - ut)$

Divide both sides by t^2, we have

$$\mathbf{a} = \frac{2(v - ut)}{t^2}$$

3.2.3 Transposition involving addition/subtraction and multiplication/division

Consider the formula: $v = u + at$. If the question is to make 't' the subject then u and a must be transferred to the other side of the formula.

$v = u + at$ may be written as $u + at = v$

Transpose u to the RHS, $at = v - u$

On the LHS, a and t are multiplied, and to separate them we must do the opposite, i.e. divide.

Divide both sides by a, $\frac{at}{a} = \frac{v - u}{a}$

or $\mathbf{t} = \frac{v - u}{a}$

3.2.4 Transposition involving squares/square roots

The formula to determine the area of a circle is: $A = 0.25\,\pi d^2$. We have to deal with the square if d is to be the new subject of the formula. d^2 can be changed to d by taking square root of the whole formula.

$0.25\,\pi d^2 = A$

$$d^2 = \frac{A}{0.25\pi}$$

Take square root of both sides $\sqrt{d^2} = \sqrt{\frac{A}{0.25\pi}}$

$$d = \sqrt{\frac{A}{0.25\pi}}$$

The square root is the reverse of the square, therefore they cancel out each other.

Example 3.3

Transpose $V = \pi\, r^2\, h$ to make r the subject.

Solution:

a) $V = \pi r^2 h$ can be written as $\pi r^2 h = V$

Divide both sides by πh to have only r^2 on the LHS

$$\frac{\pi r^2 h}{\pi h} = \frac{V}{\pi h}$$

$$r^2 = \frac{V}{\pi h}$$

Take square root of both sides. The square root is the reverse of squaring, therefore, the square and the square root cancel out to give r:

$$\sqrt{r^2} = \sqrt{\frac{V}{\pi h}}$$

$$\mathbf{r} = \sqrt{\frac{V}{\pi h}}$$

Example 3.4

Transpose $r = \sqrt{\frac{A}{\pi}}$ to make A the subject.

Solution:

$r = \sqrt{\frac{A}{\pi}}$ can be written as $\sqrt{\frac{A}{\pi}} = r$

Square both sides to get rid of the square root:

$$\frac{A}{\pi} = r^2$$

Multiply both sides by π:

$$\frac{A \times \pi}{\pi} = r^2\pi$$

$A = r^2 \pi$ or $\mathbf{A = \pi r^2}$

Example 3.5

Transpose $E = \frac{p}{w} + z + \frac{v^2}{2g}$ to make v the subject.

Solution:

Transpose $\frac{p}{w}$ and z to the left hand side of the equation:

$$E - \frac{p}{w} - z = \frac{v^2}{2g} \text{ or } \frac{v^2}{2g} = E - \frac{p}{w} - z$$

$v^2 = 2g(E - \frac{p}{w} - z)$ (multiply both sides by 2g)

$$\mathbf{v} = \sqrt{2g\left(E - \frac{p}{w} - z\right)}$$

Example 3.6

Transpose $z = \frac{bd^3}{6}$ to make d the subject.

Solution:

$z = \frac{bd^3}{6}$ can be written as $\frac{bd^3}{6} = z$

After transposing 6 to the RHS we have, $bd^3 = 6z$

$$d^3 = \frac{6z}{b}$$

Take cube root of both sides to change d^3 into d

$$(d^3)^{1/3} = (\frac{6z}{b})^{1/3}$$

$d = (\frac{6z}{b})^{1/3}$ (cube and cube root cancel out leaving just d)

This can also be written as $\mathbf{d} = \sqrt[3]{\frac{6z}{b}}$ ($\sqrt[3]{\ }$ is the symbol for cube root)

Example 3.7

Transpose $N = 10\log_{10}\left(\frac{I_2}{I_1}\right)$ to make I_2 the subject.

Solution:

$$10\log_{10}\left(\frac{I_2}{I_1}\right) = N$$

$$\log_{10}\left(\frac{I_2}{I_1}\right) = \frac{N}{10}$$

To remove $\log_{10}$ from the LHS, take antilogarithm (antilog) of both sides.

Antilogarithm is the reverse of logarithm and cancels it out.

$$\text{antilog}\,(\log_{10}\left(\frac{I_2}{I_1}\right)) = \text{antilog}\left(\frac{N}{10}\right)$$

$\left(\frac{I_2}{I_1}\right) = \text{antilog}\left(\frac{N}{10}\right)$ (antilog cancels $\log_{10}$)

$$\mathbf{I_2 = I_1 \times antilog\left(\frac{N}{10}\right)}$$

3.3 Evaluation of formulae

The process of finding the value of one term of a formula (usually the subject) by replacing the other terms by their respective values, is known as the **evaluation of formulae.**

For example, the area of a rectangle, A = Length × Width

If the length and the width of a rectangle are 15 cm and 12 cm respectively, the area of the rectangle is:

$A = 15 \times 12 = 180 \text{ cm}^2$

It is important that the units of different quantities are compatible so that the answer has got the right units, otherwise the answer will be wrong. For instance, the length of a rectangle could be in millimetres and the width in centimetres. In this case it is necessary to convert either the length into centimetres or the width into millimetres.

Example 3.8

The area of a triangle (A) is given by: $A = \frac{\text{Base} \times \text{Height}}{2}$. If the base and the height of a triangle measure 10 cm and 15 cm respectively, calculate the area of the triangle.

Solution:

$$\text{Area, } A = \frac{\text{Base} \times \text{Height}}{2}$$
$$= \frac{10 \times 15}{2} = \mathbf{75\,cm^2}$$

Example 3.9

Evaluate R in the formula $R = \sqrt{\frac{L}{2\pi}}$, if $L = 250 \text{ cm}^2$.

Solution:

$$R = \sqrt{\frac{L}{2\pi}}$$
$$R = \sqrt{\frac{250}{2\pi}}$$
$$R = \sqrt{39.7887} = \mathbf{6.31cm}$$

Example 3.10

Evaluate I_2 in the formula $I_2 = I_1 \times \text{antilog}\left(\frac{N}{10}\right)$, if N = 85 and $I_1 = 1 \times 10^{-12}$.

Solution:

$$I_2 = I_1 \times \text{antilog}\left(\frac{N}{10}\right)$$

$$= 1 \times 10^{-12} \times \text{antilog}\left(\frac{85}{10}\right)$$

$$= 1 \times 10^{-12} \text{ antilog } 8.5$$

$$= \mathbf{3.162 \times 10^{-4}}$$

3.4 Binomial theorem

3.4.1 Introduction

The binomial theorem is a formula that allows us to expand a binomial expression to any power. A binomial expression consists of two terms, for example, $1 + x$, $x + 2y$, $a + b^2$ etc.

The general binomial expression of $(a + x)^n$ is

$$(a+x)^n = a^n + na^{n-1}x + \frac{n(n-1)}{2!}a^{n-2}x^2 + \frac{n(n-1)(n-2)}{3!}a^{n-3}x^3 + \ldots\ldots\ldots\ldots\ldots + x^n$$

where n may be a fraction, a decimal fraction or a positive/negative integer

In the above expression the fourth term is: $\frac{n(n-1)(n-2)}{3!}a^{n-3}x^3$

Similarly the fifth term will be: $\frac{n(n-1)(n-2)(n-3)}{4!}a^{n-4}x^4$

The r'th term is given by $\frac{n(n-1)(n-2) \quad \text{to} \, (r-1)}{(r-1)!}a^{n-(r-1)}x^{r-1}$

3.4.2 Pascal's triangle

If we expand the expression $(a + x)^n$ for values of n ranging between 0 and 7, the following results are obtained:

$(a + x)^0 = 1$

$(a + x)^1 = a + x$

$(a + x)^2 = a^2 + 2ax + x^2$

$(a + x)^3 = a^3 + 3a^2 x + 3ax^2 + x^3$

$(a + x)^4 = a^4 + 4a^3 x + 6a^2 x^2 + 4ax^3 + x^4$

$(a + x)^5 = a^5 + 5a^4 x + 10a^3 x^2 + 10a^2 x^3 + 5ax^4 + x^5$

$(a + x)^6 = a^6 + 6a^5 x + 15a^4 x^2 + 20a^3 x^3 + 15a^2 x^4 + 6ax^5 + x^6$

$(a + x)^7 = a^7 + 7a^6 x + 21a^5 x^2 + 35a^4 x^3 + 35a^3 x^4 + 21a^2 x^5 + 7ax^6 + x^7$

It is evident from the above results:

i) The power of 'a' decreases as we move from left to right

ii) 'x' increases in power as we move from left to right

The coefficients are shown in Table 3.1; this arrangement is known as Pascal's triangle. The coefficient of a term may be obtained by adding the two adjacent coefficients immediately above in the previous row.

Table 3.1

$(a + x)^0 =$	1
$(a + x)^1 =$	1 1
$(a + x)^2 =$	1 2 1
$(a + x)^3 =$	1 3 3 1
$(a + x)^4 =$	1 4 6 4 1
$(a + x)^5 =$	1 5 10 10 5 1
$(a + x)^6 =$	1 6 15 20 15 6 1
$(a + x)^7 =$	1 7 21 35 35 21 7 1

Example 3.11

Expand $(a + x)^8$ by using Pascal's triangle.

Solution:

From Table 3.1 the row of Pascal's triangle corresponding to $(a + x)^7$ is

1 7 21 35 35 21 7 1

Coefficients of $(a + x)^8$ are: 1 8 28 56 70 56 28 8 1

Therefore, $(a + x)^8 = a^8 + 8\,a^7\,x + 28\,a^6\,x^2 + 56\,a^5\,x^3 + 70\,a^4\,x^4 + 56\,a^3\,x^5 + 28\,a^2\,x^6 + 8ax^7 + x^8$

Example 3.12

Use the binomial theorem to expand the term $(2x + 2y)^3$.

Solution:

The binomial expression of $(a + x)^n$ is

$$(a+x)^n = a^n + na^{n-1}x + \frac{n(n-1)}{2!}a^{n-2}x^2 + \frac{n(n-1)(n-2)}{3!}a^{n-3}x^2 + \frac{n(n-1)(n-2)}{3!}a^{n-3}x^3 + \dots\dots + x^n$$

Using the above, the expansion of $(2x + 2y)^3$ is:

$$(2x+2y)^3 = (2x)^3 + 3(2x)^{3-1}(2y) + \frac{3(3-1)}{2!}(2x)^{3-2}(2y)^2 + (2y)^3$$

$$= 8x^3 + 3(4x^2)(2y) + \frac{3\times 2}{2\times 1}(2x)^1(4y^2) + 8y^3$$

$$= 8x^3 + 24x^2y + 24xy^2 + 8y^3$$

Example 3.13

Expand $(x - 3y)^3$.

Solution:

The expression can be written as $(x + (-3y))^3$. Comparing it to the standard expression of $(a + x)^n$, we have x in place of a, $-3y$ in place of x and 3 in place of n.

$$\text{Therefore, } (x+(-3y))^3 = x^3 + 3(x)^{3-1}(-3y) + \frac{3(3-1)}{2!}x^{3-2}(-3y)^2 + (-3y)^3$$

$$= x^3 + 3(x)^2(-3y) + 3x^1(9y^2) - 27y^3$$

$$= x^3 - 9x^2y + 27xy^2 - 27y^3$$

Example 3.14

Without expanding $(2p + q)^5$ determine the fourth term.

Solution:

The r'th term of the expansion $(a + x)^n$ is given by:

$$\frac{n(n-1)(n-2)\text{.........} \text{ to } (r-1)}{(r-1)!}a^{n-(r-1)}x^{r-1}$$

Here $n = 5$; $a = 2p$; $x = q$; $r = 4$ and $r - 1 = 3$

Substituting these values, the fourth term of $(2p + q)^5$ is:

$$\frac{5(5-1)(5-2)}{3!}(2p)^{5-(4-1)}q^{4-1} = 10(2p)^2q^{4-1}$$

$$= 10\times 4p^2q^3 = \mathbf{40p^2q^3}$$

3.4.3 Binomial series for fractional or negative indices

The binomial theorem is true if the index n is a positive integer (whole number):

$$(a+x)^n = a^n + na^{n-1}x + \frac{n(n-1)}{2!}a^{n-2}x^2 + \frac{n(n-1)(n-2)}{3!}a^{n-3}x^3 + \text{.........} + x^n$$

For negative and fractional values of n the series of the terms goes on for ever and is called an infinite series, for example the expansion of $(1 + x)^{-2}$

$$(1+x)^{-2} = 1 + (-2)x + \frac{(-2)(-3)}{2!}x^2 + \frac{(-2)(-3)(-4)}{3!}x^3 + \frac{(-2)(-3)(-4)(-5)}{4!}x^4 + \text{.......}$$

$$= 1 - 2x + 3x^2 - 4x^3 + 5x^4 - \text{................}$$

This would result into an approximate answer which is only valid if the sequence converges. For a convergent series, x should be less than a [for $(1 + x)^{-2}$, x should be less than 1]. This is written as $|x| < 1$ numerically, or $-1 < x < 1$.

Example 3.15

Use the binomial theorem to expand $\frac{1}{(1+3x)^3}$ to four terms.

Solution:

$$\frac{1}{(1+3x)^3} = (1+3x)^{-3}$$

$$\text{Expansion of } (1+3x)^{-3} = 1 + (-3)(1)^{-3-1}(3x) + \frac{(-3)(-3-1)}{2!}(1)^{-3-2}(3x)^2 + \frac{(-3)(-3-1)(-3-2)}{3!}(1)^{-3-3}(3x)^3$$

$$= \mathbf{1 - 9x + 54x^2 - 270x^3}$$

This is true provided $3x < 1$ numerically

or $x < \frac{1}{3}$; i.e. $-\frac{1}{3} < x < \frac{1}{3}$

3.4.4 Approximations

We can find the value of $(1 + x)^n$ to any degree of accuracy by the binomial expansion, if x is small.

Since $(1+x)^n = 1 + nx + \frac{n(n-1)}{2!}x^2 + \frac{n(n-1)(n-2)}{3!}x^3 + ...$, a rough approximation is obtained by rejecting all powers of x above x^1.

Therefore, $(1 + x)^n :\approx 1 + nx$

For accuracy, the term containing x^2 or higher powers can be retained.

When x and y are very small fractions, then $(1 + x)^m (1 + y)^n$ becomes

$(1 + x)^m (1 + y)^n = 1 + mx + ny$ approximately.

Example 3.16

Using the binomial theorem, find the approximate as well as the value correct to 4 decimal places of $(1.004)^6$.

Solution:

a) $(1.004)^6 = (1 + 0.004)^6$, Here $x = 0.004$ and $n = 6$

Approximate value $= 1 + nx = 1 + 6(0.004) = \mathbf{1.024}$

b) $(1+x)^n = 1 + nx + \frac{n(n-1)}{2!}x^2 + \frac{n(n-1)(n-2)}{3!}x^3 + ...,$

$$(1.004)^6 = (1+0.004)^6 = 1+6(0.004)+\frac{6(6-1)}{2\times 1}(0.004)^2$$

$$= 1 + 0.024 + 0.00024 = 1.02424$$

$$= \mathbf{1.0242} \qquad \text{(correct to 4 d.p.)}$$

3.4.5 Practical problems

Examples 3.17 and 3.18 show how the binomial theorem can be used to find the approximate effect of small numerical changes on a calculation.

Example 3.17

The cross-sectional area (A) of a pipe is given by

$A = \frac{1}{4}\pi d^2$, where d is the diameter of the pipe.

Find the approximate error in calculating the area if the diameter has been measured 1% too small.

Solution:

$A = \frac{\pi}{4}d^2$, and the new diameter is $\left(d - d\times\frac{1}{100}\right)$

New area $A' = \frac{\pi}{4}\left(d - d\times\frac{1}{100}\right)^2$

$A' = \frac{\pi}{4}d^2\left(1 - 1\times\frac{1}{100}\right)^2$

The calculation can be approximated as $\frac{1}{100}$ is small compared with 1

$A' \approx \frac{\pi}{4}d^2\left(1 - 2\times\frac{1}{100}\right)$

$\approx A\left(1 - \frac{2}{100}\right)$

From $1 - \frac{2}{100}$ it can be concluded that the **new area is 2% too small.**

Example 3.18

In the formula $Q = \frac{km^2t}{L}$ find the percentage error in Q due to an error of +2% in t, –1% error in m and +1% error in L.

Solution:

Let Q' be the new quantity.

$$Q' = \frac{k[m(1-\frac{1}{100})]^2\, t(1+\frac{2}{100})}{L(1+\frac{1}{100})}$$

$$= \frac{km^2t(1-\frac{1}{100})^2(1+\frac{2}{100})}{L(1+\frac{1}{100})}$$

$$= \frac{km^2t}{L}\left(1-\frac{1}{100}\right)^2\left(1+\frac{2}{100}\right)^1\left(1+\frac{1}{100}\right)^{-1}$$

$$= \frac{km^2t}{L}\left(1-\frac{2}{100}+\frac{2}{100}+\frac{1\times(-1)}{100}\right)$$

$$= \frac{km^2t}{L}\left(1-\frac{2}{100}+\frac{2}{100}-\frac{1}{100}\right) = \frac{km^2t}{L}\left(1-\frac{1}{100}\right)$$

$= Q\left(1-\frac{1}{100}\right)$, therefore the **percentage error in Q is – 1%**

Exercise 3.1

1. Transpose $f = c + d - e$, to make e the subject
2. Transpose $y = mx + c$, to make x the subject
3. a) The thermal conductivity (k) of a material is given by, $k = \frac{Qd}{AT}$. Transpose the formula to make Q the subject.

   b) Evaluate Q, if $k = 0.8$, $A = 10$, $T = 40$ and $d = 0.5$
4. If $P = \frac{V^2}{R}$, transpose the formula to make V the subject.
5. a) The surface area (A) of an object is given by, $A = 2\pi r^2$. Transpose the formula to make r the subject.

   b) Evaluate r, if $A = 8000\ cm^2$
6. The area of a trapezium (A) is given by $A = \frac{1}{2}(a+b)\times d$. Transpose the formula:

   to make d the subject.

   to make b the subject.
7. a) The velocity (v) of water flow in an open channel is given by, $v = c\sqrt{RS}$. Transpose the formula to make R the subject.

   b) Calculate R if $v = 2.5$ m/s, $c = 50$ and $S = 0.02$
8. The relationship between porosity (n) and void ratio (e) of a soil is given by:

   $n = \frac{e}{1+e}$. Transpose the formula to make **e** the subject.

9. The deflection of a beam (d) is given by, $d = \frac{5wL^3}{384EI}$. Transpose the formula to make L the subject.

10. a) The velocity (v) of water flow in an open channel is given by, $v = \frac{1}{n}R^{\frac{2}{3}}S^{\frac{1}{2}}$. Transpose the formula to make R the subject.

    b) Find the value of R, if: v = 2.75 m/s, n = 0.01 and S = 0.02

11. Transpose the equation U = B(T + A(S – T)) to make T the subject.

12. Transpose $D = \frac{(G+e)d}{1+e}$ to make e the subject.

13. a) Transpose $A_2 = \frac{A_1}{1-\frac{h}{H}}$ to make h the subject.

    b) Evaluate h if $A_2 = 10$, $A_1 = 5$ and H = 5

14. Transpose $\frac{w}{G} = w - \frac{R}{1000}$ to make w the subject.

15. a) Transpose $w_1 - w_2 = C \times \log_{10}\left(\frac{n_2}{n_1}\right)$ to make n_2 the subject

    b) Evaluate n_2 if $w_1 = 25$, $w_2 = 24$, $n_1 = 20$ and C = 4.34

16. Calculate I_2 in the formula $I_2 = I_1 \times \text{antilog}\left(\frac{N}{10}\right)$, if N = 97 and $I_1 = 1\times10^{-12}$

Exercise 3.2

1. Use Pascal's triangle method to determine the expansion of $(2m - 3n)^4$
2. Expand $(3 + x)^3$ using the binomial theorem:
3. Use the binomial theorem to expand $(2p + q)^5$
4. Without expanding $(3 + 2x)^8$ determine the fifth term
5. Without expanding $(2 - 3x)^7$ determine the fourth term
6. Expand $\frac{1}{\sqrt[3]{1+2x}}$ to 4 terms
7. Expand $(3 - 2x)^{3/2}$ using the binomial theorem
8. In the formula $Q = \frac{km^2t}{L}$ find the percentage error in Q due to an error of +3% in t, –2% error in m and +2% error in L.
9. The maximum deflection (δ) in a beam is given by $\delta = \frac{KwL^3}{d^4}$. Find the approximate percentage change in deflection if W increases by 2%, L decreases by 2% and d decreases by 1.5%.
10. The area of steel (A) in a reinforced concrete beam is given by $A = \frac{M}{tjd}$. Find the approximate percentage change in the area of steel if M increases by 3% and t decreases by 2%.

Answers – Exercise 3.1

1. $e = c + d - f$
2. $x = \frac{y-c}{m}$
3. a) $Q = \frac{kAT}{d}$
   b) 640
4. $V = \sqrt{PR}$
5. a) $r = \sqrt{\frac{A}{2\pi}}$
   b) 35.68 cm
6. a) $d = \frac{2A}{a+b}$
   b) $b = \frac{2A}{d} - a$
7. a) $R = \frac{v^2}{c^2 S}$
   b) 0.125
8. $e = \frac{n}{1-n}$
9. $L = \sqrt[3]{\frac{384dEI}{5w}}$
10. a) $R = \left(\frac{vn}{S^{\frac{1}{2}}}\right)^{\frac{3}{2}}$
    b) $R = 0.0857$
11. $T = \frac{U-BAS}{B-BA}$
12. $e = \frac{Gd-D}{D-d}$
13. a) $h = H\left(1-\frac{A_1}{A_2}\right)$
    b) 2.5
14. $w = \frac{R}{1000} \times \frac{G}{G-1}$
15. a) $n_1 \times \text{antilog}\left(\frac{w_1 - w_2}{C}\right)$
    b) 34
16. 5.01×10^{-3}

Answers – Exercise 3.2

1. $16m^4 - 96m^3 n + 216m^2 n^2 - 216mn^3 + 81n^4$
2. $27 + 27x + 9x^2 + x^3$
3. $32p^5 + 80p^4 q + 80p^3 q^2 + 40p^2 q^3 + 10pq^4 + q^5$
4. $90720 x^4$
5. $-15120 x^3$
6. $1 - \frac{2}{3}x + \frac{8}{9}x^2 - \frac{112}{81}x^3$

   Provided $x < \frac{1}{2}$; i e. $-\frac{1}{2} < x < \frac{1}{2}$
7. $(3)^{3/2} - (3)^{3/2}x + \frac{3^{1/2}}{2}x^2 + \frac{3^{-3/2}}{2}x^3$

   Provided $x < \frac{3}{2}$; i.e. $-1.5 < x < 1.5$
8. – 3%
9. + 2%
10. +5%

CHAPTER 4

Simultaneous and quadratic equations

Topics covered in this chapter:

- Simultaneous equations: elimination method and substitution method
- Quadratic equations: factorisation, quadratic formula and completing the squares

4.1 Simultaneous equations

As explained in chapter 2, an equation is a mathematical statement that shows the equality of 2 expressions. Generally an equation has 1 unknown, but in some cases equations may have 2 or more unknowns. For example, the equation $2x - 3y = 8$, has two unknowns, x and y. The equation does not have a unique solution as several values of x and y satisfy the equation:

$x = 1, y = -2$ $\quad$ $2x - 3y = 2 \times 1 - 3 \times (-2)$

$= 2 + 6 = 8$

$x = 2.5, y = -1$ $\quad$ $2x - 3y = 2 \times 2.5 - 3 \times (-1)$

$= 5 + 3 = 8$

If we have two equations involving two unknowns, say x and y, then there is a unique solution that satisfies both equations. These equations are known as simultaneous equations.

Simultaneous equations can be solved by:

a) The elimination method.

b) The substitution method.

c) The graphical method.

The graphical method is explained in chapter 5.

4.1.1 The elimination method

This method involves the elimination of one unknown so that the other unknown can be determined. For example, if we are given the equations $x + 2y = -3$ and $x - y = 3$, the solution can be found by eliminating x first and then y, or y first and then x. The elimination process involves:

a) Multiplication of the equations by suitable numbers to make the coefficients of x or y the same, if they are not equal initially.
b) Either subtraction or addition, depending on whether the x or y terms have the opposite sign or the same sign.

This technique is explained in Example 4.1.

Example 4.1

Solve the equations, $x + 2y = -3$ and $x - y = 3$

Solution:

$$x + 2y = -3 \quad (1)$$

$$x - y = 3 \quad (2)$$

In both equations the x terms are positive and their coefficients are equal, i.e. 1 and 1. To eliminate x, subtract 1 equation from the other. It doesn't matter whether the first equation is subtracted from the the second or the second from the first. Subtraction also involves the change of signs; from + to – and from – to +

Subtract equation (2) from equation (1):

$$\begin{array}{l} x + 2y = -3 \\ x - y = 3 \\ \mathbf{-} \quad \mathbf{+} \quad \mathbf{-} \\ \hline \quad 3y = -6 \\ \text{or } y = -6/3 = -2 \end{array}$$

(the new signs of the second equation are shown in bold; they supersede the original signs)

$(x - x = 0; 2y + y = 3y; -3 - 3 = -6)$

To eliminate y from the two equations multiply equation (2) by 2 so that the coefficients of y terms are the same in both equations:

$$2 \times (x - y = 3) = 2x - 2y = 6 \quad (3)$$

The signs of **2y** are different in equations (1) and (3). In this situation the equations are added. The addition of numbers/letters does not require the change of signs.

Add equations (1) and (3)

$$\begin{array}{l} x + 2y = -3 \\ 2x - 2y = 6 \\ \hline 3x \quad = 3 \\ \text{or } x = 3/3 = 1 \end{array}$$

$(x + 2x = 3x; +2y - 2y = 0; -3 + 6 = 3)$

The solution is $x = 1$ and $y = -2$

4.1.2 The substitution method

In this method the value of x (or y) is calculated in terms of the other unknown from 1 equation and substituted in the other equation. This method is easy to use when the coefficients of x and y are small numbers. It can also be used in conjunction with the elimination method, as shown in example 4.3.

Example 4.2

Solve the equations, $x + 3 = -2y$ and $x - y = 3$ by the substitution method.

Solution:

After rearranging $x + 3 = -2y$, we have:

$x + 2y = -3$ (1)

After rearranging $x - y = 3$, we have: $x = 3 + y$ (2)

Substitute the value of x from equation (2) into equation (1)

$3 + y + 2y = -3$

Transpose 3 to the RHS, $3y = -3 -3$

$3y = -6$, or $\mathbf{y} = -6/3 = \mathbf{-2}$

Substitute $y = -2$ into either equation (1) or equation (2)

Let us use equation (2)

$x = 3 + (-2)$

or, $x = 3 - 2 = 1$

The solution is $\mathbf{x} = \mathbf{1}$ and $\mathbf{y} = \mathbf{-2}$

Example 4.3

Solve simultaneously the equations, $2x + 4y = 16$ and $7 - 3y = -x$

Solution:

$2x + 4y = 16$ (1)

$x - 3y = -7$ (2) (Rearrange the equation)

Multiply equation (2) by 2 so that both equations have 2x, and subtract

$2x + 4y = 16$

$2x - 6y = -14$ (Change the signs of the second equation)

$-\ \ +\ \ \ \ +$

$10y = 30$

or, $y = 30/10 = 3$

Substitute the value of y into equation (1)

$2x + (4 \times 3) = 16$

$2x + 12 = 16$

$2x = 16 - 12 = 4$

$x = 4/2 = 2$

The solution is $\mathbf{x} = \mathbf{2}$ and $\mathbf{y} = \mathbf{3}$.

4.1.3 Application of simultaneous equations

Example 4.4

3 bricklayers and 2 general operatives complete a building extension in 6 days and earn £2520. On similar work, 5 bricklayers and 3 general operatives earn £2720 and complete it in 4 days. Calculate the daily earnings of 1 bricklayer, and 1 general operative.

Solution:

The question can be written in the form of 2 equations that can be solved to determine the earnings of a bricklayer and a general operative. In the first project, £2520 was earned in 6 days whereas in the second project the amount earned in 4 days was £2720. It is important to have the earnings for the same duration in both cases. Let us first calculate the earnings per day:

3 bricklayers and 2 general operatives earn £420 in 1 day (2520 ÷ 6 = 420)

5 bricklayers and 3 general operatives earn £680 in 1 day (2720 ÷ 4 = 680)

Let b and g represent the bricklayers and the general operatives respectively.

$$3b + 2g = 420 \quad (1)$$

$$5b + 3g = 680 \quad (2)$$

Multiply equation (1) by 3 and equation (2) by 2 and subtract

$$\begin{array}{r} 9b + 6g = 1260 \\ 10b + 6g = 1360 \\ - \quad - \quad - \\ \hline -b \quad = -100 \end{array} \quad \text{or} \quad b = 100$$

To find g, substitute the value of b in equation (1) or equation (2).

Let us use equation (1): $3 \times 100 + 2g = 420$

$$2g = 420 - 300 = 120$$

Therefore, $g = 60$.

A **bricklayer** earns **£100** per day, and a **general operative** earns **£60** per day.

Example 4.5

The demand and supply of housing in a region can be represented by the equations

P + 15 Q = 406 000, and P – 13 Q = 58800 respectively. P represents the price in Pounds and Q represents the quantity, i.e. the number of houses. Find the price and quantity at which the supply equals the demand.

Solution:

The solution of the given equations will give us the price and quantity at which the supply of houses equals demand. The supply curve (Figure 4.1) shows that the builders will construct more houses as the prices go up. However, the demand for houses increases as their prices fall. Point E gives the equilibrium price at which the quantity demanded equals the quantity supplied. The solution of the equations will yield the equilibrium price.

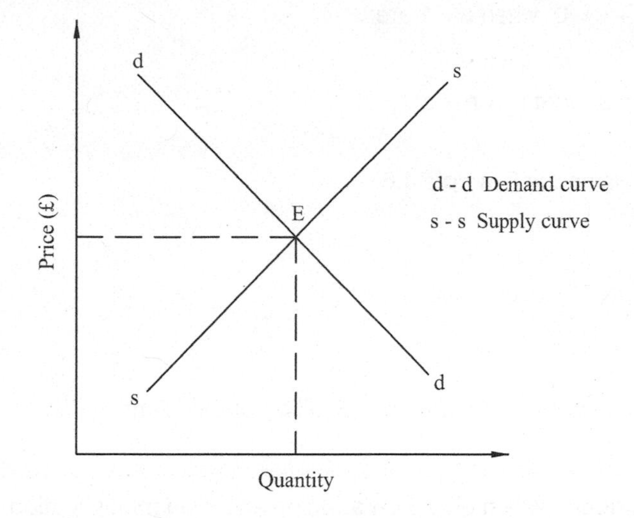

Figure 4.1

$$P + 15\,Q = 406000 \quad (1)$$
$$P - 13\,Q = 58800 \quad (2)$$
$$- \; + \qquad -$$
$$28\,Q = 347200 \qquad \text{or } Q = 12400$$

Substitute the value of Q in equation (2)

$$P - (13 \times 12\,400) = 58\,800$$
$$P = £220\,000$$

At a Price of **£220 000**, quantity supplied = quantity demanded = **12 400**

4.2 Quadratic equations

An equation of the form $ax^2 + bx + c = 0$, where a, b and c are constants (numbers) and, $a \neq 0$ is known as the quadratic equation. Constants a, b and c are known, whilst x is unknown which can be determined by a range of methods:

a) Factorising.
b) The quadratic formula.
c) Completing the square.
d) The graphical method (explained in chapter 5).

As the highest power of x in a quadratic equation is 2, therefore there are 2 values of x, which satisfy the equation. These values are also known as the roots of the equation.

4.2.1 Factorising

The factors of $ax^2 + bx + c = 0$, when $a = 1$ are:

$(x + m)(x + n) = 0$

m and n are numbers so that $m \times n = c$

and $m + n = b$

The procedure is explained in Example 4.6.

Example 4.6

Solve the equation $x^2 + 5x + 6 = 0$

Solution:

As explained earlier, the factors of $x^2 + 5x + 6 = 0$ will be of the form:

$(x + m)(x + n) = 0$ (1)

We need to select 2 numbers, which give 5 on addition and 6 on multiplication. Numbers 2 and 3 satisfy these conditions:

$2 + 3 = 5$ and $2 \times 3 = 6$ ($m = 2$ and $n = 3$)

Replacing m and n in equation (1) with 2 and 3:

$(x + 2)(x + 3) = 0$

Since the product of the 2 factors is equal to 0, then

Either $x + 2 = 0$ (2)

Or $x + 3 = 0$ (3)

If $x + 2 = 0$, then $x = -2$

If $x + 3 = 0$ then $x = -3$

The solution is $\mathbf{x = -2}$ **or** $\mathbf{x = -3}$

Example 4.7

Solve the equation $x^2 - 7x + 12 = 0$

Solution:

As explained earlier, we need to select 2 numbers, which give −7 on addition and +12 on multiplication. The numbers are −3 and −4.

$-3 + (-4) = -3 - 4 = -7$

$-3 \times -4 = 12$

The factors of $x^2 - 7x + 12 = 0$ will be of the form $(x + m)(x + n) = 0$

$(x - 3)(x - 4) = 0$ (replace m with −3 and n with −4)

Either $x - 3 = 0$, therefore $x = 3$

Or $x - 4 = 0$, hence $x = 4$

The solution is $\mathbf{x = 3}$**, or** $\mathbf{x = 4}$

Example 4.8

Solve the equation $x^2 - x - 6 = 0$

Solution:

Comparing $x^2 - x - 6 = 0$ to the general form $ax^2 + bx + c = 0$ we find that in the given equation, $a = 1$, $b = -1$ and $c = -6$. In this case the product of the two numbers should be -6 and their algebraic sum -1. We can only get -1 as the sum if the bigger number is negative; -3 and 2 are such numbers that will satisfy these conditions.

$-3 + 2 = -1$

$-3 \times 2 = -6$

The factors will be of the form:

$(x + m)(x + n) = 0, \quad (1)$

Replacing m and n with -3 and 2 in equation (1) we have:

$(x - 3)(x + 2) = 0$

Either $x - 3 = 0$, therefore $x = 3$

Or $\quad x + 2 = 0$, therefore $x = -2$

The solution is $\mathbf{x = 3}$ **or** $\mathbf{x = -2}$

Example 4.9

Solve the equation $5x^2 - 8x - 4 = 0$

Solution:

In this equation the coefficient of x^2 is greater than 1, which makes the solution more complex. The factors of 5 are 1 and 5:

$1 \times 5 = 5$

The factors of 4 are 1, 2 and 4: $\quad 1 \times 4 = 4$ and $2 \times 2 = 4$

Several factors are possible:

$(5x \pm 4)(x \pm 1)$ or $(5x \pm 2)(x \pm 2)$ or $(5x \pm 1)(x \pm 4)$

Only one of these factors will give the right answer. This can be checked by multiplying the brackets out:

$(5x + 4)(x - 1) = 5x^2 + 4x - 5x - 4 = 5x^2 - x - 4$

$(5x - 4)(x + 1) = 5x^2 - 4x + 5x - 4 = 5x^2 + x - 4$

$(5x - 2)(x + 2) = 5x^2 - 2x + 10x - 4 = 5x^2 + 8x - 4$

$(5x + 2)(x - 2) = 5x^2 + 2x - 10x - 4 = 5x^2 - 8x - 4$

$(5x + 2)$ and $(x - 2)$ are the right factors of $5x^2 - 8x - 4$

Therefore $5x^2 - 8x - 4 = (5x + 2)(x - 2) = 0$

Either $5x + 2 = 0$, or $x - 2 = 0$

If $\quad 5x + 2 = 0$, then $x = -2/5$ or -0.4

If $\quad x - 2 = 0$, then $x = 2$

The solution is $x = -0.4$, or $x = 2$

4.2.2 The quadratic formula

If $ax^2 + bx + c = 0$, where $a \neq 0$, the solutions of the equation are given by the formula:

$$x = \frac{-b \pm \sqrt{(b)^2 - 4ac}}{2a}$$

This formula can be used to solve all types of quadratic equations, especially those which cannot be solved by the other methods.

Example 4.10

Solve the equation $2x^2 = 6x + 8$

Solution:

The first step is to rearrange and write the equation to match the general form, i.e.

$ax^2 + bx + c = 0$:

$2x^2 - 6x - 8 = 0$

In this equation, $a = 2$, $b = -6$ and $c = -8$

Put the values of a, b and c in the quadratic formula:

b b a c

$$x = \frac{-(-6) \pm \sqrt{(-6)^2 - 4 \times 2 \times (-8)}}{2 \times 2}$$

a

$$\frac{6 \pm \sqrt{36 + 64}}{4} \qquad [-(-6) = 6; (-6)^2 = 36]$$

$$= \frac{6 \pm \sqrt{100}}{4} = \frac{6 \pm 10}{4}$$

Either $x = \frac{6 + 10}{4}$ or $x = \frac{6 - 10}{4}$

Solve the above to get the solution. The solution is **$x = 4$ or $x = -1$**

4.2.3 Completing the square

Quadratic equations can be rearranged by transposing the constant to the RHS. If the constant is greater than zero, it is possible to solve the equation by completing the square. A number is added to the LHS to make it a perfect square. For example, $(x + 1)^2$, $(x + 2)^2$, $(x - 1)^2$, $(x - 2)^2$ are perfect squares, and we must try to get the LHS of the equation written in this form. The next example explains the procedure.

Example 4.11

Solve the equation $x^2 + 2x - 5.25 = 0$

Solution:

Rearrange the equation by transposing 5.25 to the RHS:

$x^2 + 2x = 5.25$

We know that $(x + 1)^2 = x^2 + 2x + 1$, therefore $x^2 + 2x = (x + 1)^2 - 1$

$(x + 1)^2 - 1 = 5.25$

$(x + 1)^2 = 5.25 + 1 = 6.25$

$(x + 1)^2 = (2.5)^2$

Take square root of both sides. The square root cancels out the square

$x + 1 = \pm 2.5$

$x + 1 = +2.5,\ x = 2.5 - 1 = 1.5$

or $x + 1 = -2.5,\ x = -2.5 - 1 = -3.5$

The solution is $\mathbf{x = 1.5}$ **or** $\mathbf{x = -3.5}$

4.2.4 Application of quadratic equations

Example 4.12

A rectangular lawn measuring 15 m by 10 m is surrounded by a path. The total area of the lawn and the path is 218.75 m^2. Calculate the width of the path.

Solution:

Let x metres be the width of the path.

Length of the lawn (including path) = $15 + 2x$; Width = $10 + 2x$

$218.75 = (15 + 2x) \times (10 + 2x)$

$218.75 = 150 + 50x + 4x^2$

$0 = 4x^2 + 50x - 68.75$

This is a quadratic equation, and can be written as: $4x^2 + 50x - 68.75 = 0$

Although the quadratic formula will be used to solve the equation here, other methods may also be used:

$$x = \frac{-b \pm \sqrt{b^2 - 4ac}}{2a} = \frac{-50 \pm \sqrt{50^2 - 4(4)(-68.75)}}{2 \times 4}$$

$$= \frac{-50 \pm \sqrt{3600}}{8} = \frac{-50 \pm 60}{8}$$

$$\text{Either } x = \frac{-50 + 60}{8} = 1.25\,\text{m}$$

$$\text{or, } x = \frac{-50 - 60}{8} = -13.75\,\text{m}$$

Rejecting 13.75 m, as the width cannot be negative, the **width of the path is 1.25 m.**

Example 4.13

The length of a rectangular building plot is $(x + 20)$ metres and its width $(x + 5)$ metres. Calculate the length and the width of the building plot if its area is 2700 m^2.

Solution:

Area = length × width

$2700 = (x + 20) \times (x + 5)$

$2700 = x^2 + 20x + 5x + 100$

or, $2700 = x^2 + 25x + 100$

$0 = x^2 + 25x - 2600$ or $x^2 + 25x - 2600 = 0$

$$x = \frac{-b \pm \sqrt{b^2 - 4ac}}{2a} = \frac{-25 \pm \sqrt{25^2 - 4(1)(-2600)}}{2 \times 1}$$

$$= \frac{-25 \pm \sqrt{11025}}{2} = \frac{-25 \pm 105}{2}$$

Either $x = \frac{-25 + 105}{2} = 40\,m$

or $x = \frac{-25 - 105}{2} = -65\,m$

Rejecting the negative value, **x = 40 m**

Length = 40 + 20 = 60m; Width = 40 + 5 = 45 m

Exercise 4.1

1. Solve the following simultaneous equations by substitution:
   a) $x + y = 3$ and $x - y = -1$
   b) $2x + y = -1$ and $x + 2y = 1$
2. Solve the following equations by elimination/substitution:
   a) $x + y = 5$ and $3x - 2y = -5$
   b) $x + y = 1$ and $3x - y = -5$
   c) $2x - 1 = y$ and $x - 2y = -2.5$
   d) $5y = 0.5 - 4x$ and $2x = 4 - 4y$
   e) $1.5a + 2.5b = 3$, and $2a + 3.5b = 5$
3. 2 doors and 1 window cost £260 at B & B DIY store. At another DIY store, where prices are similar to those at B & B store, 3 doors and 2 windows cost £430. Find the cost of 1 door, and 1 window.
4. The demand and supply of housing in a region can be represented by simultaneous equations, P + 29 Q = 800 000, and P – 27 Q = 100 000 respectively. P represents the price in Pounds Sterling and Q represents the quantity, i.e. the number of houses. Find the price and quantity at which the supply equals the demand.
5. The shear strength of soils is given by the equation, $s = c + nR$, where
   s = shear strength (kN/m^2); c = cohesion (kN/m^2); n = normal stress (kN/m^2);
   R = shearing resistance component. Two tests were performed on samples of a particular soil to yield the following data:

Test 1: $s = 320$ kN/m^2 at $n = 310$ kN/m^2

Test 2: $s = 200$ kN/m^2 at $n = 110$ kN/m^2

Form 2 equations from the above data and calculate the values of c and R.

6. The number of bricks (N) at a site (existing stock + delivered) is represented by the equation: $N = a D + e$
   where D = number of days; a = constant; e = existing stock before the delivery program started.
   After 2 days the number of bricks at the site was 22 000 and after 5 days the number of bricks was 40 000. Assuming that the bricks are delivered to the site regularly, calculate the values of a and e, and hence calculate the number of bricks after 8 days.

Exercise 4.2

1. Solve the following equations by factorisation:
   a) $x^2 - 4x + 3 = 0$
   b) $x^2 + x - 20 = 0$
   c) $x^2 + 8x + 12 = 0$
   d) $4x^2 - 4x - 3 = 0$
2. Use the quadratic formula to solve the following equations:
   a) $3x^2 + 10x - 8 = 0$
   b) $6x^2 + 9x - 6 = 0$
   c) $2x^2 - 3x - 5 = 0$
   d) $4x^2 + 8x = -3$
   e) $x^2 = -5x - 6$
3. Solve the following equations by completing the square
   a) $x^2 + 4x = 5$
   b) $x^2 + 2x = 3$
   c) $x^2 - 6x = 7$
4. The perimeter of a rectangular plot of land is 84 m and the length of its diagonal is 30 m. Calculate the dimensions of the plot.
5. The perimeter of a rectangle is 21 m and its area 24.5 m^2. Calculate the dimensions of the rectangle.
6. The owner of a house requires 121 m long fencing on three sides of the back garden. Find the length and the width of the garden if its area is 605 m^2.
7. The bending moment (M) at any point on one particular beam is given by:
   $M = 9x - 1.5x^2$
   where x is the distance between the point and the left support.
   Find the distance, from the left support, where the bending moment is 12 kN-m.
8. The cross-section of an irrigation canal is shown in Figure 4.2. If the shaded portion shows the cross-sectional area of the water flow:
   a) obtain an expression, in terms of x, for the cross-sectional area of the water flow.
   b) calculate the value of x if the cross-sectional area of flow is 5.4 m^2.

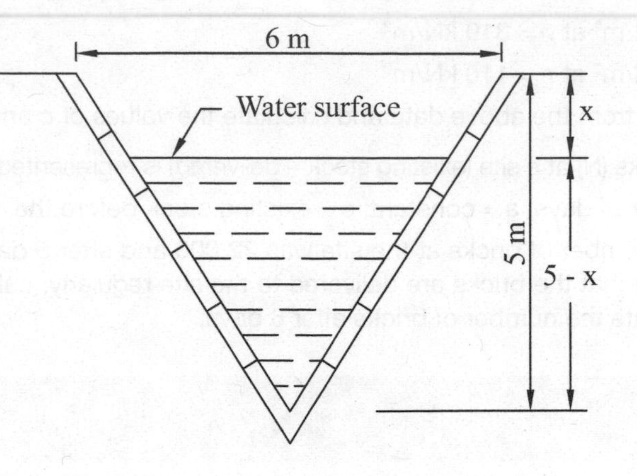

Figure 4.2 Irrigation canal

Answers – Exercise 4.1

1. a) $x = 1, y = 2$ b) $x = -1, y = 1$
2. a) $x = 1, y = 4$ b) $x = -1, y = 2$ c) $x = 1.5, y = 2$
   d) $x = -3, y = 2.5$ e) $x = -8, y = 6$
3. Door = £90, Window = £80
4. Price = £437 500, Quantity = 12 500
5. $c = 134$ kN/m^2, $R = 0.6$
6. $a = 6000$, $e = 10\,000$, $N = 58\,000$

Answers – Exercise 4.2

1. a) $x = 1$ or $x = 3$ b) $x = -5$ or $x = 4$ c) $x = -6$ or $x = -2$
   d) $x = 1.5$ or $x = -0.5$
2. a) $x = 2/3$ or $x = -4$ b) $x = 0.5$ or $x = -2$ c) $x = 2.5$ or $x = -1$
   d) $x = -0.5$ or $x = -1.5$ e) $x = -2$ or $x = -3$
3. a) $x = 1$ or $x = -5$ b) $x = 1$ or $x = -3$ c) $x = -1$ or $x = 7$
4. Length = 24 m, width = 18 m
5. Length = 7 m, width = 3.5 m
6. Length = 55 m, width = 11 m
7. $x = 4$ m or 2 m
8. a) $0.6x^2 - 6x + 15$ b) $x = 2$ m

CHAPTER 5

Graphical solutions

Topics covered in this chapter:

- Straight line graphs and determination of the law of a straight-line
- Solution of simultaneous, quadratic and cubic equations by graphical technique
- Plotting of experimental data and production of the 'best-fit' line

5.1 Introduction

A graph represents the relationship between two quantities and basically shows how 1 quantity varies relative to the other quantity. A graph consists of 2 axes, the x-axis and the y-axis, drawn at right angles to each other. The 2 axes cross at the origin, and are known as rectangular or cartesian axes (Figure 5.1). As a graph is used to show the relationship between 2 variables, 1 variable is represented along the x-axis and the other variable represented along the y-axis.

If a set of values, x, is connected to another set of values, y, and for each value of x there is only 1 value of y, then y is known as a function of x. If the values of x are continuous, a graph can be produced to represent the function.

The data is collected and plotted as points, each point having 2 values: the 'x' value and the 'y' value. These values are known as the x-coordinate and the y-co-ordinate; if the co-ordinates of a point are (2, 1), then:

i) The first number, i.e. 2, is known as the 'x co-ordinate'.
ii) The second number, i.e. 1, is known as the 'y co-ordinate'.

This is shown as point A in Figure 5.1. Similarly, if the x co-ordinate of point B is –4 and its y co-ordinate is 3, these are written as (–4, 3). For establishing the x co-ordinate on the graph, we move horizontally. Positive values of x are shown on the right hand side of the origin whereas the negative values of x are shown on the left hand side of the origin. The positive values of y are plotted above the origin and its negative values plotted below the origin. Figure 5.1 also shows points C and D having co-ordinates (–3, –3) and (3, –3) respectively.

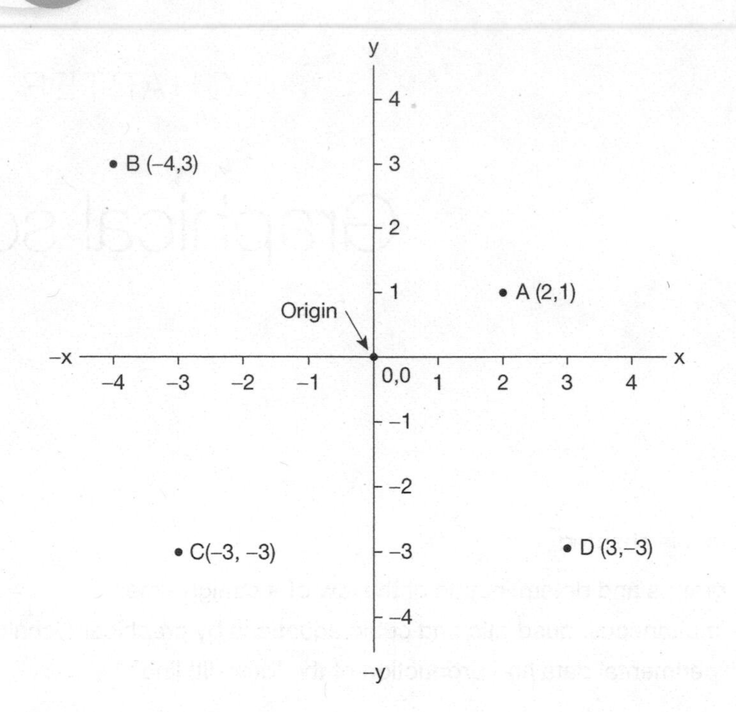

Figure 5.1

Example 5.1

The air temperatures on a typical day, last autumn were:

Time	0800	1000	1200	1400	1600	1800	2000	2200
Temperature (°C)	–1	1	3	7	6	3	0	–1

Plot a graph to show the air temperature against time.

Solution:

On a graph paper the axes are drawn as shown in Figure 5.2. The x-axis is normally used to represent the independent variable, i.e. time, and the y-axis used to represent the dependent variable, i.e. temperature. The first point on the graph should represent a temperature of –1°C at 0800 hours. To plot this point draw a vertical line at 0800 hours and a horizontal line at –1°C, and wherever the lines meet mark that point with a dot or a cross. This point represents a temperature of – 1°C at 0800 hours, and similarly the other points are established on the graph paper. Finally, join the points to show how the air temperature varies with time.

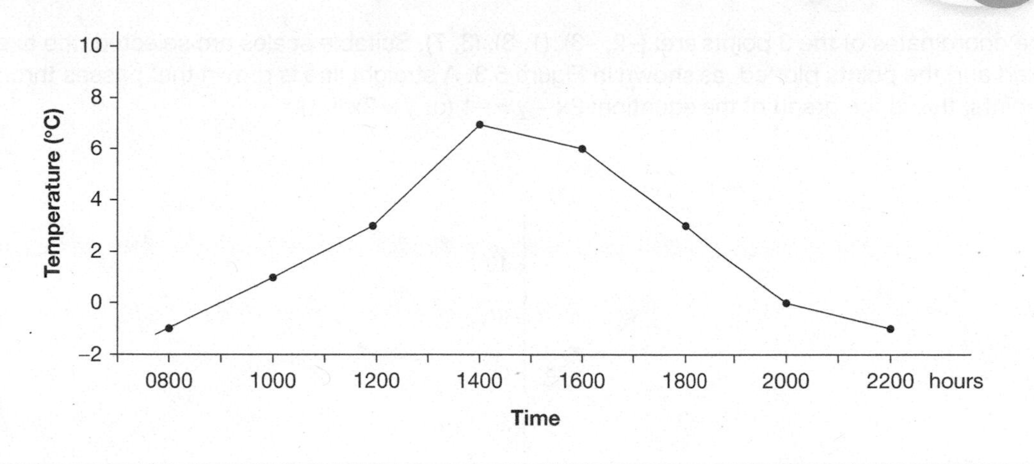

Figure 5.2

5.2 Linear equations

The general form of a linear equation is $ax + by = c$, where a, b and c are constants. The equation is rearranged so that the values of y can be determined corresponding to assumed values of x. Since the graph will be a straight line, only 3 points are required, as illustrated in example 5.2.

Example 5.2

Draw the graphs of: a) $2x - y = -1$, from $x = -2$ to $x = 3$
b) $2x + y = 4$, from $x = -2$ to $x = 4$

Solution:

Before the points are plotted on a graph paper and the graph produced, it is necessary to use at least 3 values of x from the given range, and find the corresponding values of y. To make this process easier the equations may be written as:

$y = 2x + 1$ (equation a) and $y = 4 - 2x$ (equation b)

a) If $x = -2$, $y = (2 \times -2) + 1$
$= -4 + 1 = -3$

Similarly for $x = 1$ and 3, the corresponding values of y are determined, which are shown in Table 5.1.

Table 5.1

x	−2	1	3
y	−3	3	7

The coordinates of the 3 points are: (–2, –3), (1, 3), (3, 7). Suitable scales are selected, the axes marked and the points plotted, as shown in Figure 5.3. A straight line is drawn that passes through all the points; this is the graph of the equation: $2x - y = -1$ (or $y = 2x + 1$)

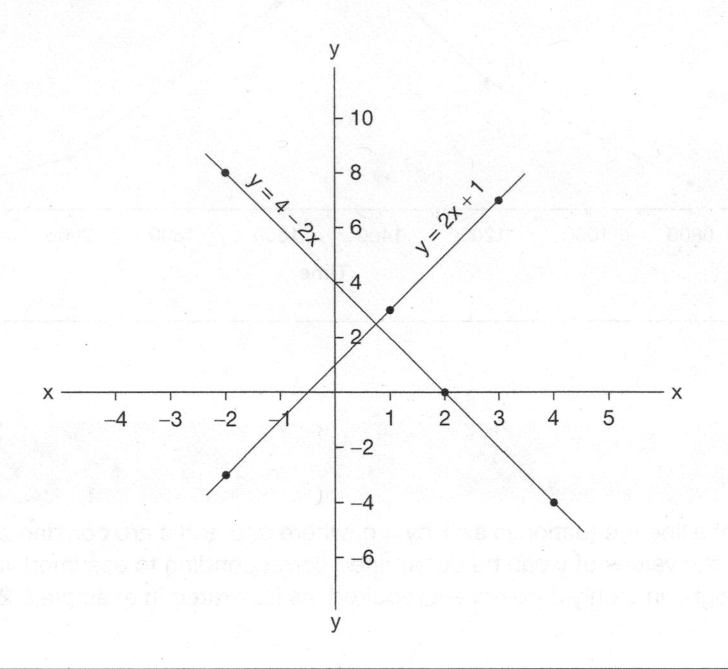

Figure 5.3

b) The process explained above is used for producing the graph of $y = 4 - 2x$.

If $x = -2$, $\quad y = 4 - (2 \times -2)$

$\quad = 4 + 4 = 8$

The other values of x and y are as shown in Table 5.2.

Table 5.2

x	–2	2	4
y	8	0	–4

These points are plotted to get a straight line graph, as shown in Figure 5.3.

5.3 Linear simultaneous equations

The analytical methods of solving linear simultaneous equations are explained in chapter 4; in this section the graphical method of solving the equations will be described. The graphs of linear simultaneous equations are drawn in the same way as explained in section 5.2. The graphs are plotted on the same axes and the co-ordinates of the point of intersection are determined. The point of intersection is common to both lines and hence is the solution.

Example 5.3

Solve graphically the simultaneous equations:

$x - 3 = -2y$ and $x + y = 4$

Solution:

The equations may be given in any form, but it is necessary to rearrange each equation into the form of the straight line law, i.e. $y = mx + c$:

Equation 1: $2y = -x + 3$ or $y = \dfrac{-x+3}{2}$

Equation 2: $y = -x + 4$

At least three values of x are assumed in each case and the corresponding values of y determined:

Equation 1: If $x = 1$, $y = \dfrac{-1+3}{2} = \dfrac{2}{2} = 1$

If $x = 3$, $y = \dfrac{-3+3}{2} = \dfrac{0}{2} = 0$

Similarly when $x = 5$, $y = -1$

The co-ordinates for equation 2 are calculated in a similar manner. The following tables (Tables 5.3a, 5.3b) summarise the calculations:

Table 5.3a

Equation 1	x	1	3	5
	$y = \dfrac{-x+3}{2}$	1	0	−1

Table 5.3b

Equation 2	x	1	3	5
	$y = -x + 4$	3	1	−1

The graphs are plotted as shown in Figure 5.4. The solution of the 2 equations is the point of intersection (5, –1). Therefore **$x = 5$ and $y = -1$.**

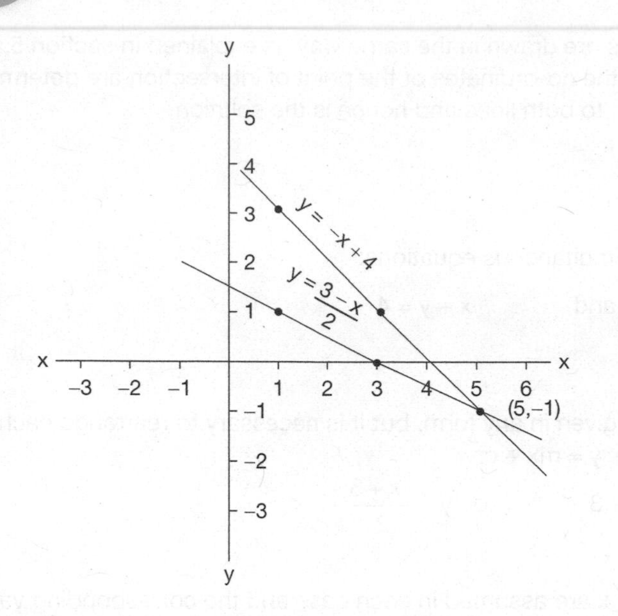

Figure 5.4

5.4 The law of straight line

The general equation of the straight line (or linear function) is given by:

$y = mx + c$, where x and y are independent and dependent variables, respectively.
m is the slope or gradient of the straight line.
c is the intercept made by the straight line on the y-axis.

Any equation that conforms to this form will produce a straight-line graph, for example, the graphs shown in Figures 5.3 and 5.4.

Figure 5.5 shows two straight-line graphs, one with positive slope and the other with negative slope. Graph (a) slopes upwards from left to right; its slope (or gradient) is taken as positive. Graph (b) slopes downwards from left to right, therefore its gradient is considered to be negative. The gradient (m) of a line graph can be determined by drawing a right-angled triangle, of any size, as shown in Figure 5.5 and is given by:

$$m = \frac{\text{Vertical side}}{\text{Horizontal side}} = \frac{AC}{BC} = \frac{1.5}{3} = 0.5$$

In some cases when data from experimental work is plotted, a straight-line graph is produced. If the task involves the determination of the equation of the straight line, then it is necessary to determine the gradient (m) and the intercept on the y-axis (c) due to the straight-line graph. The law of the straight line is determined by incorporating the values of m and c into the equation $y = mx + c$.

Example 5.4

A 10 cm long sample of a ductile material was stretched by a tensile force. The lengths of the rod at different levels of the tensile force were:

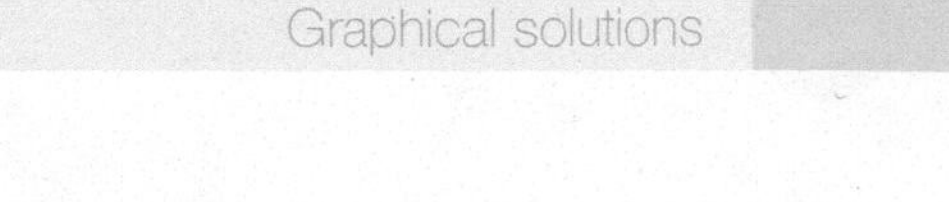

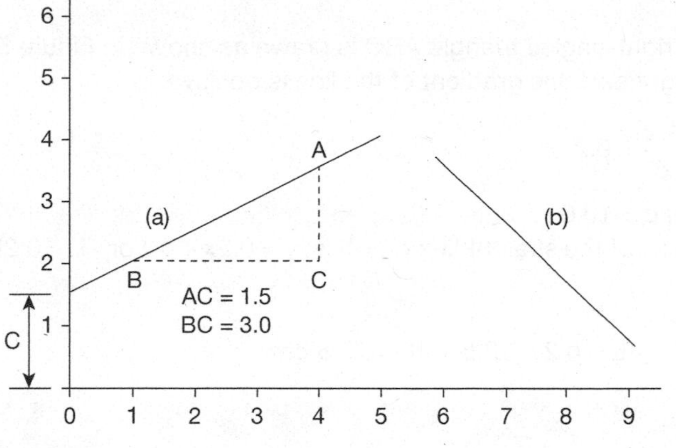

Figure 5.5

Length, L (cm)	10	10.4	10.8	11.2	11.6
Force, F (Newtons)	0	2	4	6	8

a) Show that the equation connecting length (L) and force (F) is of the form:
   $L = mF + c$

b) Find the equation of the graph.

c) Find the length of the rod if a force of 12.5 N acts on it

Solution:

a)

Length (L) is plotted on the y-axis and the force (F) plotted on the x-axis.

Suitable scales are selected and marked on the axes as shown in Figure 5.6. It is evident on plotting the points that they follow a straight line. Therefore, it can be concluded that the equation connecting L and F is of the form: $L = mF + c$.

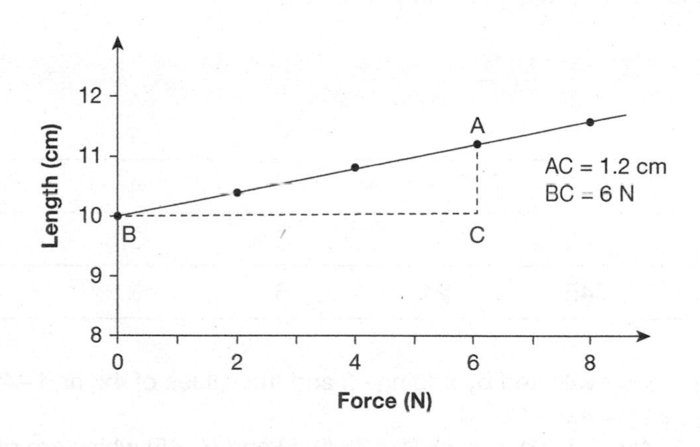

Figure 5.6

b)

To find the gradient, a right-angled triangle ABC is drawn as shown in Figure 5.6. As the left end of the line is lower than the right end, the gradient of the line is positive.

$$\text{Gradient } m = \frac{AC}{BC} = \frac{1.2}{6} = 0.2$$

Intercept on the y-axis, $c = 10.0$

Therefore the equation of the straight line graph is: $y = 0.2x + 10$ or, $L = 0.2F + 10$

c)

When $F = 12.5$ N, $L = 0.2 \times 12.5 + 10 =$ **12.5 cm**

5.5 Quadratic equations

The graphical method, although time consuming, can be used to solve any quadratic equation. There is only 1 unknown (e.g. x) in a quadratic equation, but for plotting a graph we need the values of y co-ordinates as well. Therefore, for graphical solution the quadratic is equated to y. For example, if we are asked to solve the equation

$4x^2 - 4x - 3 = 0$, then as a starting point we say that:

$y = 4x^2 - 4x - 3 = 0$

At least 6 or 7 values of x are assumed and the corresponding values of y are determined. The points are plotted on a graph and joined by a smooth curve. As $y = 0$, the solution of the equation is determined by finding the values of x co-ordinates where the curve meets the x-axis.

Example 5.5

Solve the equation $4x^2 - 4x - 3 = 0$ by graphical method.

Solution:

Assume any reasonable values of x, for example we can start off by assuming –3, –2, –1, 1, 2, 4 and then the corresponding values of y are calculated as shown below. Depending on the shape of the graph further points may be necessary to find the solution.

x	–3	–2	–1	1	2	4
$4x^2$	36	16	4	4	16	64
$-4x$	12	8	4	–4	–8	–16
–3	–3	–3	–3	–3	–3	–3
$y = 4x^2 - 4x - 3$	**45**	**21**	**5**	**–3**	**5**	**45**

The equation $4x^2 - 4x - 3$ is evaluated by adding –3 and the values of $4x^2$ and $-4x$. These are also the values of y coordinate, as $y = 4x^2 - 4x - 3$.

We have 6 points (–3, 45), (–2, 21), (–1, 5), (1, –3), (2, 5) and (4, 45) which are plotted as shown in Figure 5.7. A smooth curve is drawn passing through all the points.

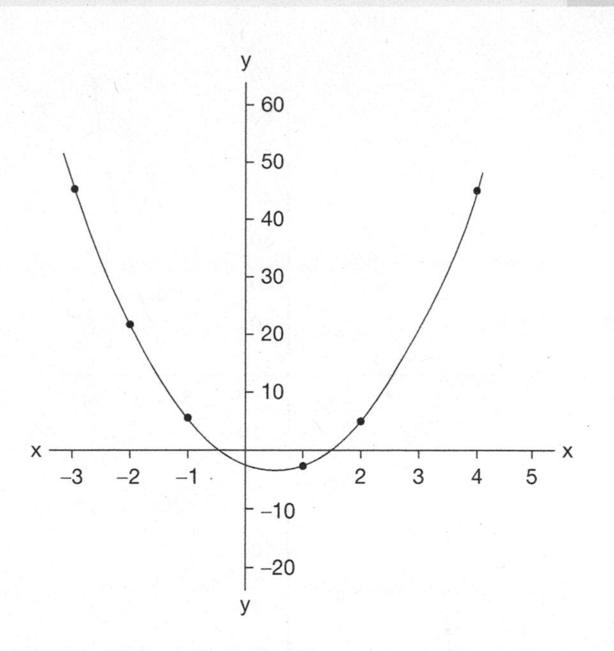

Figure 5.7

The curve crosses the x-axis at $x = -0.5$ and $x = 1.5$, therefore the solution of equation $4x^2 - 4x - 3$ is: $\mathbf{x = -0.5}$ **and** $\mathbf{x = 1.5}$

5.6 Cubic equations

The general form of a cubic equation is $ax^3 + bx^2 + cx + d = 0$, where a, b, c and d are constants. The values of b, c, and d may be equal to 0, but constant a must have a value other than 0. A cubic equation can have 1, 2 or 3 real roots. If a cubic equation has 3 real roots, then their values will be different. However, there are cases where we could have only 1 or 2 solutions. A range of methods are available for solving a cubic equation, but here the graphical method will be used.

Example 5.6

Solve the equation $2x^3 - 2x^2 - 8x + 4 = 0$, graphically.

Solution:

Let $y = 2x^3 - 2x^2 - 8x + 4 = 0$

$x = -3, \quad y = 2(-3)^3 - 2(-3)^2 - 8(-3) + 4 = -54 - 18 + 24 + 4 = -44$

$x = -2, \quad y = 2(-2)^3 - 2(-2)^2 - 8(-2) + 4 = -16 - 8 + 16 + 4 = -4$

$x = 0, \quad y = 2(0)^3 - 2(0)^2 - 8(0) + 4 = 4$

$x = 2, \quad y = 2(2)^3 - 2(2)^2 - 8(2) + 4 = 16 - 8 - 16 + 4 = -4$

$x = 4, \quad y = 2(4)^3 - 2(4)^2 - 8(4) + 4 = 128 - 32 - 32 + 4 = 68$

The points are plotted and joined by a smooth curve as shown in Figure 5.8.

The solution of the equation is: $x = -1.8$, or $x = 0.5$, or $x = 2.35$

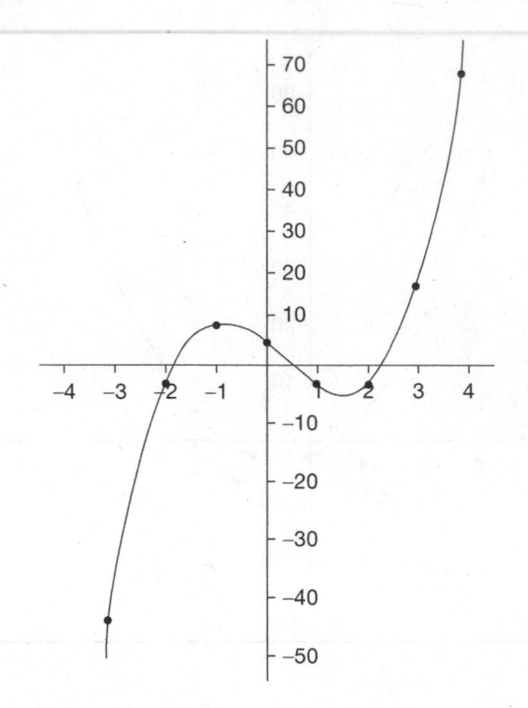

Figure 5.8

5.7 Curve fitting

When data is collected from experimental work or surveys and plotted on a graph, very often a relationship is found to exist between the variables. The data may not lie exactly on a straight line or a curve, even if there is a relationship between the variables. This may be visualised by drawing an approximating curve; Figure 5.9 shows some of the possibilities.

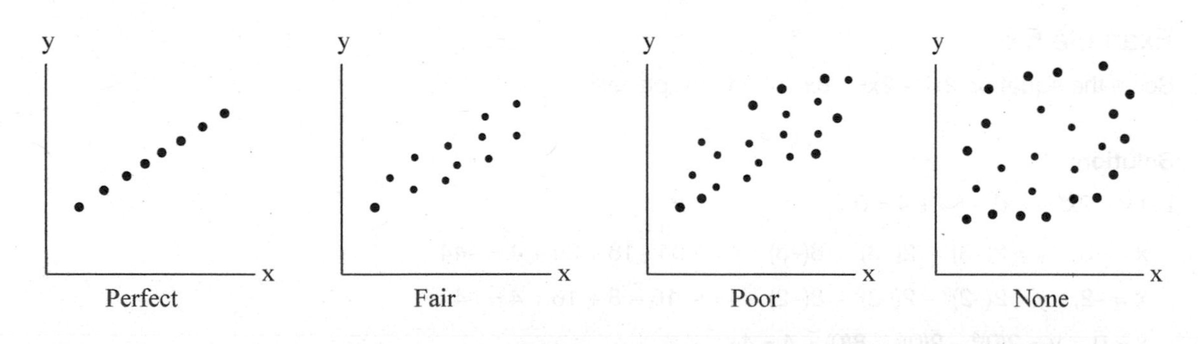

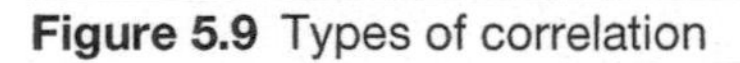

Figure 5.9 Types of correlation

Curve fitting is the process of finding equations of the curves which fit the given data. The method of least squares is a technique which avoids individual judgement in drawing lines, parabolas and other curves to fit the given data. If X and Y are the independent and the dependent variables, the quantity: $D_1^2 + D_2^2 + \dots\dots\dots + D_n^2$ must be minimum for a best fitting curve; see Figure 5.10.

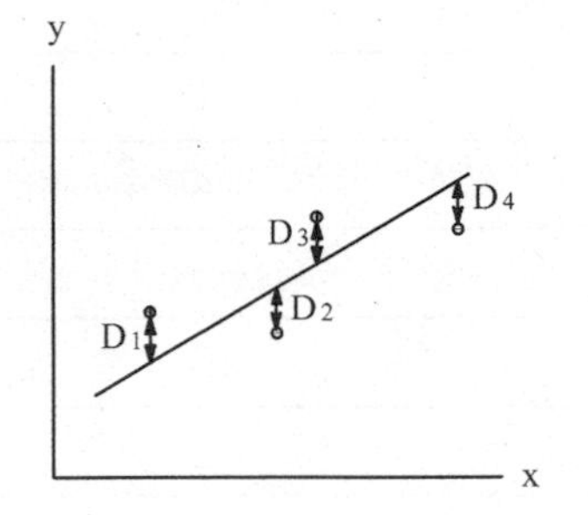

Figure 5.10

The least squares line approximating the points (X_1, Y_1), (X_2, Y_2), ………, (X_n, Y_n) has the equation:

$Y = a + bX$

This is same as the equation $y = mx + c$. Here, a is the intercept on the y-axis, and b is the slope of the straight line. The values of a and b are determined from the equations: $\Sigma Y = a\,N + b\,\Sigma\,X$ and $\Sigma\,XY = a\,\Sigma\,X + b\,\Sigma\,X^2$, and substituted into

$Y = a + bX$ to produce the 'best fit' straight line. The line is called the linear regression of y on x. Similarly y can be taken as the independent variable and the line known as the regression of x on y produced.

Example 5.7

In an experiment to prove Hooke's law the following data shown in Table 5.4 was obtained by stretching a spring:

Table 5.4

Applied force (Newtons)	0	1	2	3	4	5
Extension (mm)	0	10	19	23	35	39

a) Plot the data showing the force on the x-axis and the extension on the y-axis.
b) Find the equation of the least squares line.
c) Plot the least squares line.

Solution:

a) The points are plotted by taking force on the x-axis and extension on the y-axis, as shown in Figure 5.11.
b) Table 5.5 is prepared first to solve the equations already explained in section 5.7,

i.e. $\Sigma Y = a\,N + b\,\Sigma\,X$ (1)

$\Sigma\,XY = a\,\Sigma\,X + b\,\Sigma\,X^2$ (2)

Table 5.5

X	Y	X^2	XY
0	0	0	0
1	10	1	10
2	19	4	38
3	23	9	69
4	35	16	140
5	39	25	195
$\Sigma X = 15$	$\Sigma Y = 126$	$\Sigma X^2 = 55$	$\Sigma XY = 452$

The above values are substituted in equations 1 and 2:

$126 = a \times 6 + b \times 15$

$452 = a \times 15 + b \times 55$

Solving these equations simultaneously, a = 1.43 and, b = 7.83

The equation of the least squares line (or the best fit line) is:

$\mathbf{Y = 1.43 + 7.83\,X}$

c) We need the co-ordinates of two points to draw the best fit line:

When X = 1.5, Y = 1.43 + 7.83 × 1.5 = 13.18 (1.5, 13.18)

When X = 4, Y = 1.43 + 7.83 × 4 = 32.75 (4, 32.75)

The 2 points are plotted on the graph (shown as crosses in Figure 5.11) and a straight line drawn through these points. This line is the least squares line or the regression line.

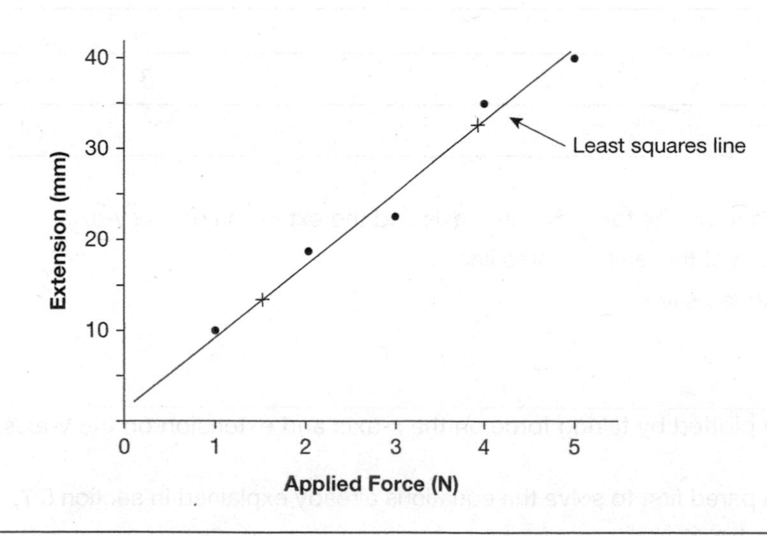

Figure 5.11

Exercise 5.1

1. A test was conducted to determine the effect of water content on the density of compacted clay. The results are shown in Table 5.6:

Table 5.6

Density (kg/m^3)	1800	1840	1870	1870	1840
Water content (%)	9	11	13	15	17

Draw a graph between density and water content taking water content on the x-axis.

2. Draw the graphs of the following equations:
   a) $y = 2x - 3$, taking values of x between –2 and 4
   b) $y = 4 - 0.5x$, taking values of x between –2 and 4

3. In an experiment to prove Hooke's law, the following data (Table 5.7) was obtained by stretching a spring:

Table 5.7

Stretching force, F (Newton)	0	1	2	3	4	5	6
Extension, L (mm)	0	8.5	16	24.5	33	41.5	50

   a) Plot a graph of stretching force (y-axis) against extension (x-axis) and find the gradient of the straight line.
   b) Show that the law connecting F and L is of the form $F = mL$, and find the law.

4. Solve the following simultaneous equations graphically:
   a) $x + y = 5$, and $3x - 2y = -5$
   b) $x + y = 1$, and $3x - y = -5$

5. Solve graphically: a) $x^2 + x - 20 = 0$
   b) $3x^2 + 10x - 8 = 0$

6. Solve, graphically, the equation $0.5x^3 - 2x^2 - 6x + 6 = 0$

7. A soil sample was tested to determine its shear strength and the data shown in Table 5.8 was obtained:

Table 5.8

Shear stress (kN/m^2)	110	190	220	280	380	430
Normal stress (kN/m^2)	0	100	200	300	400	500

a) Plot the data showing the shear stress on the y-axis and the normal stress on the x-axis.

b) Find the equation of the least squares line.

c) Plot the least squares line.

Answers – Exercise 5.1

1. Refer to Figure S5.1 (Appendix 3)
2. Refer to Figure S5.2 (Appendix 3)
3. a) $m = 0.12$

   b) $F = 0.12L$
4. a) $x = 1$ and $y = 4$

   b) $x = -1$ and $y = 2$
5. a) $x = -5$ or $x = 4$

   b) $x = -4$ or $x = 0.7$
6. $x = -2.5$ or 0.8 or 5.7
7. a) Refer to Figure S5.7 (Appendix 3)

   b) $Y = 109.08 + 0.637X$

   c) Refer to Figure S5.7 (Appendix 3)

CHAPTER 6

Geometry, areas and volumes

Topics covered in this chapter:

- Properties of angles, triangles, quadrilaterals and circles
- Pythagoras' theorem and its application
- Area of regular and irregular shapes
- Volume of regular and irregular objects

6.1 Geometry

Geometry is a branch of mathematics that studies the properties of points, lines, angles, shapes and space. Shapes are studied in 2 dimensions and 3 dimensions. 2-dimensional shapes have length and width, such as rectangles, triangles etc., whereas 3 dimensional shapes (cubes, cylinders, pyramids etc.) have length, width and height. Geometry plays a very important role in the study of several areas of construction/civil engineering, for example land surveying, quantity surveying and construction technology cannot be studied without a knowledge of lines, angles, areas and volumes.

6.1.1 Angles

An angle is formed when two straight lines meet at a point. Figure 6.1 shows two lines PQ and PR meeting at point P, and hence producing angle QPR, which may be denoted as ∠QPR or ∠P. The size of the angle depends on the amount of rotation of line PQ from line PR. A complete revolution by line PQ will produce an angle of 360 degrees, denoted as 360°.

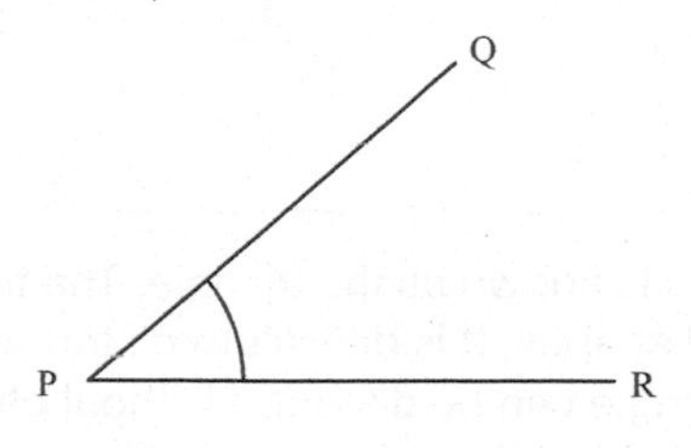

Figure 6.1

The basic unit for measuring angles is a degree, but for more accurate work, minutes and seconds are also used:

$$1° = \frac{1}{360} \text{ of a complete revolution}$$

$$1° = 60 \text{ minutes } (60')$$

$$1' = 60 \text{ seconds } (60'')$$

If an angle measures 20 degrees, 30 minutes and 50 seconds, then in the calculation it is written as 20°30′50″.

An angle may be measured in radians as well:

$$\pi \text{ radians} = 180°$$

$$1 \text{ radian} = \frac{180°}{\pi} = 57.30° (\text{correct to 2 d.p.})$$

$$1° = \frac{\pi}{180} \text{ radians}$$

Example 6.1

Convert a) 60° 45′ 30″ into radians

b) $\frac{2\pi}{3}$ radians into degrees.

Solution:

a) $30'' = \frac{30}{60} = 0.5'$

$45' + 0.5' = 45.5'$

$45.5' = \frac{45.5}{60} = 0.7583°$, Therefore $60°45'30'' = 60.7583°$

$60.7583° = 60.7583° \times \frac{\pi}{180} = \mathbf{1.06\ radians}$

The question can also be solved by converting 60° 45′ 30″ into decimal format with a scientific calculator, and then multiplying it with $\frac{\pi}{180}$

b) $1 \text{ radian} = \frac{180}{\pi} \text{ degrees}$

$$\frac{2\pi}{3} \text{ radians} = \frac{2\pi}{3} \times \frac{180}{\pi} = 120°$$

6.1.2 Triangles

A figure enclosed by 3 straight lines is known as the triangle. The triangular shape is more stable as compared to the other geometrical shapes. It is difficult to distort a triangle without changing the length of one of its sides, whereas a rectangle can be distorted without changing the length of its sides. A roof truss is triangular in shape as the slope of the sides can drain rain water quickly, but also, the triangular shape provides stability to the roof trusses and hence the roof structure. The stability of the triangular shape is also used in the fabrication of steel frames for multi-storey buildings. The rectangular grids of the selected steel frames of a building are changed into triangles by fixing extra structural sections – a process known as 'triangulation'. The use of extra structural sections is called 'wind bracing'.

The main types of triangles are:

a) **Acute angled triangle:** Each angle of the triangle is less than 90°.
b) **Obtuse angled triangle**: One angle is more than 90°.
c) **Right-angled triangle**: One of the angles is equal to 90°.
d) **Scalene triangle:** All angles and all sides of the triangle are unequal. Acute angled, obtuse angled and right-angled triangles can be scalene triangles as well. The 3-4-5 triangle is a right-angled as well as a scalene triangle.
e) **Equilateral triangle:** All sides and all angles are equal. Each angle is a 60° angle.
d) **Isosceles triangle:** Two sides and two angles equal.

The conventional method of denoting the angles of a triangle is to use capital letters, e.g. A, B, C, P, Q, R etc. The sides are denoted by lower case letters; side **a** is opposite angle A; side **b** is opposite angle B; side **p** is opposite angle P and so on, as illustrated in Figure 6.2.

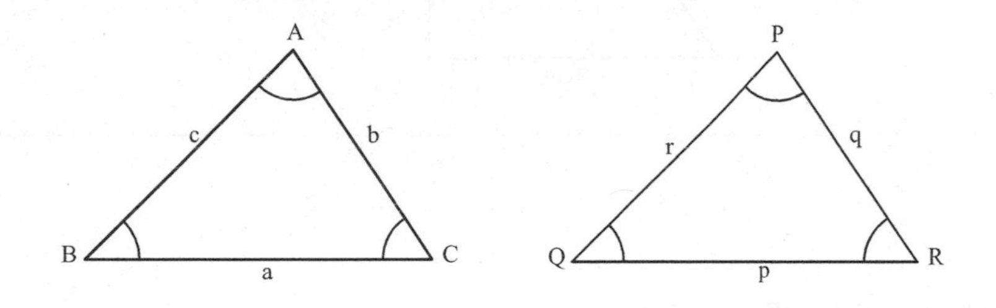

Figure 6.2

The sum of the 3 internal angles of a triangle is 180°:

$\angle A + \angle B + \angle C = 180°$

$\angle P + \angle Q + \angle R = 180°$

6.1.3 Similar triangles

2 triangles are similar if the angles of 1 triangle are equal to the angles of the other triangle. The sides of similar triangles are unequal, but proportional because of the same shape, as shown in Figure 6.3.

$$\frac{a}{d} = \frac{b}{e} = \frac{c}{f}$$

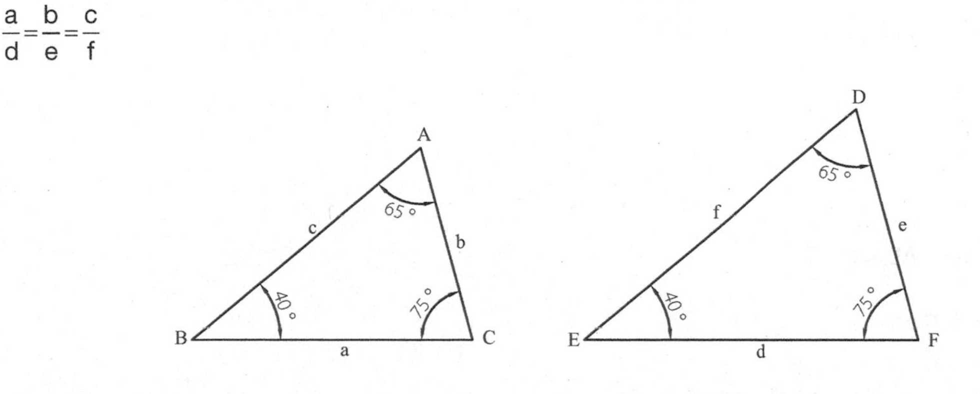

Figure 6.3

Example 6.2

Figure 6.4 shows triangle ABC with a line DE forming another triangle. If lines BC and DE are parallel:

a) Show that triangles ADE and ABC are similar.

b) Calculate AD and AE.

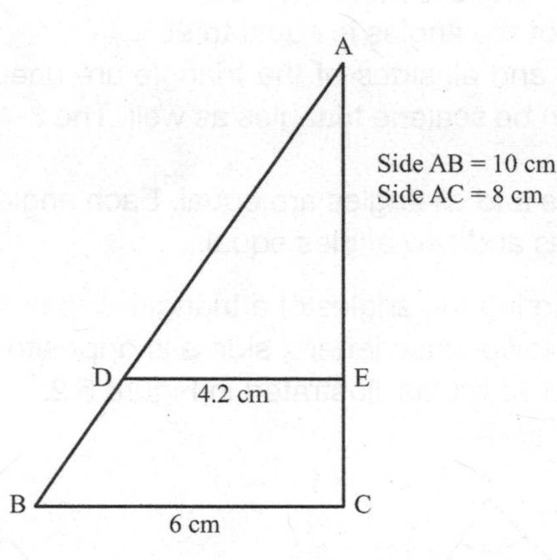

Figure 6.4

Solution:

a) Line BC is parallel to line DE, therefore:

$\angle D = \angle B$, and $\angle E = \angle C$ (corresponding angles)

$\angle A$ is common to both triangles.

As the angles in the two triangles are equal, ΔADE and ΔABC are similar.

b) As ΔABC and ΔADE are similar, their sides should be proportional.

$$\frac{BC}{DE} = \frac{AB}{AD}$$

$$\frac{6}{4.2} = \frac{10}{AD}$$

Transposing, $6 \times AD = 4.2 \times 10$

$$AD = \frac{4.2 \times 10}{6} = \mathbf{7.0\ cm}$$

Also, $\frac{AB}{AD} = \frac{AC}{AE}$

$$\frac{10}{7} = \frac{8}{AE}$$

Transposing, $10 \times AE = 8 \times 7$

$$AE = \frac{8 \times 7}{10} = \mathbf{5.6\ cm}$$

6.1.4 Pythagoras' theorem

Pythagoras' theorem states that in a right-angled triangle the area of the squares on the hypotenuse is equal to the sum of the areas of the squares on the other sides.

$c^2 = a^2 + b^2$

$(\text{Hypotenuse})^2$ = Sum of the squares of the other two sides

Hypotenuse, the longest side of a right-angled triangle, is always opposite the right angle (Figure 6.5a).

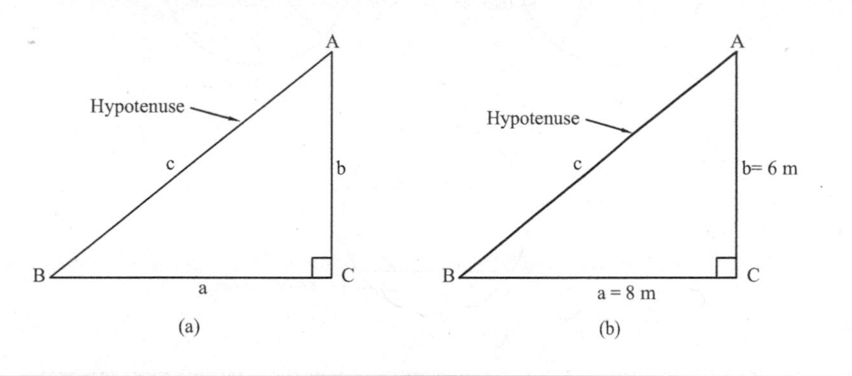

Figure 6.5

The theorem is based on the areas of the squares on the sides of right-angled triangles, but is used mainly for calculating the sides of such triangles. The principle behind the theorem is used in setting out small buildings by making use of the 3-4-5 triangle. In this triangle, if 2 sides that enclose the right-angle, measure 3 m and 4 m, the third side (hypotenuse) will measure 5 m. This is true for all units of measurement.

Example 6.3

Figure 6.5b shows a right-angled triangle with sides 6 m and 8 m. Show that the length of the hypotenuse (side AB) is 10 m.

Solution:

From Pythagoras' theorem:

$(\text{Hypotenuse})^2$ = Sum of the squares of the other 2 sides

$c^2 = 6^2 + 8^2 = 36 + 64 = 100$

$c^2 = 100$, therefore, $c = \sqrt{100} = \mathbf{10\ m}$

6.1.5 The circle

A plane figure enclosed by a curved line so that every point on the curve is equidistant from the centre is known as a circle (Figure 6.6a). Line OA is called the radius (r); all lines drawn this way are equal:

OA = OB = OC = OD (Figure 6.6a)

Any straight line that passes through the centre and touches the curved line at 2 points is known as the diameter (d). The diameter is twice the length of the radius:

d = 2 × radius

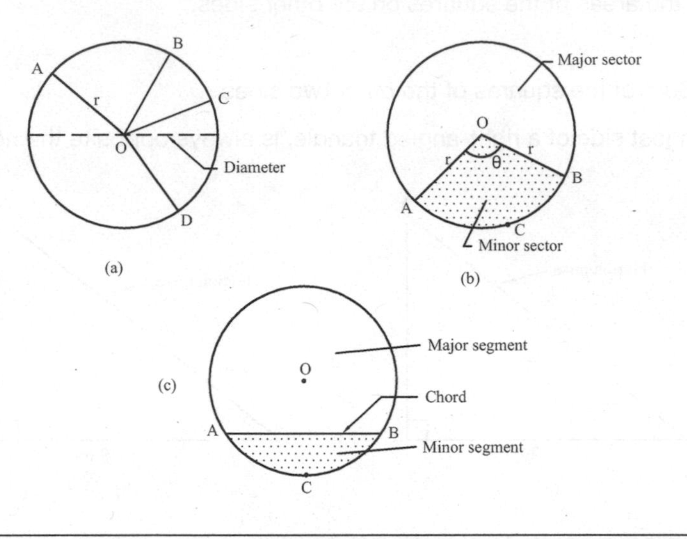

Figure 6.6

The length of the curved line is known as the circumference (c) of the circle. The ratio of the circumference to the diameter of a circle is a constant and denoted by the Greek letter π (pronounced as pie):

$$\frac{c}{d} = \pi \text{ or } c = \pi \times d$$

As $d = 2r$, therefore $c = \pi \times 2r = 2\pi r$

The value of π is 3.14159 (5 d.p.)

As explained in section 6.1, an angle may be measured in either radians or degrees. In Figure 6.6b, if l is the length of arc ACB and r the radius of the circle, then angle θ is given by:

$$\theta\,(\text{in radians}) = \frac{l}{r}$$

The part of a circle between 2 radii is known as the sector. If the sector is smaller than the semi-circle then it is called a minor sector, otherwise it is called a major sector (Figure 6.6b). ∠AOB in a **minor sector** is less than 180°, whereas in a **major sector** this angle is greater than 180°.

$$\text{Area of minor sector} = \text{Area of the circle} \times \frac{\theta}{360^\circ} \qquad (\text{angle } \theta \text{ in degrees})$$

$$= \text{Area of the circle} \times \frac{\theta}{2\pi} \qquad (\text{angle } \theta \text{ in radians})$$

$$= \pi r^2 \times \frac{\theta}{2\pi} = \frac{1}{2} r^2 \theta \qquad (\text{angle } \theta \text{ in radians})$$

Any straight line that divides a circle into 2 parts is called a chord. In Figure 6.6c line AB is the chord, which divides the circle into minor and major segments. An arc is a portion of the circumference of a circle, for example portion ACB in Figure 6.6c.

Area of the shaded segment (Figure 6.7a) is:

= Area of sector OACB – Area of triangle OAB

$$= \frac{1}{2}r^2\theta - \frac{1}{2}r^2\sin\theta$$

$$= \frac{1}{2}r^2(\theta - \sin\theta)\ (\text{angle } \theta \text{ is in radians})$$

A line that meets the circle at 1 point, without cutting it, is known as the tangent (Figure 6.7b). The radius drawn from point A, where the tangent meets the circle, is at right angles to the tangent.

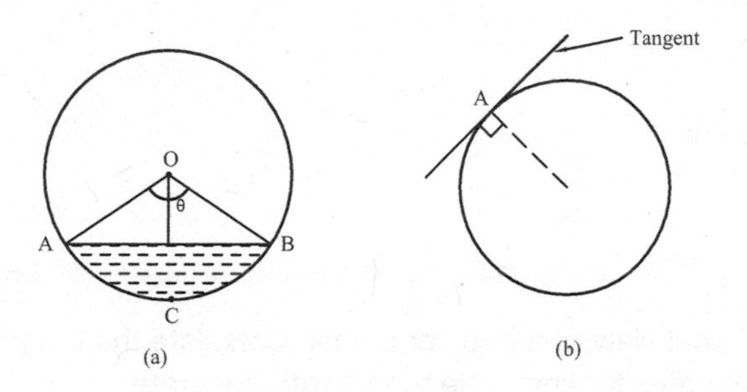

Figure 6.7

Example 6.4

The circle shown in Figure 6.8 has a radius of 15.0 cm. Calculate:

a) the circumference.
b) area of the minor sector.
c) length of arc ACB.

Solution:

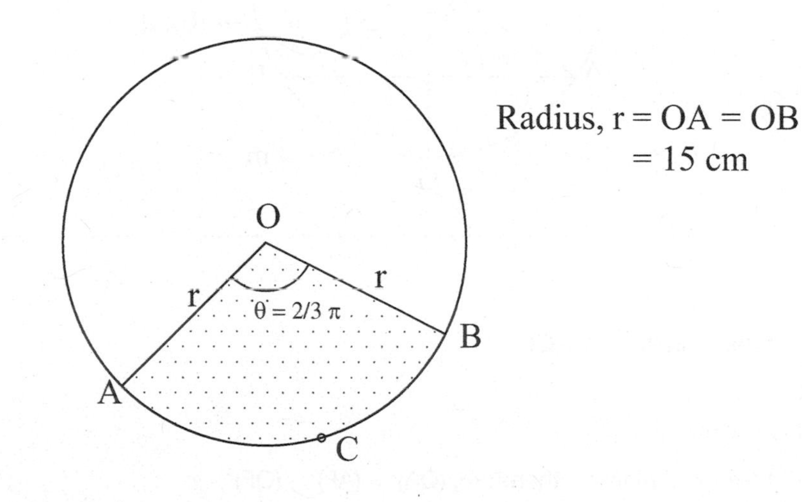

Figure 6.8

a) Radius, r = 15.0 cm

The circumference of the circle = $2\pi r$

$= 2\pi \times 15.0 =$ **94.25 cm.**

b) Area of the minor sector $= \pi r^2 \times \frac{\frac{2\pi}{3}}{2\pi}$

$= \pi \times 15^2 \times \frac{1}{3} =$ **235.62 cm²**

c) Length of arc ACB $= 2\pi r \times \frac{\theta}{2\pi}$

$= 2\pi \times 15 \times \frac{\frac{2\pi}{3}}{2\pi} =$ **31.42 cm**

Example 6.5

Figure 6.9 shows the sectional view of a circular tunnel. Calculate the length of arc ABCDE so the engineers can prepare estimates for lining the tunnel with concrete.

Solution:

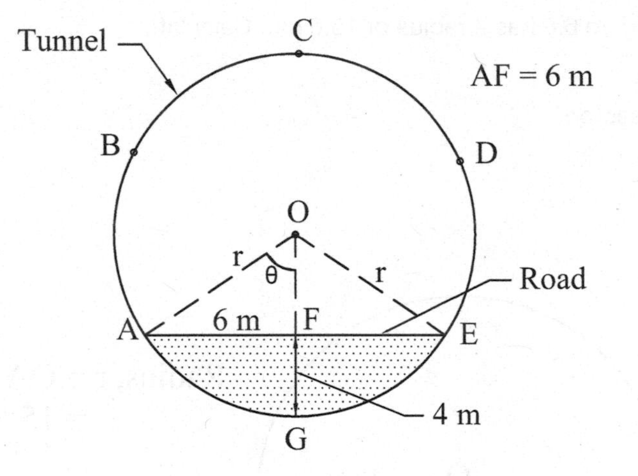

Figure 6.9

Let r be the radius of the tunnel; OA = OE = r

OF = OG – FG

$= r - 4$ (FG = 4 m)

Triangle OAF is a right-angled triangle, therefore, $(OA)^2 = (AF)^2 + (OF)^2$

$r^2 = 6^2 + (r - 4)^2$

$= 36 + r^2 - 8r + 16$

$r^2 - r^2 + 8r = 52$

$8r = 52,$ therefore, $r = \frac{52}{8} = 6.5\,\text{m}$

In triangle OAF, $\sin\theta = \frac{AF}{OA} = \frac{6}{6.5} = 0.9231$

$\angle\theta = \sin^{-1} 0.9231 = 67.3836°$

$\angle AOE = \angle 2\theta = 134.77°$

Length of arc EGA $= 2\pi r \times \frac{134.77°}{360°}$

$= 2\pi \times 6.5 \times \frac{134.77°}{360°} = 15.289\,\text{m}$

Length of arc ABCDE = Circumference of the circle – 15.289

$= 2\pi \times 6.5 - 15.289 = 40.84 - 15.289 =$ **25.552 m**

6.2 Area

Area is defined as the amount of space taken up by a 2-dimensional figure or the surface of a 3-dimensional object. A summary of the formulae used in calculating the areas, and other properties of some regular shapes is given in Table 6.1.

The main units of area used in the metric systems are: mm^2, cm^2, m^2, km^2 and hectares.

Table 6.1

Shape	Area and other properties
Triangle	Area $= \frac{B \times H}{2}$
Rectangle	Area = L × B Perimeter = 2(L + B)
TRAPEZIUM	Area $= \frac{1}{2}(A + B) \times H$

(continued)

Table 6.1 *(continued)*

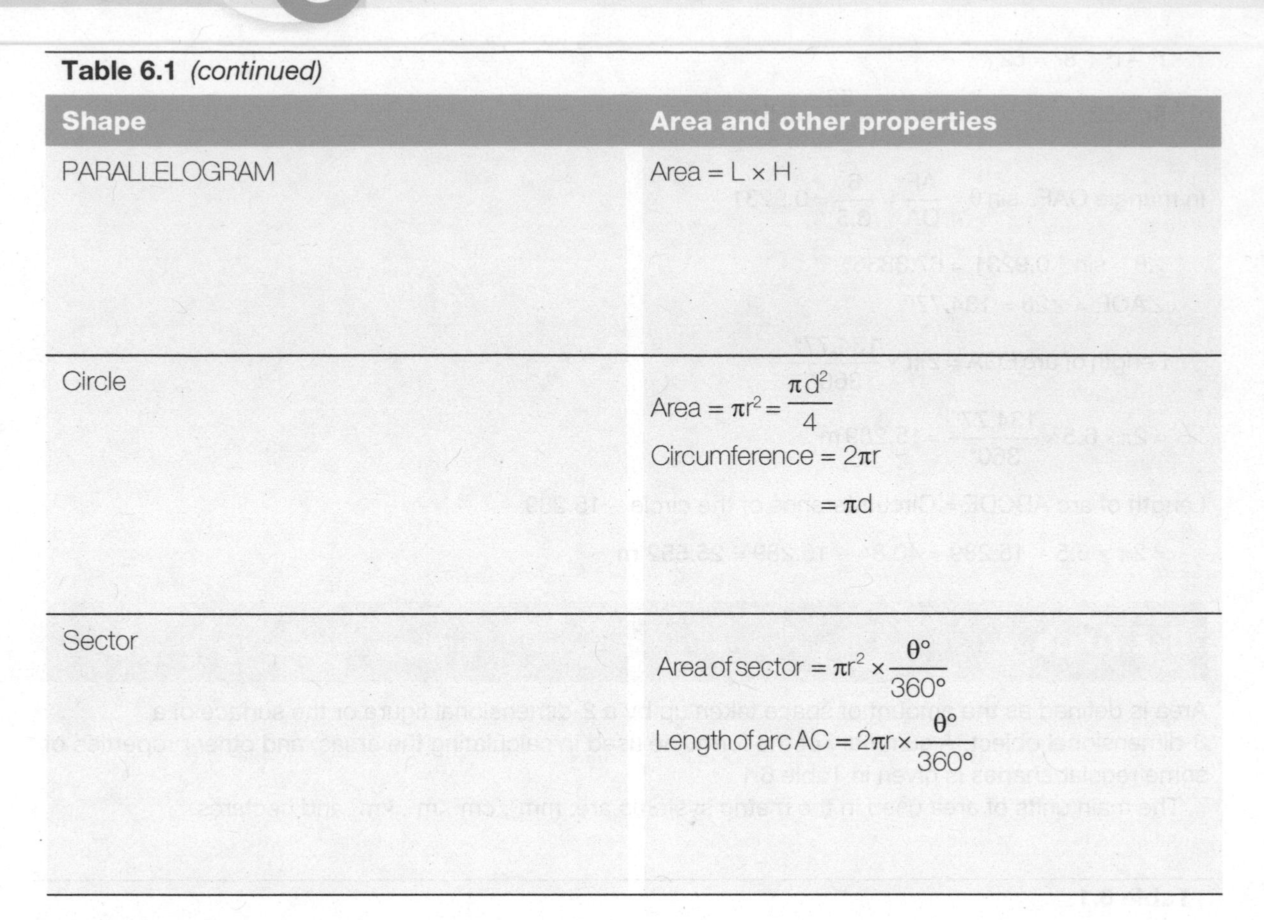

Shape	Area and other properties
PARALLELOGRAM	Area = L × H
Circle	$\text{Area} = \pi r^2 = \frac{\pi d^2}{4}$ Circumference = $2\pi r$ $= \pi d$
Sector	$\text{Area of sector} = \pi r^2 \times \frac{\theta°}{360°}$ $\text{Length of arc AC} = 2\pi r \times \frac{\theta°}{360°}$

6.2.1 Area of regular shapes

(a) Triangles

Several techniques are available to determine the area of triangles, which depend on the available information. In this section, the formulae given in Table 6.1 and other formulae will be considered to find the area of a right-angled triangle.

Example 6.6

Find the area of the triangle shown in Figure 6.10 by 3 methods.

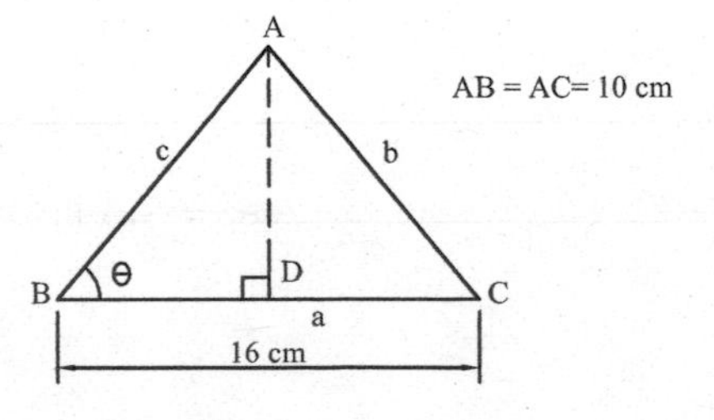

Figure 6.10

Solution:

i) **Method 1**: Area of triangle ABC $= \frac{\text{base} \times \text{height}}{2}$

(This method can be used if the vertical height of the triangle is either known or can be calculated)

Base BC = 16 cm; Height AD $= \sqrt{10^2 - 8^2} = 6$ cm

Area of triangle ABC $= \frac{16 \times 6}{2} =$ **48 cm²**

ii) **Method 2**: Area of triangle ABC $= \sqrt{s(s-a)(s-b)(s-c)}$

where, $s = \frac{a+b+c}{2}$

(This method is used where all 3 sides (sides a, b and c) of a triangle are known. s is known as the average perimeter of the triangle)

$s = \frac{16+10+10}{2} = 18$

Area of triangle ABC $= \sqrt{18(18-16)(18-10)(18-10)}$

$= \sqrt{18(2)(8)(8)} = \sqrt{2304} =$ **48 cm²**

iii) **Method 3**: Area of triangle ABC $= \frac{1}{2} ac \sin\theta$

$= \frac{1}{2} \times 16 \times 10 \times \frac{6}{10} (a = 16; c = 10; \sin\theta = \frac{AD}{AB} = \frac{6}{10})$

$= 48 \text{ cm}^2$

(Also refer to chapter 7: trigonometry)

6.2.2 Area of quadrilaterals

A polygon formed by 4 sides is called a **quadrilateral**, typical examples being rectangle, square, trapezium, parallelogram and rhombus. The calculation of areas of some of the quadrilaterals is explained in this section.

Example 6.7

Figure 6.11 shows the section of a steel beam made from 2 channel sections. Calculate its cross-sectional area in mm² and cm².

Solution:

The area of the section (or cross-section) of an object is called its cross-sectional area.

Each channel section may be divided into 3 parts as shown in Figure 6.12.

Area of the heavier channel = Area A + Area B + Area C

Area A $= 101.6 \times 10.2 = 1036.32$ mm²

Area B $= 284.4 \times 14.8 = 4209.12$ mm²

Area C $= 101.6 \times 10.2 = 1036.32$ mm²

Sub total $= 1036.32 + 4209.12 + 1036.32 = 6281.76$ mm²

Area of the lighter channel = Area D + Area E + Area F

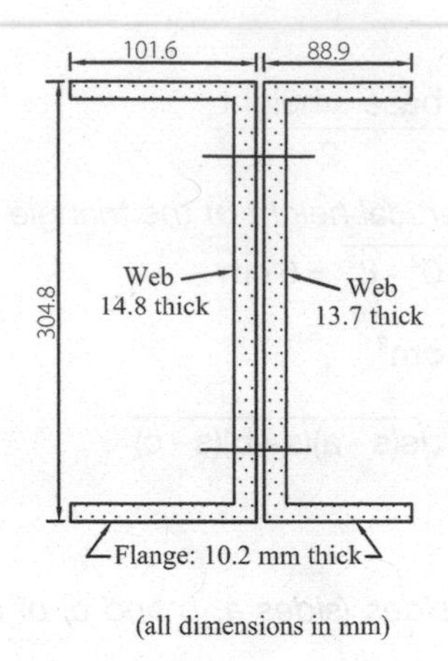

Figure 6.11

Area D = 88.9 × 10.2 = 906.78 mm²

Area E = 284.4 × 13.7 = 3896.28 mm²

Area F = 88.9 × 10.2 = 906.78 mm²

Sub total = 906.78 + 3896.28 + 906.78 = 5709.84 mm²

Total = 6281.76 + 5709.84 = **11991.6 mm²**

= 11991.6 ÷ 100 = **119.92 cm²** (1 cm² = 100 mm²)

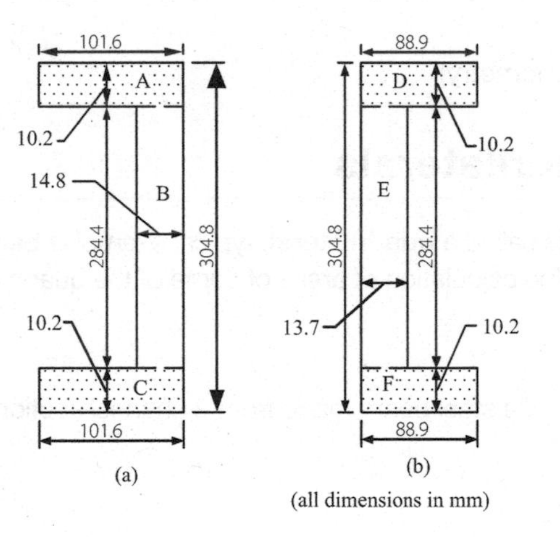

Figure 6.12

6.2.3 Area of circle

Some of the important properties of the circle have been explained in section 6.1.5.

There are two formulae that can be used to determine the area of a circle:

$$\text{Area} = \pi r^2 = \frac{\pi}{4} d^2$$

where r and d are the radius and the diameter of the circle, respectively.

Example 6.8

The depth of water flow in a drain is shown in Figure 6.13. If the diameter of the drain is 300 mm, calculate the cross-sectional area of the water flow.

Solution:

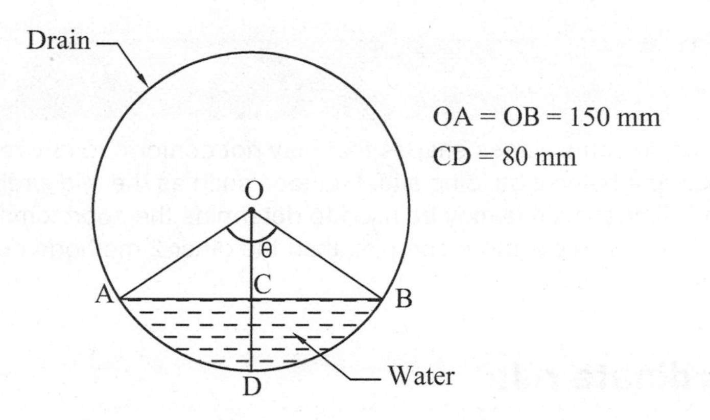

Figure 6.13

Cross-sectional area of water flow = Area of sector OAB – Area of triangle OAB

$$=\frac{1}{2}r^2(\theta-\sin\theta) \text{ (see section 6.1.5)}$$

Angle θ is in radians for the above formula.

Radius of the drain = 150 mm.

For calculating angle θ, consider right-angled triangle OAC:

$$\cos\frac{\theta}{2}=\frac{OC}{OA}=\frac{70}{150}=0.466667\ (OC=150-80=70\text{ mm})$$

$$\frac{\theta}{2}=\cos^{-1}0.466667=1.08528\text{ radians}$$

Therefore, $\angle\theta$ = 2.17056 radians

$$\text{Cross-sectional area of water flow}=\frac{1}{2}\times150^2\times(2.17056-\sin 2.17056)$$

$$= 11250 \times (2.17056 - 0.82547)$$

$$= 15132.26\text{ mm}^2$$

Example 6.9

A hot water cylinder has a cylindrical body but hemispherical top. If the total height of the cylinder is 900 mm and its diameter is 400 mm, find the surface area of the cylinder in m^2.

Solution:

Surface area of hemispherical part = $4\pi r^2 \div 2 = 2\pi r^2$

$= 2\pi \times 0.2^2$ (diameter = 0.4 m; radius = 0.2 m)

$= 0.25133\text{ m}^2$

Surface area of cylindrical part = $2\pi rh$

$= 2\pi \times 0.2 \times 0.7$ (h = 0.9 – 0.2 = 0.7 m)
$= 0.87965\ m^2$

Area of the base = $\pi r^2 = \pi \times 0.2^2 = 0.12566\ m^2$
Total surface area = 0.25133 + 0.87965 + 0.12566 = **1.257 m²**

6.3 Area of irregular shapes

In construction we often come across shapes that may not conform to any regular geometrically shape, a typical example being a building site. Methods such as the mid-ordinate rule, the trapezoidal rule and Simpson's rule may be used to determine the approximate area of any irregular shape. Simpson's rule is more complex than the other 2 methods but will give a more accurate answer.

6.3.1 Mid-ordinate rule

The irregular shape is divided into a number of strips of equal width (d) by vertical lines, known as the ordinates (Figure 6.14a). Each strip is assumed to be an approximate rectangle, as shown in Figure 6.14b. The mid-ordinates of the strips (y_1, y_2, etc.) shown in Figure 6.14a, represent the approximate average length of the strips or the assumed rectangles. Mathematically, the mid-ordinate of each strip can be determined by calculating the average of its ordinates, and the area of a strip is calculated by multiplying the length of the mid-ordinate by the width of the strip. Finally, the area of the shape is determined by adding the areas of all the strips.

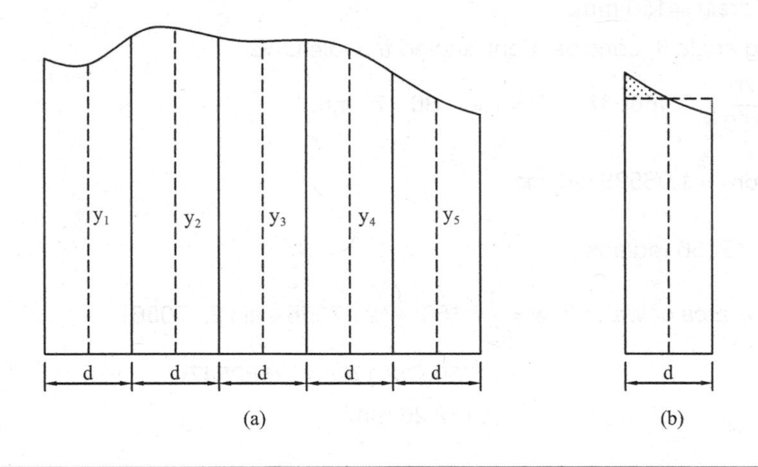

Figure 6.14

$$\text{Area} = dy_1 + dy_2 + dy_3 + dy_4 + dy_5$$
$$= d\,(y_1 + y_2 + y_3 + y_4 + y_5)$$

Area = width of strip × sum of mid-ordinates
The number of strips may vary from one task to another.

6.3.2 Trapezoidal rule

In this method the irregular shape is divided into a number of strips of equal width (Figure 6.15a), and each strip is assumed to be a trapezium, as shown in Figure 6.15b. Consider the first strip ABJK, its area is given by:

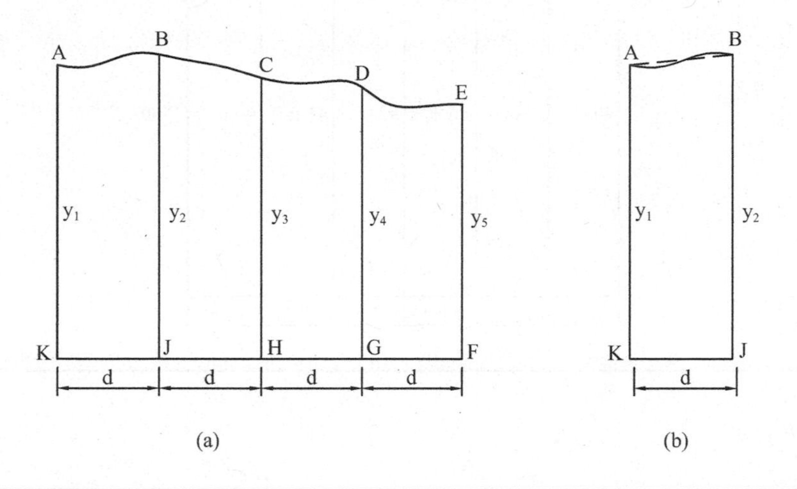

Figure 6.15

$$\text{Area of strip ABJK} = \frac{1}{2}(y_1 + y_2) \times d$$

$$\text{Similarly, area of strip BCHJ} = \frac{1}{2}(y_2 + y_3) \times d$$

The area of other strips can be determined in the same way. Finally, the area of the whole shape is determined by adding the area of the individual strips:

$$\text{Total area of the shape} = \frac{1}{2}(y_1 + y_2)d + \frac{1}{2}(y_2 + y_3)d + ... + \frac{1}{2}(y_4 + y_5)d$$

$$= d(\frac{y_1 + y_5}{2} + y_2 + y_3 + y_4)$$

$$= \text{Width of strip} \times [\frac{1}{2}(\text{sum of the first and last ordinates}) + (\text{sum of the remaining ordinates})]$$

6.3.3 Simpson's rule

Simpson's rule gives a more accurate answer as compared to the other 2 methods. The figure is divided into an even number of vertical strips of equal width, giving an odd number of ordinates. Simpson's rule cannot be used if a figure is divided into an odd number of strips.

$$\text{Area} = \frac{1}{3}(\text{width of strip})[(\text{first} + \text{last ordinate}) + 4(\text{sum of the even ordinates}) + 2(\text{sum of the remaining odd ordinates})]$$

Example 6.10

Calculate the area of the building plot shown in Figure 6.16 by 3 methods and compare the answers.

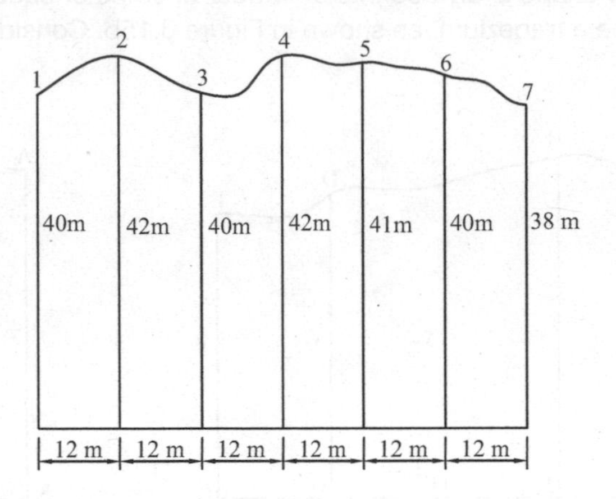

Figure 6.16

Solution:

Mid-ordinate rule:

Area = width of strip × sum of mid-ordinates

= 12 (41 + 41 + 41 + 41.5 + 40.5 + 39)

= 12 (244) = **2928 m²**

Trapezoidal rule:

$$\text{Area} = \text{Width of strip} \times [\frac{1}{2}(\text{sum of the first and last ordinates}) + (\text{sum of the remaining ordinates})]$$

$$= 12 \times [\frac{1}{2}(40 + 38) + 42 + 40 + 42 + 41 + 40]$$

$$= 12\ [244] = \mathbf{2928\ m^2}$$

Simpson's rule: The numbers at the top of Figure 6.16 show whether an ordinate is even or odd.

$$\text{Area} = \frac{1}{3}(\text{width of strip})\,[(\text{first} + \text{last ordinate}) + 4\ (\text{sum of the even ordinates (i.e. 2, 4 and 6)}) + 2\left(\text{sum of the remaining odd ordinates}\left(\text{i.e. 3 and 5}\right)\right)]$$

$$= \frac{1}{3}(12)\,[(40 + 38) + 4\,(42 + 42 + 40) + 2\,(40 + 41)]$$

$$= \frac{1}{3}(12)\,[78 + 496 + 162]$$

$$= \frac{1}{3}(12)\left[736\right] = \mathbf{2944\ m^2}$$

The answer produced by the mid-ordinate rule as well as the trapezoidal rule is 2928 m²; the answer produced by the Simpson's rule is slightly higher at 2944 m². Although the result obtained by the Simpson's rule is considered to be more accurate, the accuracy of these results can only be determined if we know the actual answer.

6.4 Volume

Volume may be defined as the space occupied by a 3-dimensional object, for example, the volume of a concrete block is the amount of concrete used to make the block. Similarly, the amount of fluid that a container can hold is known as the **capacity** of the container. The area of a surface is 2-dimensional, but the volume of an object is 3-dimensional, therefore, in some cases the volume of an object may be determined by multiplying its surface area by the thickness/height. The volume of a cuboid is given by:

$$\text{Volume, V} = \text{length}(\text{l}) \times \text{width}(\text{w}) \times \text{height}(\text{h})$$

Surface area of the cuboid = l × w

As the volume = area × height, therefore, V = (l × w) × h

Units of volume: mm^3, cm^3, m^3

Units of capacity: millilitres (ml), centilitres (cl), litres (*l*)

6.4.1 Volume of a sphere

A **sphere** is a solid generated by a semi-circle revolving through 360° about its diameter as an axis.

$$\text{Volume of a sphere} = \frac{4}{3}\pi r^3, \text{where r is the radius of the sphere.}$$

$$\text{Volume of a hemi} - \text{sphere} = \frac{2}{3}\pi r^3, \text{ where r is the radius of the hemi} - \text{sphere.}$$

A summary of the formulae used in calculating the volume, and other properties of some regular objects is given in Table 6.2.

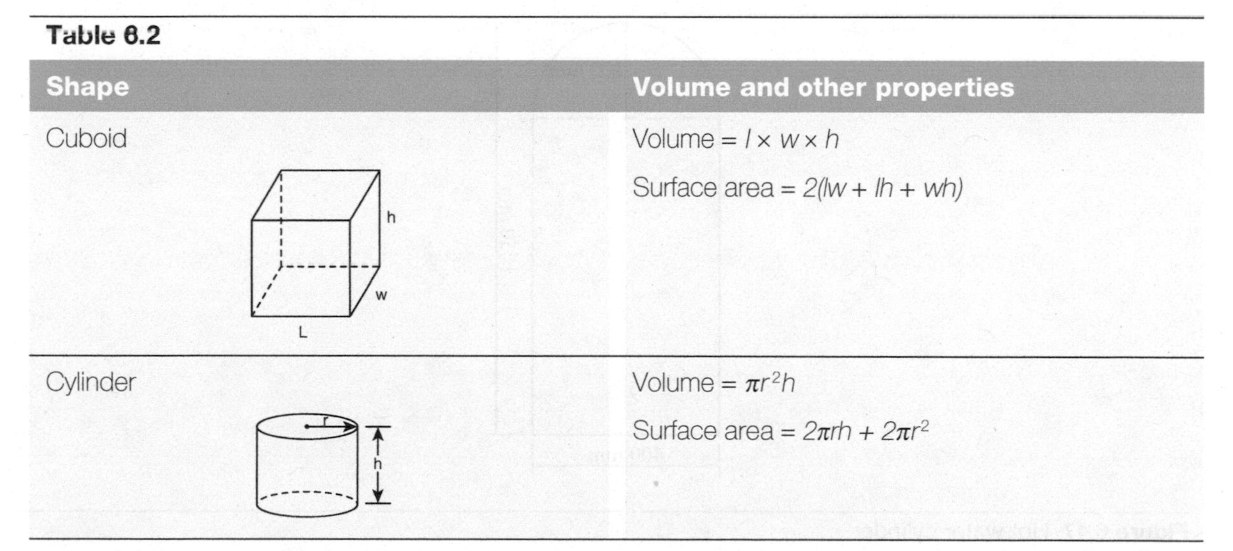

Table 6.2

Shape	Volume and other properties
Cuboid	Volume = $l \times w \times h$ Surface area = $2(lw + lh + wh)$
Cylinder	Volume = $\pi r^2 h$ Surface area = $2\pi rh + 2\pi r^2$

(continued)

Table 6.2 *(continued)*

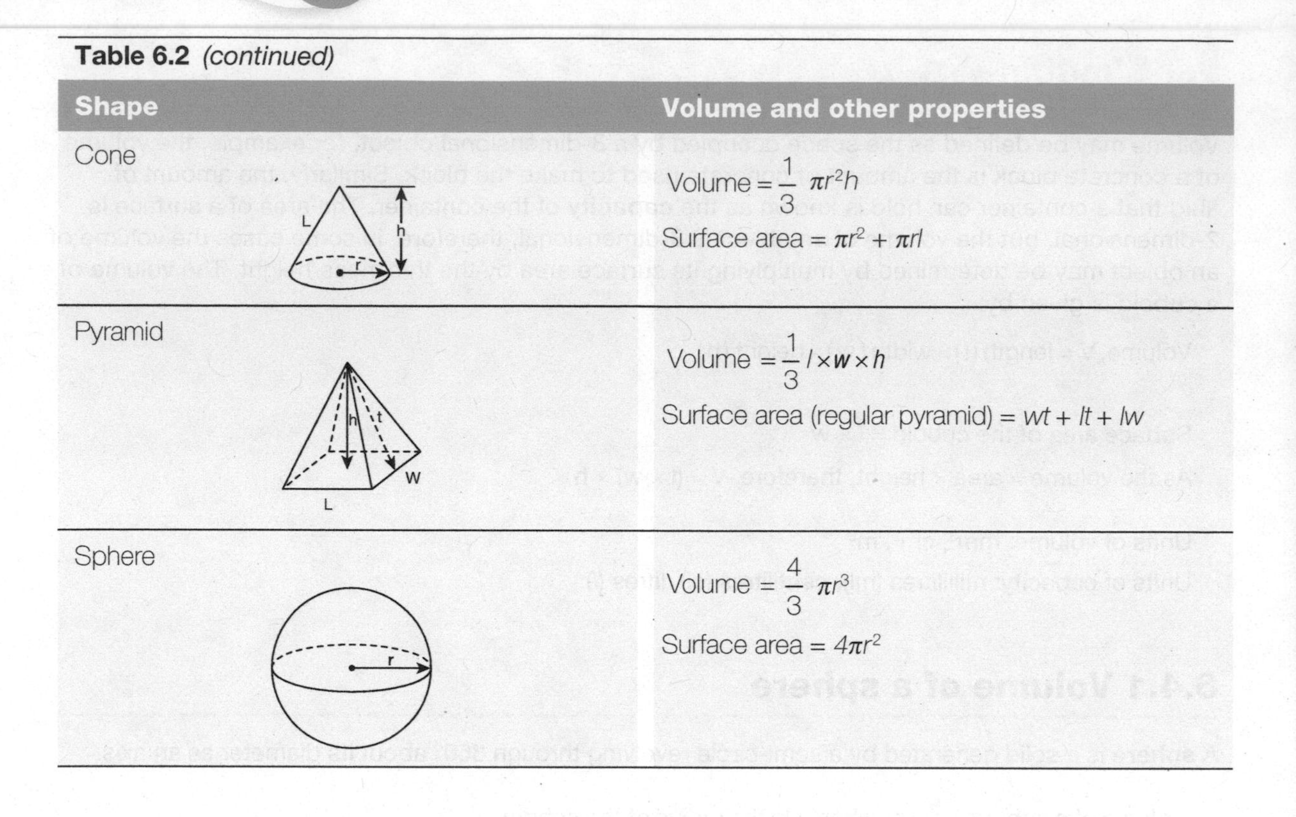

Shape	Volume and other properties
Cone	Volume $= \dfrac{1}{3}\pi r^2 h$ Surface area $= \pi r^2 + \pi r l$
Pyramid	Volume $= \dfrac{1}{3} l \times w \times h$ Surface area (regular pyramid) $= wt + lt + lw$
Sphere	Volume $= \dfrac{4}{3}\pi r^3$ Surface area $= 4\pi r^2$

Example 6.11

A hot water cylinder has a cylindrical body but hemispherical top. If the total height of the cylinder is 1050 mm and its diameter is 400 mm, find its capacity in cubic metres and litres.

Solution:

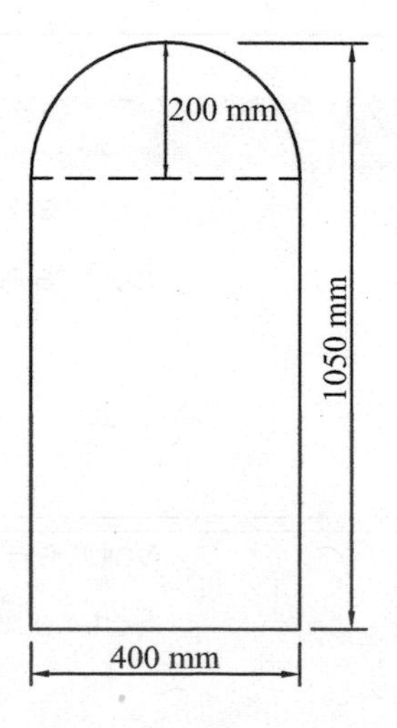

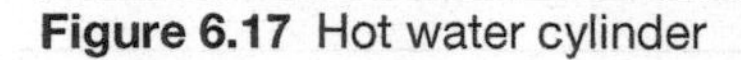

Figure 6.17 Hot water cylinder

The cylinder may be divided into 2 parts (Figure 6.17) for calculating the total volume/capacity:

a) cylindrical body b) hemi-spherical top

a) Volume of the cylindrical part $= \pi r^2 h$

$$= \pi \times (0.2)^2 \times 0.850 \quad (r = 0.200\ m;\ h = 0.850\ m)$$

$$= 0.1068\ m^3$$

b) Volume of the hemi – spherical part $= \frac{2}{3}\pi r^3,$

$$= \frac{2}{3}\pi(0.200)^3$$

$$= 0.0168\ m^3$$

Total volume = 0.1068 + 0.0168 = 0.1236 m^3

1 m^3 = 1000 litres, therefore 0.1236 m^3 = 123.6 litres

Therefore, capacity of the hot water cylinder is **0.1236 m^3** or **123.6 litres**

Example 6.12

A concrete retaining wall is 0.6 m wide at the top and 1.30 m wide at the base. If the height of the wall is 3.0 m, find the volume of concrete used:

a) per metre length
b) to construct a 15 m long wall.

Solution:

a) Figure 6.18 shows the sectional details of the wall. The volume of the retaining wall is:

Volume = Cross-sectional area × length

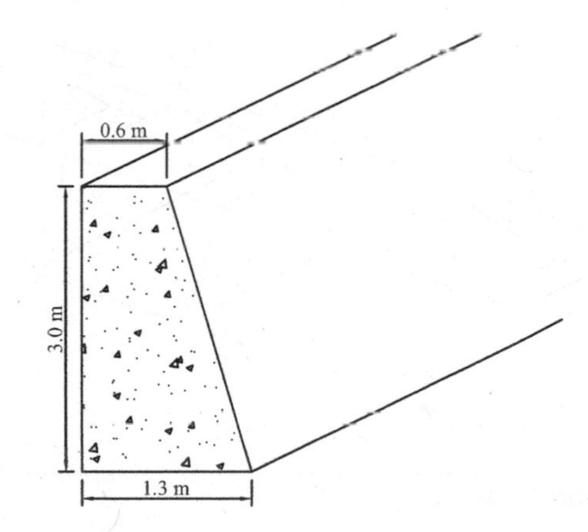

Figure 6.18 Retaining wall

The wall has trapezoidal cross-section, therefore:

$$\text{Cross – sectional area} = \frac{0.6 + 1.30}{2} \times 3.0 = 2.85\ m^2$$

Volume per metre length = Cross-sectional area × 1 m

$= 2.85 \times 1 = \mathbf{2.85\ m^3}$

b) Volume of 15 m long wall = $2.85 \times 15 = \mathbf{42.75\ m^3}$

6.5 Volume of irregular objects

The volume of irregular objects may be determined by any of the 3 methods, i.e. the mid-ordinate rule, the trapezoidal rule and Simpson's rule. In section 6.3, these methods have been used to determine the areas of irregular surfaces. Their use can be extended to the determination of the volume of irregular objects if the ordinates that were used in area calculation are replaced by the sectional areas.

Simpson's rule, as discussed before, gives more accurate results.

a) **Mid-ordinate rule:**

Volume = Width of strip × [sum of the areas of mid-sections]

b) **Trapezoidal rule:**

$$\text{Volume} = \text{Width of strip} \times [\frac{1}{2}(\text{first area} + \text{last area}) + (\text{sum of the remaining areas})]$$

c) **Simpson's rule:**

$$\text{Volume} = \frac{1}{3}(\text{width of strip})[(\text{first area} + \text{last area}) + 4(\text{sum of the even areas}) + 2(\text{sum of the remaining odd areas})]$$

Example 6.13

The cross-sectional areas of an embankment, at 10 m intervals, are shown in Figure 6.19. Use the trapezoidal rule and Simpson's rule to calculate the volume of the soil used.

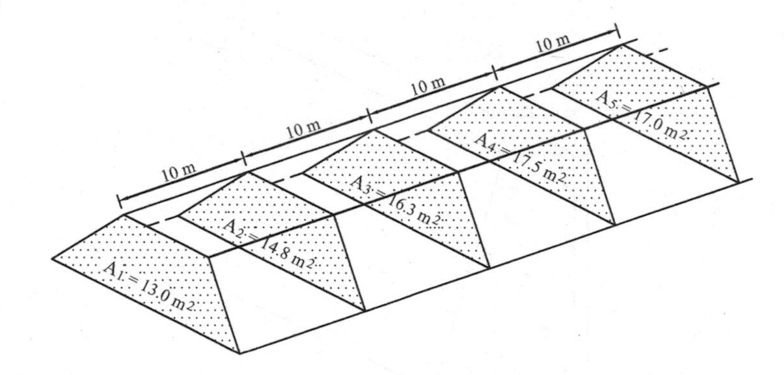

Figure 6.19 Sections of the embankment

Solution:

Trapezoidal rule:

$A_1 = 13.0\ m^2$, $A_2 = 14.8\ m^2$, $A_3 = 16.3\ m^2$, $A_4 = 17.5\ m^2$, $A_5 = 17.0\ m^2$

$$\text{Volume} = 10 \times [\frac{1}{2}(13.0 + 17.0) + (14.8 + 16.3 + 17.5)]$$

Volume = 10 × [(15.0) + (48.6)]

= 10 [63.6] = **636 m³**

Simpson's rule:

$$\text{Volume} = \frac{1}{3}(\text{width of strip})[(\text{first area} + \text{last area}) + 4(\text{sum of the even areas}) + 2(\text{sum of the remaining odd areas})]$$

$$= \frac{1}{3}(10)[(13.0 + 17.0) + 4(14.8 + 17.5) + 2(16.3)]$$

$$= \frac{1}{3}(10)[(30.0) + 4(32.3) + 2(16.3)]$$

$$= \frac{1}{3}(10)[30.0 + 129.2 + 32.6]$$

$$= \frac{1}{3}(10)[191.8] = \mathbf{639.33\ m^3}$$

6.5.1 Mass-haul diagram

Mass-haul diagrams are prepared for highway projects where large quantities of soil have to be moved; these diagrams show the volume of soil to be excavated (cut) and the volume of soil required to fill the areas which are below the balance line (Figure 6.20). Mass-haul diagrams also show:

i) the distance and the direction of haul.

ii) the gradients of ground to balance cut and fill.

iii) the volume of soil to be carted away and/or the volume of soil to be borrowed.

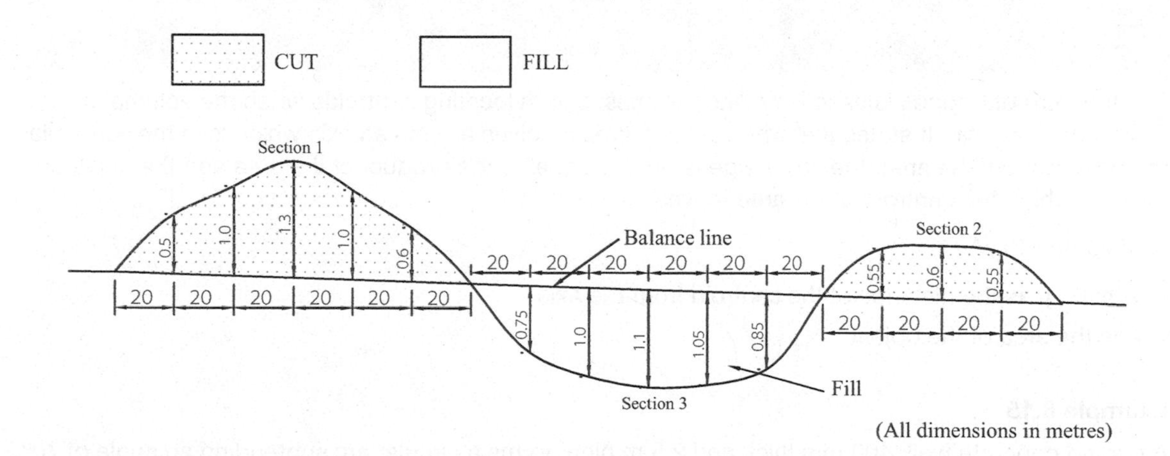

Figure 6.20 Mass-haul diagram

Example 6.14

Figure 6.20 shows the mass-haul diagram for a 320 m long section of a highway construction project. Calculate the volume of soil to be excavated, volume of soil required to fill the area that is below the balance line and the volume of surplus soil. Assume that the ordinates represent average values across the width of the highway.

Width of the highway = 40 m

Solution:

In Figure 6.20, 2 parts require excavation and removal of the excavated soil. The third part is below the balance line, therefore, requires to be filled.

Using the trapezoidal rule:

$$\text{Section 1(cut): Cross-sectional area} = 20 \times [(\frac{0+0}{2}) + 0.5 + 1.0 + 1.3 + 1.0 + 0.6]$$

$$= 20 \times [0 + 4.4] = 20 \times 4.4 = 88.0 \text{ m}^2$$

Volume of soil = 88.0 × 40 = **3520 m³**

$$\text{Section 2 (cut): Cross-sectional area} = 20 \times [(\frac{0+0}{2}) + 0.55 + 0.6 + 0.55]$$

$$= 20 \times [0 + 1.7] = 20 \times 1.7 = 34.0 \text{ m}^2$$

Volume of soil = 34.0 × 40 = **1360 m³**

Total volume of soil (cutting) = 3520 + 1360 = **4880 m³**

$$\text{Section 3(fill): Cross-sectional area} = 20 \times [(\frac{0+0}{2}) + 0.75 + 1.0 + 1.1 + 1.05 + 0.85]$$

$$= 20 \times [0 + 4.75] = 20 \times 4.75 = 95.0 \text{ m}^2$$

Volume of soil (for filling) = 95.0 × 40 = **3800 m³**

Surplus amount of soil that has to be carted away from the site = cut – fill

= 4880 – 3800

= **1080 m³**

6.6 The theorem of Pappus

The theorem of Pappus is used in finding volumes, and in locating centroids when the volume of revolution is known. It states that when a plane area revolves around an axis which is in the same plane but does not cut the area, the volume generated is equal to the product of the area and the distance through which the centroid of the area moves:

$$\text{Volume} = 2\pi\bar{y}A$$

where, $\bar{y}$ is the distance of the centroid from the axis

A is the area of the object

Example 6.15

A curved concrete wall, 400 mm thick and 2.5 m high, forms a circular arc subtending an angle of 70°, as shown in Figure 6.21. Calculate the volume of the wall.

Solution:

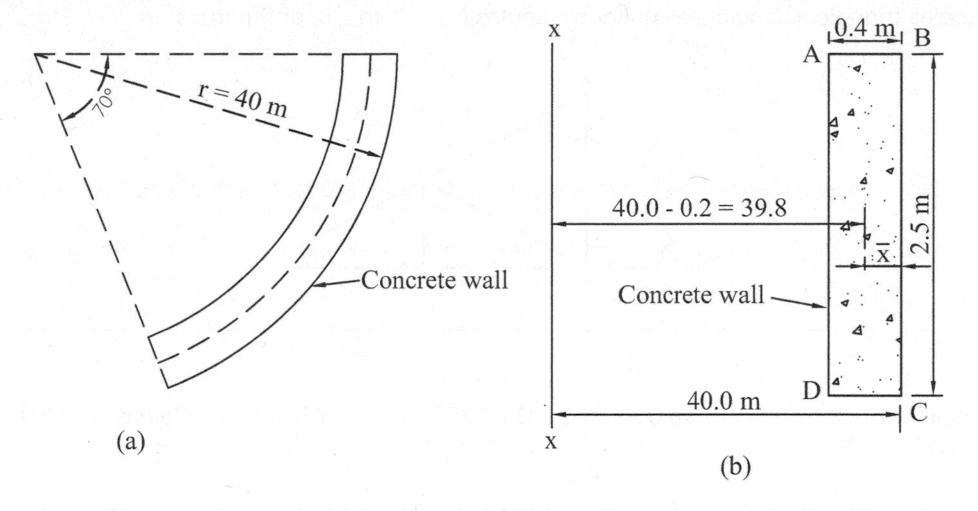

Figure 6.21

The solid wall is generated by rotating rectangle ABCD about axis x-x through 70° as shown in Figure 6.21.

The position of the centroid of the wall is situated at the centre of the wall, i.e. where the two diagonals meet. The distance between the centroid and face BC of the wall is 0.2 m.

Distance between axis x-x and the centroid of the wall = 40 – 0.2 = 39.8 m

Area of the wall = 0.4 × 2.5 = 1.0 m²

$$\text{Volume} = 2\pi\bar{y}A \times \frac{70^\circ}{360^\circ} \quad (\bar{y} = 39.8\text{ m})$$

$$= 2\pi \times 39.8 \times 1.0 \times \frac{70^\circ}{360^\circ} = \mathbf{48.625\ m^3}$$

Exercise 6.1

1. Convert a) 35.5° into radians

   b) $\frac{2\pi}{5}$ radians into degrees

2. Find angles c to g shown in Figure 6.22. Line AB is parallel to line CD.

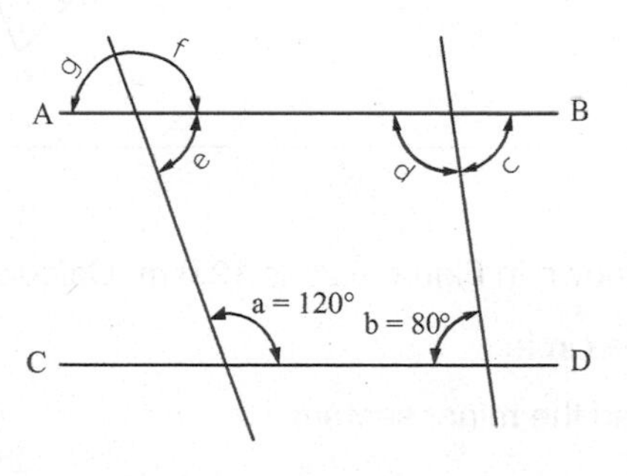

Figure 6.22

3. Figure 6.23 shows a north-light roof; angles ACB and BAF are right angles and triangle DCF is an isosceles triangle. Calculate all unknown angles (i.e. ∠1 to ∠8) of the truss.

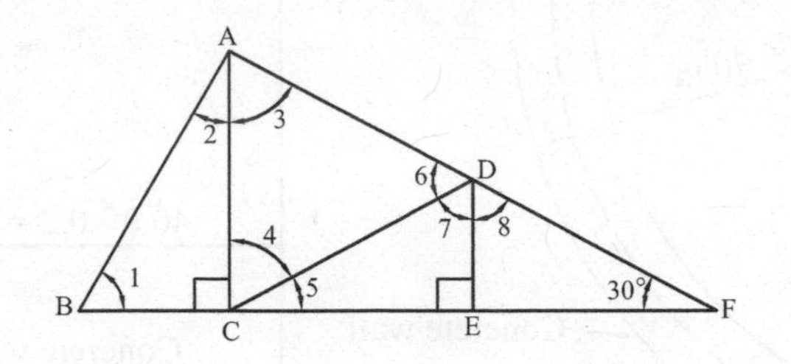

Figure 6.23

4. Triangle ABC, shown in Figure 6.24, has ∠C = 135° and side AB = 13 m. Calculate the length of sides AC and BC.

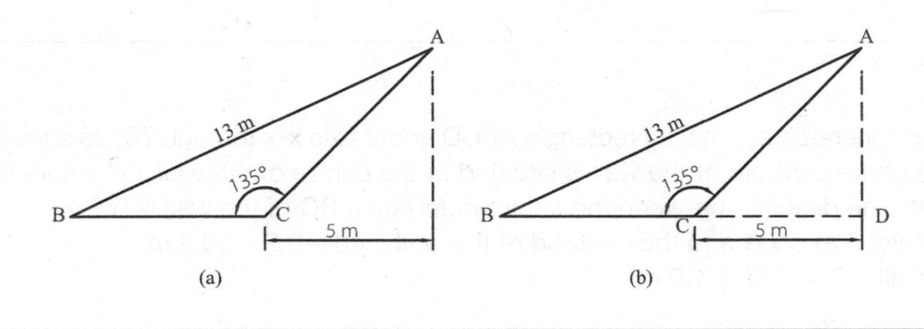

Figure 6.24

5. In Figure 6.25 side AB = side AC and side CE = side CD. Find ∠1 to ∠5.

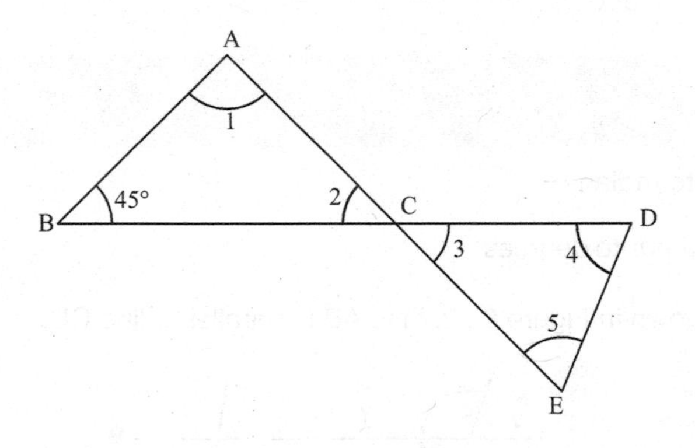

Figure 6.25

6. The diameter of the circle shown in Figure 6.26 is 12.0 m. Calculate:

   a) The circumference of the circle.

   b) The area of the major and the minor sectors.

   c) The length of arc ACB.

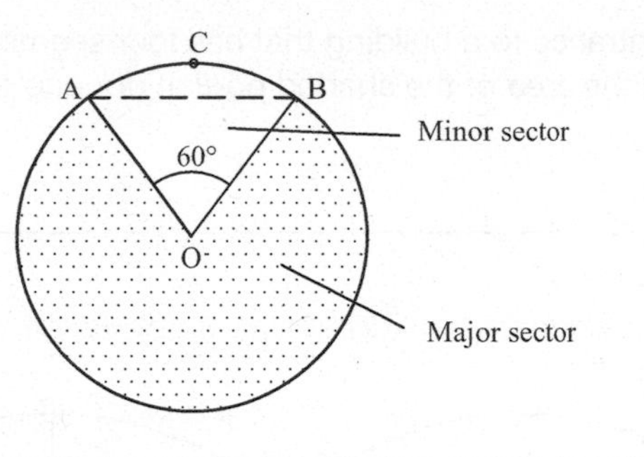

Figure 6.26

7. The base of an isosceles triangle is 12.0 cm long, and the other two sides are each 15.0 cm long. Calculate the area of the triangle by 3 methods.

8. Figure 6.27 shows a bridge beam (plate girder) made up of structural steel plates and angle sections. Find the cross-sectional area of the beam.

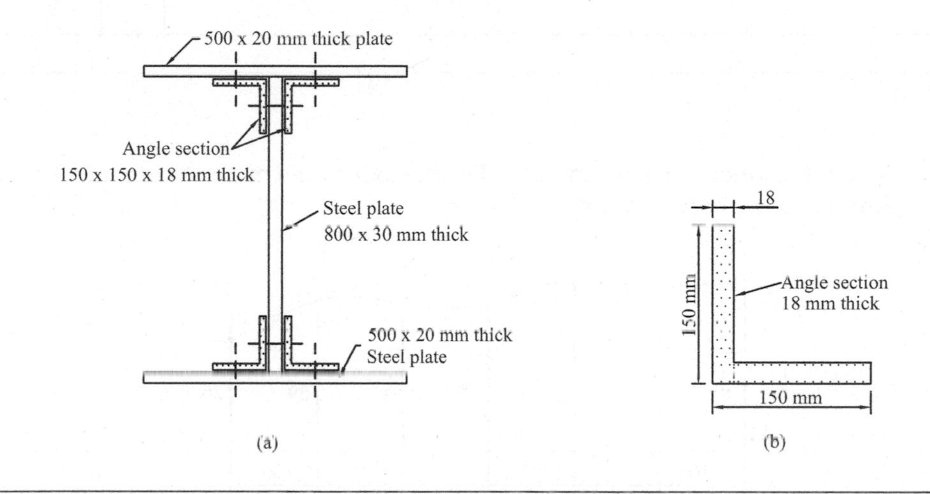

Figure 6.27 Plate girder

9. The depth of water flow in a drain is shown in Figure 6.28. If the diameter of the drain is 450 mm, calculate the cross-sectional area of the water flow.

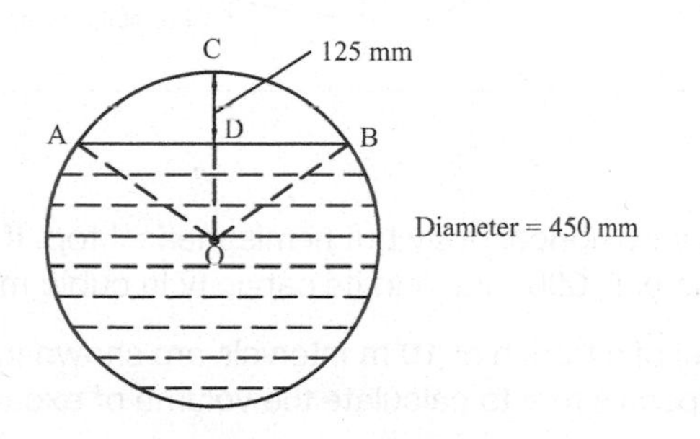

Figure 6.28

10. Figure 6.29 shows the entrance to a building that has four segmental arches, each arch having a span of 8.0 m. Calculate the area of the shaded portion (for one arch) which is to be provided with stained glass.

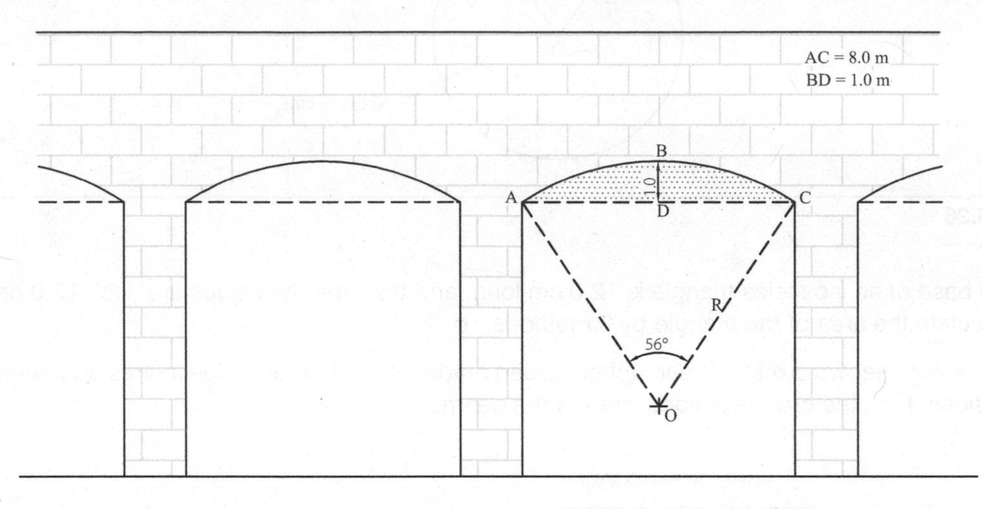

Figure 6.29

11. Find the area of the irregular shape shown in Figure 6.30, by the mid-ordinate rule, trapezoidal rule and Simpson's rule, and compare the results.

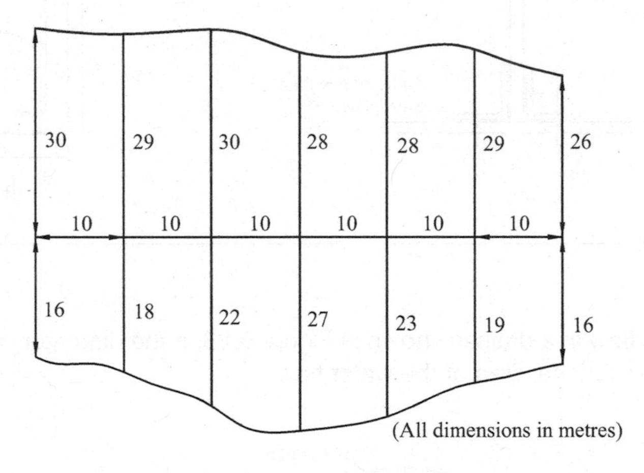

Figure 6.30

12. A hot water cylinder has a cylindrical body but hemispherical top. If the total height of the cylinder is 1200 mm and its diameter is 600 mm, find its capacity in cubic metres and litres.

13. The cross-sectional areas of a trench at 10 m intervals are shown in Figure 6.31. Use the trapezoidal rule and Simpson's rule to calculate the volume of excavated soil.

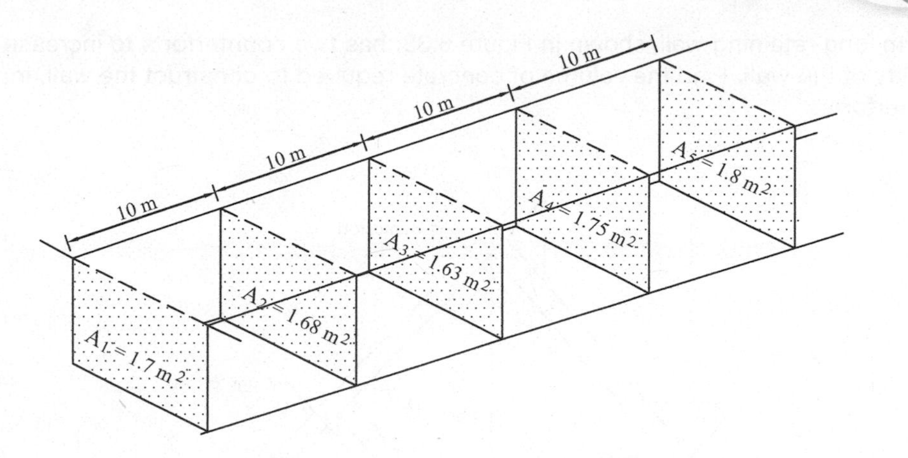

Figure 6.31 Cross-sections of a trench

14. The spot levels of a building site, on a grid of 10 m × 10 m, are shown in Figure 6.32. The earth needs to be excavated to reduce the ground level to 97.0 m. Calculate the volume of the earth to be excavated.

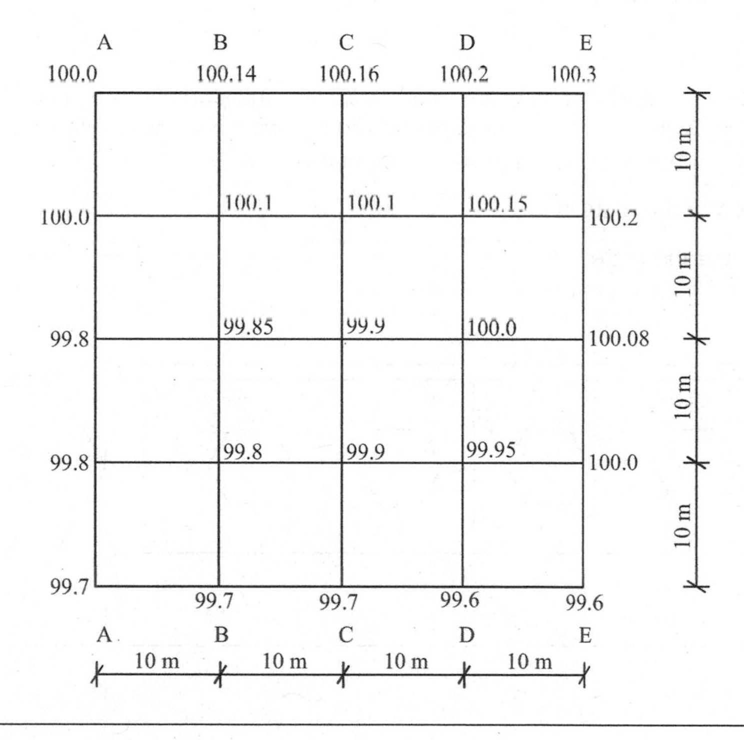

Figure 6.32

15. An 8 m long retaining wall, shown in Figure 6.33, has two counterforts to increase the stability of the wall. Find the volume of concrete required to construct the wall, including the counterforts.

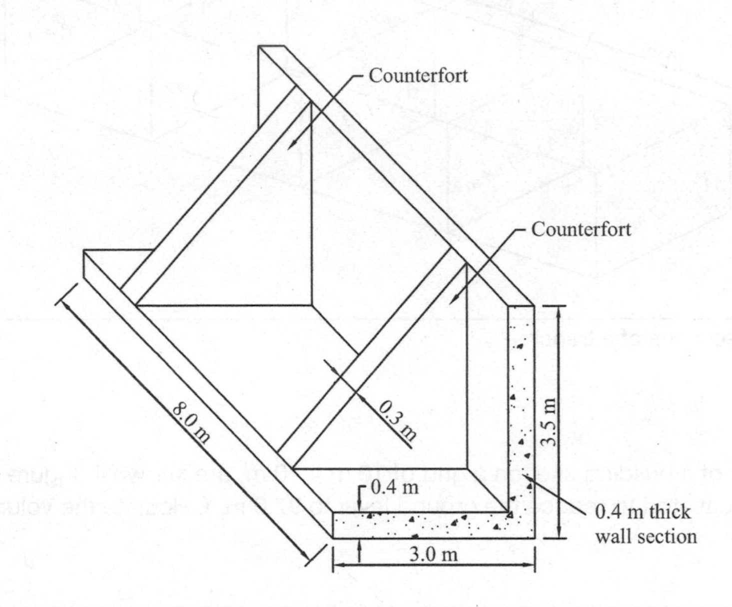

Figure 6.33 Counterfort retaining wall

16. A structural steel plate, shown in Figure 6.34, is 30 mm thick and is to be used in the fabrication of a plate girder. Some of the metal will be cut out to form trapezoidal voids for making the plate lighter in weight. If there are 18 such voids, calculate:

    a) The net area of the plate in m^2

    b) The volume of the plate in m^3

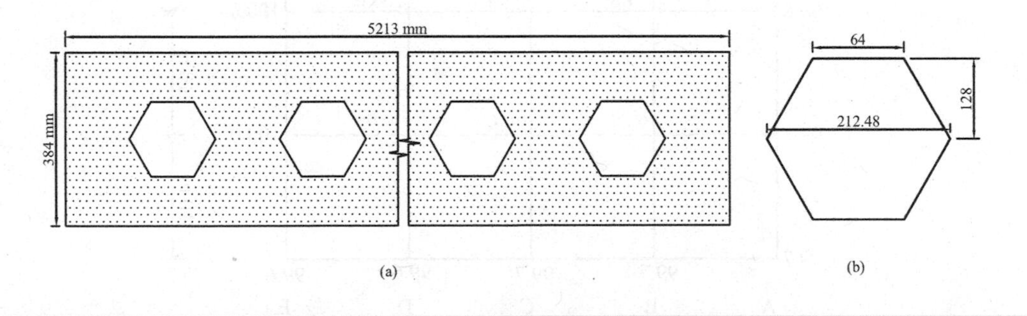

Figure 6.34

17. The dimensions of a structural steel beam section are shown in Figure 6.35. Calculate the cross-sectional area of the section and its mass per metre if the density of steel is 7850 kg/m^3.

    [Density = Mass ÷ volume]

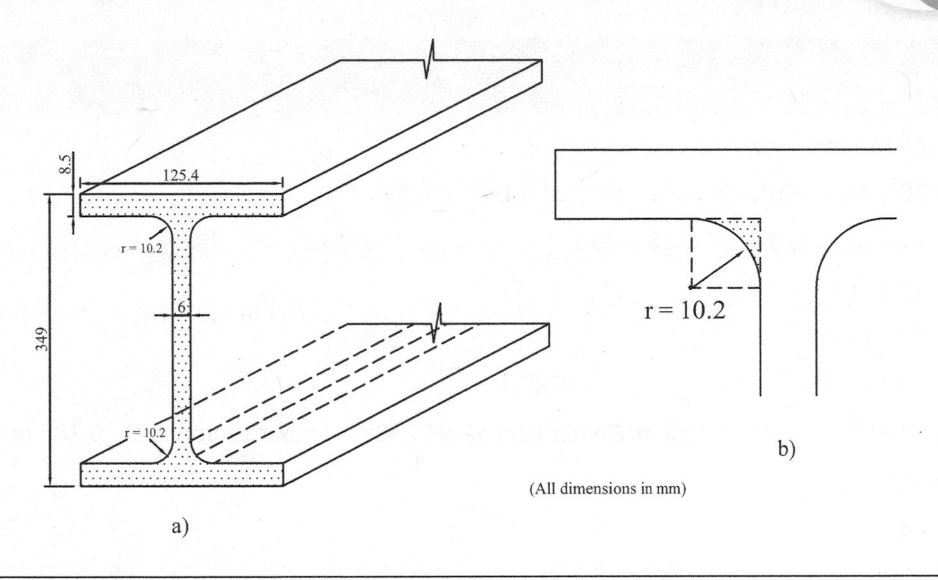

Figure 6.35 Structural steel beam

18. A concrete retaining wall, that forms a circular arc, subtends an angle of 90°. If the radius of the arc is 75.0 m and the cross-section of the wall is as shown in Figure 6.36, calculate the volume of concrete used to construct the wall.

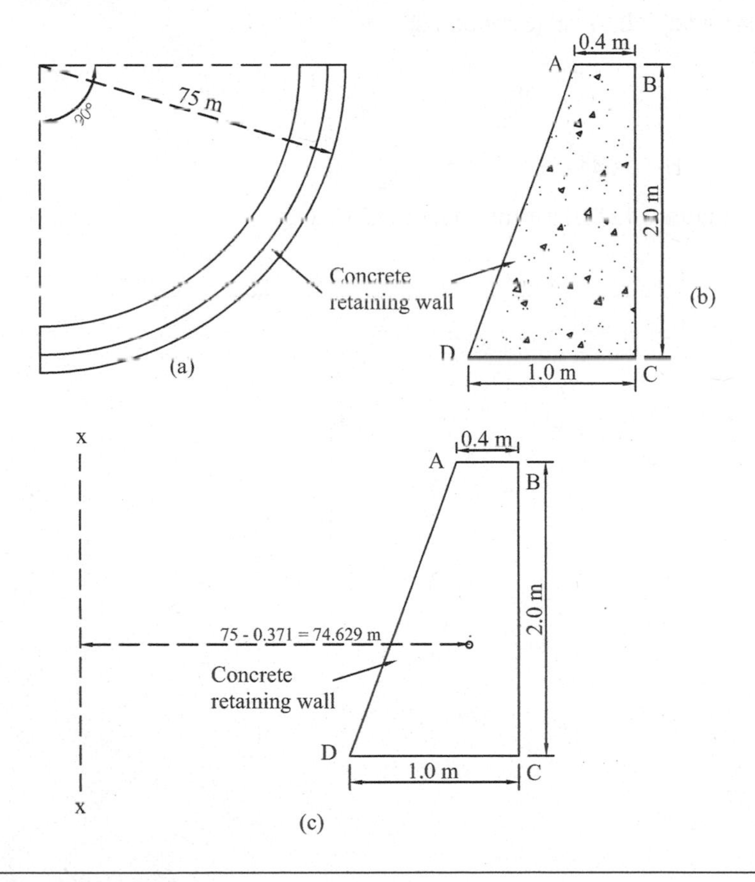

Figure 6.36 Concrete retaining wall

Answers – Exercise 6.1

1. a) 0.6196 radian b) 72°
2. $\angle c = 80°$, $\angle d = 100°$, $\angle e = \angle g = 60°$, $\angle f = 120°$
3. $\angle 1 = \angle 3 = \angle 4 = \angle 6 = \angle 7 = \angle 8 = 60°$,

   $\angle 2 = \angle 5 = 30°$
4. AC = 7.071 m; BC = 7 m
5. $\angle 1 = 90°$; $\angle 2 = \angle 3 = 45°$; $\angle 4 = \angle 5 = 67.5°$
6. a) 37.699 m, b) 18.85 m^2 (minor sector), 94.248 m^2 (major sector), c) 6.283 m
7. 82.49 m^2
8. 64304 mm^2
9. 122992.05 mm^2
10. 5.308 m^2
11. 2970 m^2 (mid-ordinate, trapezoidal); 2980 m^2 (Simpson's)
12. 0.311 m^3 or 311.02 litres
13. 68.1 m^3 (Trapezoidal); 68.27 m^3 (Simpson's)
14. 4734 m^3
15. 21.938 m^3
16. a) 1.365 m^2 b) 0.041 m^3
17. Cross-sectional area = 4213.11 mm^2; mass = 33.073 kg
18. 164.118 m^3

CHAPTER 7

Trigonometry 1

Topics covered in this chapter:

- Sine, cosine and tangent ratios
- Sine rule: $\frac{a}{\sin A} = \frac{b}{\sin B} = \frac{c}{\sin C}$
- Cosine rule: $a^2 = b^2 + c^2 - 2bc \cos A$
- Application of the sine rule and the cosine rule to solve triangles and practical problems in construction
- Area of a triangle if 2 sides and the included angle are given

7.1 Introduction

The use of trigonometry started more than 2000 years ago when astronomers started working on the measurement of angles and sides of triangles. The word trigonometry is made up of 2 Greek words 'trigonon' and 'metron', which mean triangle and measurement, respectively. Basically, trigonometry is the study of triangles (particularly right-angled triangles) and the relationship between the sides and angles. Trigonometry is of theoretical as well as practical importance as it is used in mathematics, land surveying, engineering, physics, satellite navigation and other applications.

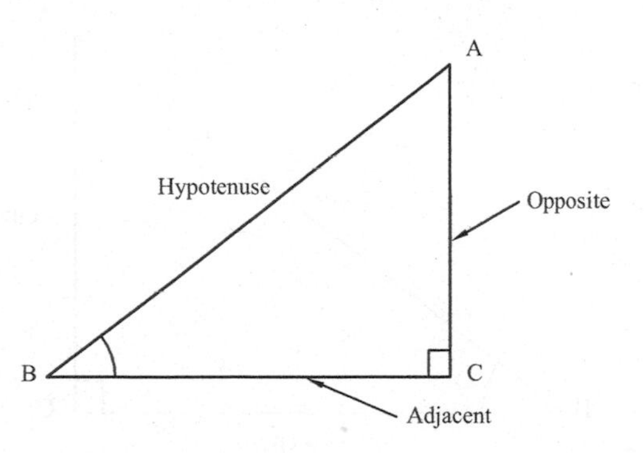

Figure 7.1

7.2 The trigonometrical ratios

Consider a right-angled triangle ABC, shown in Figure 7.1. Angle ACB (∠C) is a 90° angle or a right-angle. If ∠B is being considered for calculations, then side AC is called the 'opposite side' or just opposite. Side AB, the longest side, is called the 'hypotenuse', and the third side, BC, which is common to the right angle and ∠B is called the 'adjacent side' or adjacent.

There are 3 main trigonometric ratios, which are basically the ratios of the sides of a right-angled triangle (Figure 7.1):

i) $\frac{AC}{AB} = \frac{\text{Opposite}}{\text{Hypotenuse}} = \text{sine}$ of ∠B or **sin B** (sin is the shorter version of sine)

ii) $\frac{BC}{AB} = \frac{\text{Adjacent}}{\text{Hypotenuse}} = \text{cosine}$ of ∠B or **cos B**

iii) $\frac{AC}{BC} = \frac{\text{Opposite}}{\text{Adjacent}} = \text{tangent}$ of ∠B or **tan B**

There are also reciprocals of sine, cosine and tangent, but they are more commonly used in land surveying:

$\frac{1}{\sin B} = \text{cosecant}\,B$ or **cosec B**

$\frac{1}{\cos B} = \text{secant}\,B$ or **sec B**

$\frac{1}{\tan B} = \text{cotangent}\,B$ or **cot B**

These trigonometric ratios are only applicable if we have a right-angled triangle. For other types of triangles, sine rule and cosine rule may be used, as explained in section 7.5.

Sine, cosine and tangent work only with angles; without the angle they cannot be used.

Example 7.1

With reference to ∠B, find the sine, cosine and tangent ratios for the triangle shown in Figure 7.2.

Solution:

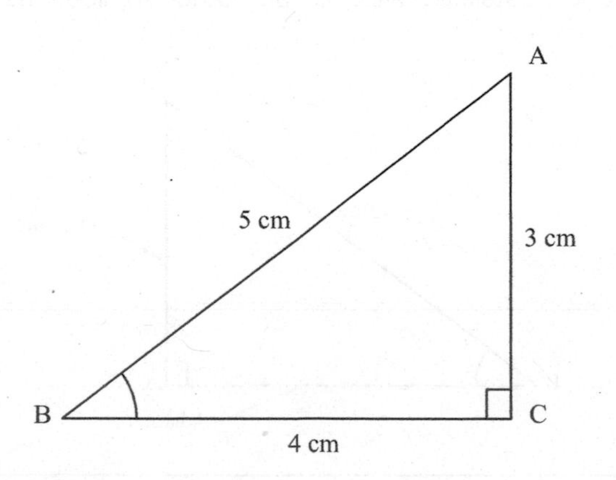

Figure 7.2

With reference to ∠B:

Side BC = adjacent; Side AC = opposite; Side AB = hypotenuse

$$\sin B = \frac{\text{Opposite}}{\text{Hypotenuse}} = \frac{AC}{AB} = \frac{3}{5} = \mathbf{0.6}$$

$$\cos B = \frac{\text{Adjacent}}{\text{Hypotenuse}} = \frac{BC}{AB} = \frac{4}{5} = \mathbf{0.8}$$

$$\tan B = \frac{\text{Opposite}}{\text{Adjacent}} = \frac{AC}{BC} = \frac{3}{4} = \mathbf{0.75}$$

Example 7.2

a) Find the value of: i) sin 30°

ii) cos 30°

b) i) If cos B = 0.6, find angle B.

ii) If tan A = 1.1, find angle A.

Solution:

a)

i) The calculator must show D (degrees) in the display area. If the calculator displays R or G then use the MODE key or the SET UP key to change the angle unit to degrees, and press the following keys:

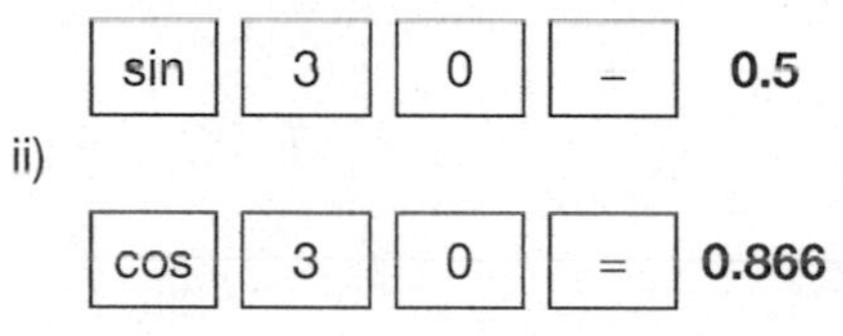

b)

i)

This question involves the determination of angles, therefore the process is the reverse of that used to solve part (a). Instead of sin, cos or tan keys, use $\sin^{-1}$, $\cos^{-1}$ and $\tan^{-1}$. Use the following sequence:

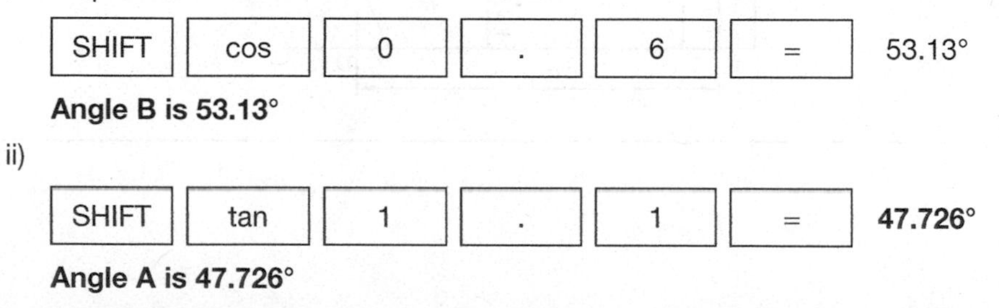

Angle B is 53.13°

ii)

Angle A is 47.726°

7.3 Angles of elevation and depression

If an object is situated above the observer, then the angle between the horizontal and the line of sight is called an angle of elevation (Figure 7.3a). If the object is situated below the observer, then the angle between the horizontal and the line of sight is known as the angle of depression, as shown in Figure 7.3b.

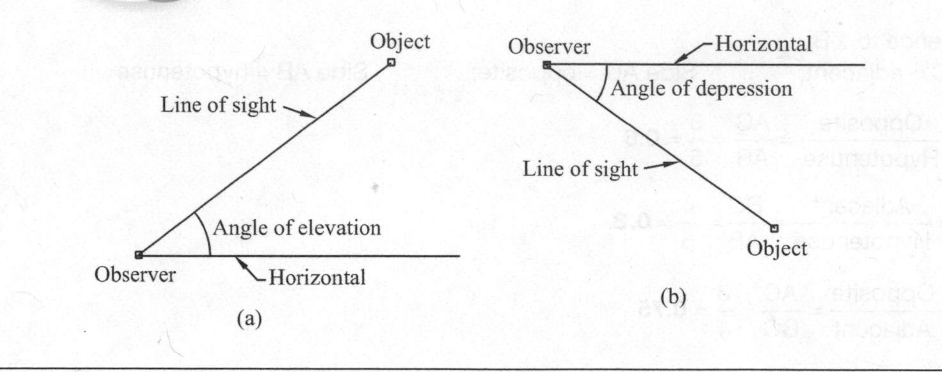

Figure 7.3

Example 7.3

From an observation point, 70.0 m from a building, the angle of elevation to the top of the building is 55°. If the height of the instrument is 1.420 m and the ground between the surveyor and the building is level, find the height of the building.

Solution:

The building and the other details are shown in Figure 7.4.

Height of the building = AC + CE = AC + 1.420 m

In triangle ABC, AC is the opposite side and BC is the adjacent side.

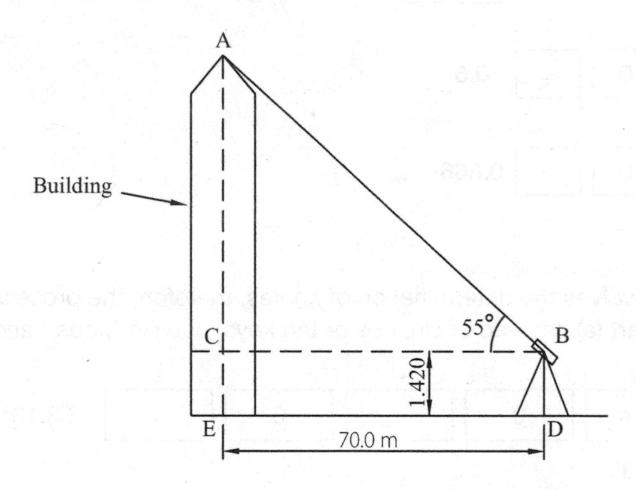

Figure 7.4

$$\frac{AC}{BC} = \tan 55°$$

$$AC = BC \times \tan 55°$$

$$= 70 \times \tan 55° = 99.97 \text{ m}$$

Height of the building = AC + 1.420 = 99.97 + 1.420 = **101.39 m**

Example 7.4

The angles of depression from the top of an 80 m high building, to points C and D, are 70° and 30°, respectively (Figure 7.5). If points C and D are on the opposite edges of a road, and A, C and D are in the same vertical plane, find the width of the road.

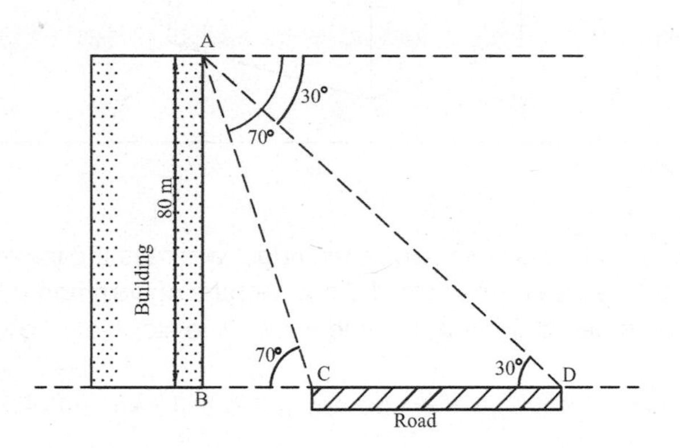

Figure 7.5

Solution:

Assuming the building is vertical and the ground is level, $\angle C = 70°$ in triangle ABC and $\angle D = 30°$ in triangle ABD.

Width of the road = CD = BD – BC

Consider right-angled triangle ABC: $\tan 70° = \frac{\text{Opp}}{\text{Adj}} = \frac{AB}{BC}$

$$BC = \frac{AB}{\tan 70°} = \frac{80}{\tan 70°} = 29.118\text{ m}$$

Consider right-angled triangle ABD: $\tan 30° = \frac{\text{Opp}}{\text{Adj}} = \frac{AB}{BD}$

$$BD = \frac{AB}{\tan 30°} = \frac{80}{\tan 30°} = 138.564\text{ m}$$

Width of the road = BD – BC

= 138.564 – 29.118 = **109.446 m**

7.4 Roofs

Trigonometry is often used in calculating the pitch of a roof, the length of common/hipped rafters and the surface area of a roof. Before embarking on the solution of problems it is important to get familiar with the main technical terms associated with roof construction. Figure 7.6 illustrates some basic terminology used in roof construction.

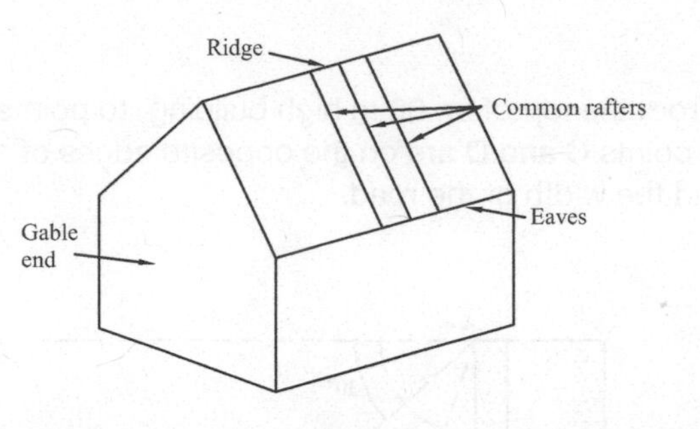

Figure 7.6 Roof terminology

The highest member of a roof truss is known as the ridge, whereas the lowest member is known as the eaves. In order to calculate the area of the roof, the true lengths of common rafters and hipped rafters are required; these can only be determined from the end views. The slope of a roof is called its pitch.

Example 7.5

The roof shown in Figure 7.7 is 10 m long, 4 m high and has a span of 12 m. Calculate:

a) Pitch of the roof.
b) True lengths of the common rafters.
c) Surface area of the roof.

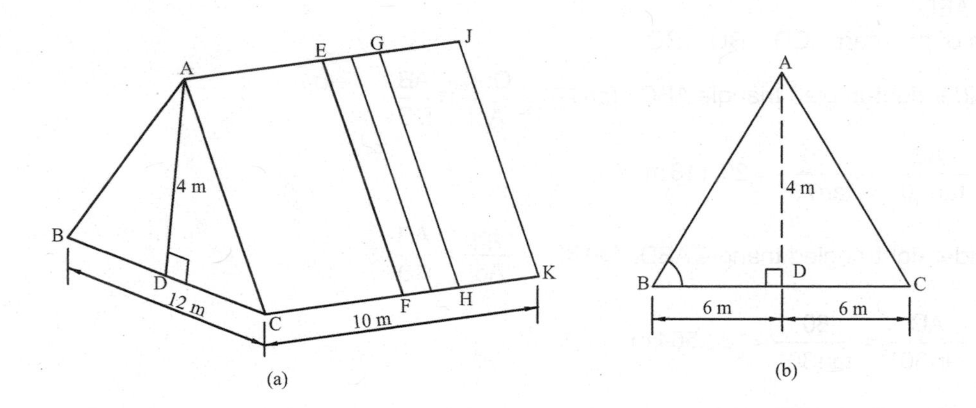

Figure 7.7

Solution:

a) As the roof is symmetrical, BD = DC, and AB = AC

$\angle B$ or $\angle C$ is the pitch of the roof

With reference to $\angle B$ in the right-angled triangle ABD:

AD = opposite side = 4m; BD = adjacent side = 6 m

$$\tan B = \frac{AD}{BD} = \frac{4}{6} = 0.6667$$

Therefore $\angle B = \tan^{-1} 0.6667 = 33.69°$

The pitch of the roof is 33.69°.

b) True length of a common rafter is equal to AB or AC

$$= \frac{6}{\cos 33.69°} = \frac{6}{0.83205} = 7.211\,\text{m}$$

Therefore, **true length of the common rafter = 7.211 m.**

(This can also be calculated by the theorem of Pythagoras.)

c) Surface area of the roof = 2 × Area of surface ACKJ

AC = KJ = 7.211 m; CK = AJ = 10 M

Therefore, surface area of the roof = 2 × (AC × CK)

= 2 × (7.211 × 10)

= **144.22 m²**

7.5 The sine rule and the cosine rule

In the previous sections the trigonometrical ratios were applied to determine the unknown angles and the sides of right-angled triangles. For triangles which do not have a right-angle, the trigonometrical ratios cannot be applied directly, instead sine and cosine rules may be used to determine the unknown sides and angles.

7.5.1 The sine rule

According to the sine rule the ratio of the length of a side to the sine of the angle opposite that side is constant in any triangle,

$$\text{or } \frac{a}{\sin A} = \frac{b}{\sin B} = \frac{c}{\sin C}$$

where a, b and c are the sides and A, B and C are the angles of ΔABC, as shown in Figure 7.8. Sides a, b and c are opposite ∠A, ∠B and ∠C respectively.

The above rule can be adapted if we have triangles DEF, PQR etc.

$$\text{In } \Delta DEF,\ \frac{d}{\sin D} = \frac{e}{\sin E} = \frac{f}{\sin F}$$

$$\text{In } \Delta PQR,\ \frac{p}{\sin P} = \frac{q}{\sin Q} = \frac{r}{\sin R}$$

The sine rule may be used for the solution of triangles when:

- 2 angles and 1 side are known.
- 2 sides and the angle opposite 1 of them are known.

Example 7.6

In ΔABC, $\angle B = 65°$, side a = 15 cm and side b = 20 cm. Find $\angle A$, $\angle C$ and side c.

Solution:

ΔABC is shown in Figure 7.8.

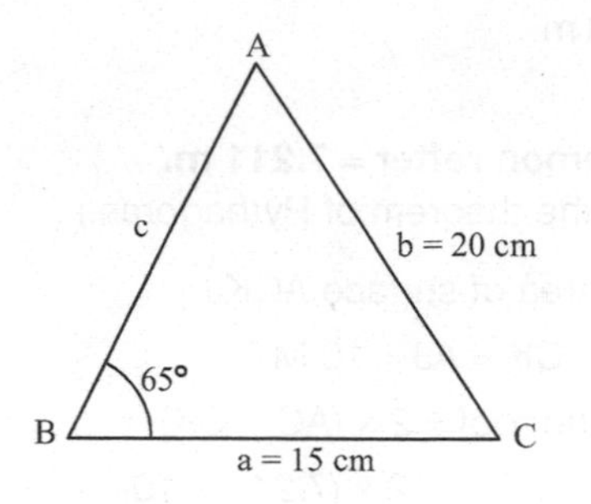

Figure 7.8

$$\frac{a}{\sin A} = \frac{b}{\sin B} = \frac{c}{\sin C}$$

$$\frac{15}{\sin A} = \frac{20}{\sin 65} = \frac{c}{\sin C}$$

Consider $\frac{15}{\sin A} = \frac{20}{\sin 65}$; After transposition, $\sin A = \frac{15 \times \sin 65°}{20}$

$\sin A = 0.679731$ or $\angle$**A** = **42.823°** or 137.177°

$\angle A$ cannot be 137.177°, as $\angle A + \angle B + \angle C$ will be more than 180°.

$\angle$**C** = 180 – 65° – 42.823° = **72.177°**

$\frac{b}{\sin B} = \frac{c}{\sin C}$; After transposition, $c = \frac{b \times \sin C}{\sin B}$

$$\text{Side } c = \frac{20 \times \sin 72.177°}{\sin 65°}$$

$$\text{Side } c = \frac{19.040}{0.9063} = \mathbf{21.01\,cm}$$

7.5.2 The cosine rule

According to the cosine rule the square of any side of a triangle is equal to the sum of the squares of the other 2 sides minus the product of those 2 sides and the cosine of their included angle:

i.e. $a^2 = b^2 + c^2 - 2bc \cos A$

$b^2 = c^2 + a^2 - 2ca \cos B$

$c^2 = a^2 + b^2 - 2ab \cos C$

The cosine rule may be used for the solution of triangles when:

- 2 sides and the included angle are known.
- 3 sides are known.

Example 7.7

In triangle ABC sides AB and BC are 80 cm and 100 cm respectively, and the angle enclosed between these 2 sides is 50°. Find the unknown side and angles.

Solution:

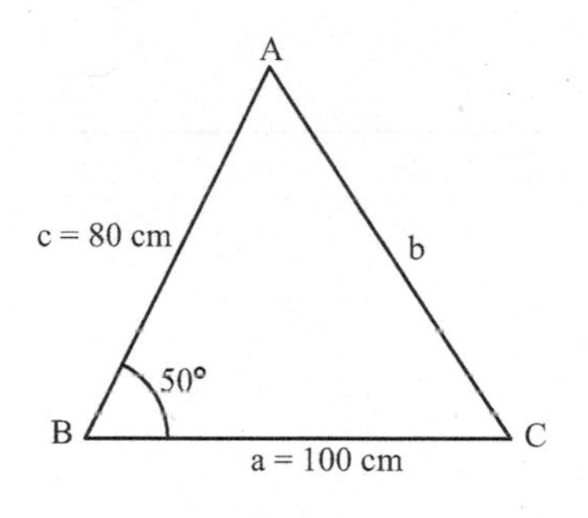

Figure 7.9

In this question 2 sides and the enclosed angle are given, therefore, the cosine rule will be used to determine the remaining side and angles (see Figure 7.9).

The equation that is appropriate to begin with, is:

$b^2 = c^2 + a^2 - 2ac \cos B$ (because $\angle B$ is known)

$= 80^2 + 100^2 - 2 \times 80 \times 100 \times \cos 50°$

$= 6400 + 10000 - 10284.602$

$= 6115.398$

Therefore, side $\mathbf{b = \sqrt{6115.398} = 78.201}$ **cm**

To find $\angle A$, use: $a^2 = b^2 + c^2 - 2bc \cos A$

$100^2 = 78.201^2 + 80^2 - 2 \times 78.201 \times 80 \times \cos A$

$10000 = 6115.398 + 6400 - 12512.16 \cos A$

$12512.16 \cos A = 12515.398 - 10\,000$

$$\cos A = \frac{2515.398}{12512.16} = 0.201036$$

$\angle\mathbf{A = \cos^{-1} 0.201036 = 78.402^\circ}$

$\angle\mathbf{C = 180^\circ - 50^\circ - 78.402^\circ = 51.598^\circ}$

Example 7.8

A building plot is shown in Figure 7.10a. Calculate ∠D, ∠DAC and the length of sides AC and DA.

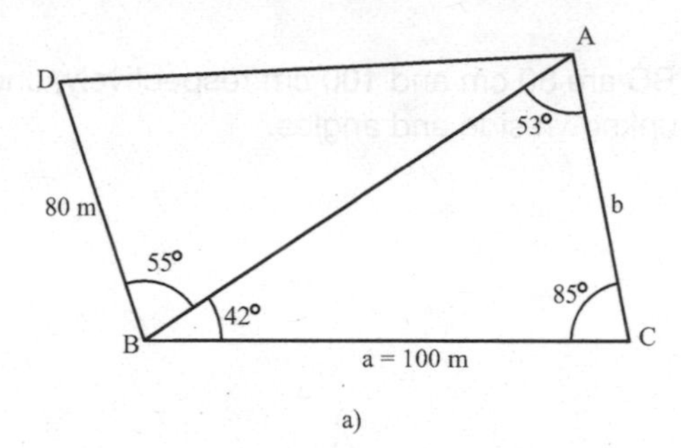

Figure 7.10a

Solution:

Consider ΔABC; $\frac{a}{\sin A} = \frac{b}{\sin B} = \frac{c}{\sin C}$

$$\frac{a}{\sin A} = \frac{b}{\sin B} \text{ or } \frac{100}{\sin 53^\circ} = \frac{b}{\sin 42^\circ}$$

Therefore, $b = \frac{100 \times \sin 42^\circ}{\sin 53^\circ} = 83.784\text{ m}$

b (side AC) = 83.784 m

Similarly, $c = \frac{a \times \sin C}{\sin A} = \frac{100 \times \sin 85^\circ}{\sin 53^\circ}$

c (side AB) = 124.737 m

Consider ΔDAB (Figure 7.10b):

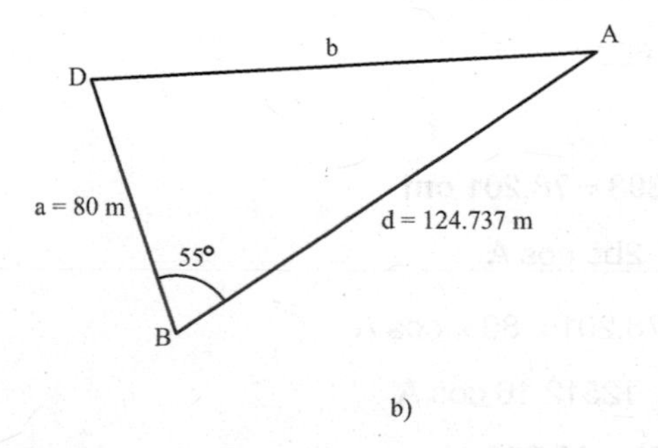

Figure 7.10b

Using cosine rule, $b^2 = d^2 + a^2 - 2da \cos B$

$$= (124.737)^2 + (80)^2 - 2 \times 124.737 \times 80 \times \cos 55°$$

$$= 15559.3192 + 6400 - 11447.393 = 10511.9266$$

b (side DA) = $\sqrt{10511.9266}$ = 102.528 m

Using sine rule, $\dfrac{80}{\sin A} = \dfrac{102.528}{\sin 55°} = \dfrac{124.737}{\sin D}$

$\dfrac{80}{\sin A} = \dfrac{102.528}{\sin 55°}$ or $\sin A = \dfrac{80 \times \sin 55°}{102.528} = 0.63916$

$\angle A = \sin^{-1} 0.63916 = 39.729°$

Total $\angle A$ ($\angle DAC$) = 53° + 39.729° = 92.729°

In ΔDBA, **$\angle D = 180° - 55° - 39.729° = 85.271°$**

7.5.3 Sine rule: the ambiguous case

If we are given 1 side and 2 angles of a triangle, there will be only 1 solution for the unknown sides. If we are given 2 sides and 1 angle, the sine rule could give us 1 or more solutions. If we know sides a and b, and $\angle A$ of ΔCAB (Figure 7.11) then side a can be swung to the left or right and produce 2 possible triangles, i.e. ΔCAB and $\Delta CAB'$. The solution will produce 2 values of $\angle B$. There are a number of conditions for a case to be ambiguous:

i) $\angle A$ is less than 90°.

ii) Side b is longer than side a.

iii) 2 sides and 1 angle are known. For example, sides a, b and $\angle A$ are known.

iv) Side a is longer than the vertical height CD.

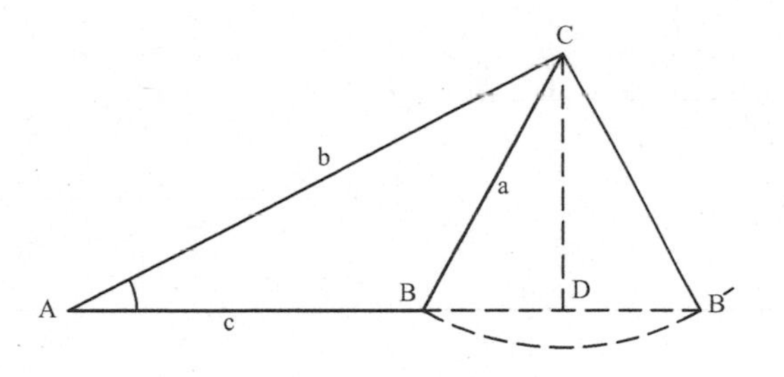

Figure 7.11 Sine rule: the ambiguous case

Example 7.9

In ΔABC, $\angle A = 30°$, side a = 10 cm and side b = 16 cm. Find $\angle B$ and $\angle C$.

Solution:

ΔABC is shown in Figure 7.12

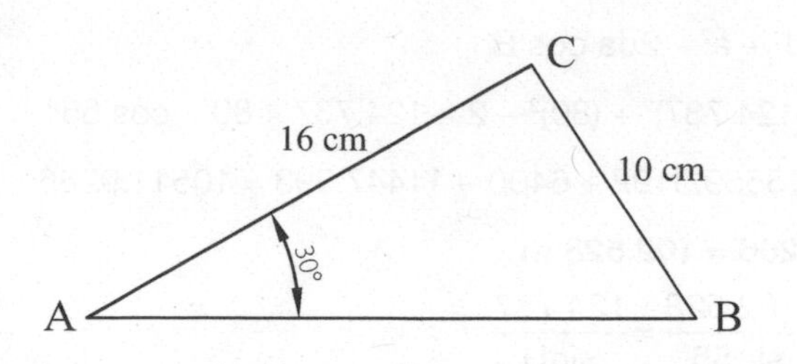

Figure 7.12

$$\frac{a}{\sin A} = \frac{b}{\sin B} = \frac{c}{\sin C}$$

$$\frac{10}{\sin 30} = \frac{16}{\sin B} = \frac{c}{\sin C}$$

Consider $\frac{10}{\sin 30} = \frac{16}{\sin B}$, After transposition $\sin B = \frac{16 \times \sin 30°}{10}$

$\sin B = 0.8$ or **$\angle B = 53.13°$ or $126.87°$**

If $\angle B = 53.13°$, $\angle C = 180° - 53.13° - 30° =$ **$96.87°$**

If $\angle B = 126.87°$, $\angle C = 180° - 126.87° - 30° =$ **$23.13°$**

This is the ambiguous case as there are 2 solutions. The solutions are shown in Figure 7.13; it is obvious that the 2 triangles are not congruent to each other.

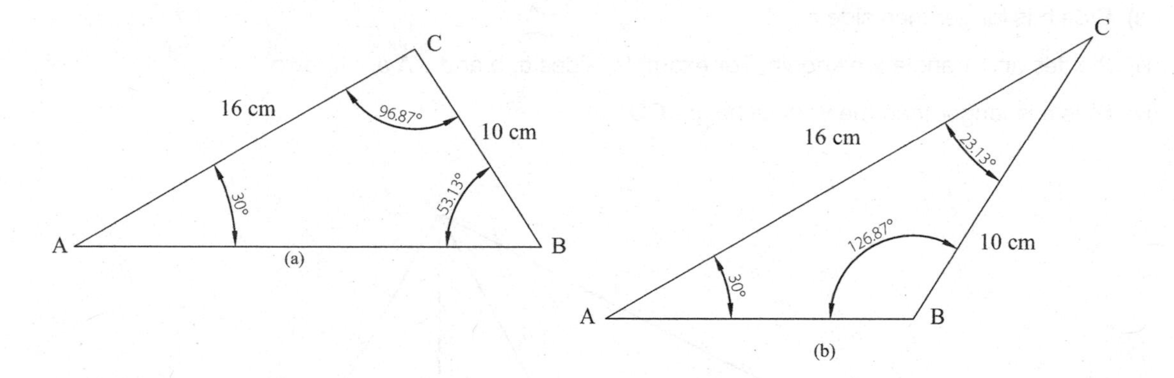

Figure 7.13

7.6 Frames

Frames are widely used in buildings and civil engineering structures to resist the applied forces and distribute them to the other parts of the structure in a safe manner. Refer to Appendix T for more details on frames.

Before a frame can be structurally designed, it is necessary to find the magnitude of force in each member of the frame. Here we are concerned with only the analysis of the forces and not the structural design. So, let us discuss the basics first:

A force which is inclined to the horizontal at an angle other than 90° or 0° can be resolved into 2 components, i.e. a horizontal component and a vertical component. Consider a force F, inclined at $\angle\theta$ to the horizontal as shown in Figure 7.14:

$$\cos\theta = \frac{\text{horizontal component}}{\text{Force F}}$$

Therefore, horizontal component = F × cos θ or F cos θ

$$\sin\theta = \frac{\text{vertical component}}{\text{Force F}}$$

Vertical component = F sin θ

The forces in the members of a frame, at any particular joint, are resolved into horizontal and vertical components. For equilibrium:

1) The sum of all horizontal forces at a joint should be equal to 0 (or $\Sigma H = 0$), and
2) The sum of all vertical forces at a joint should be equal to 0 (or $\Sigma V = 0$).

Each joint needs to be considered for determining forces in all the members.

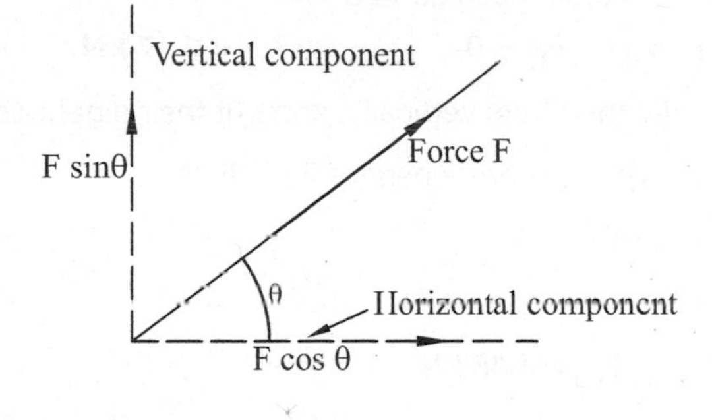

Figure 7.14

Example 7.10

Find the magnitude of forces in the members of the truss shown in Figure 7.15.

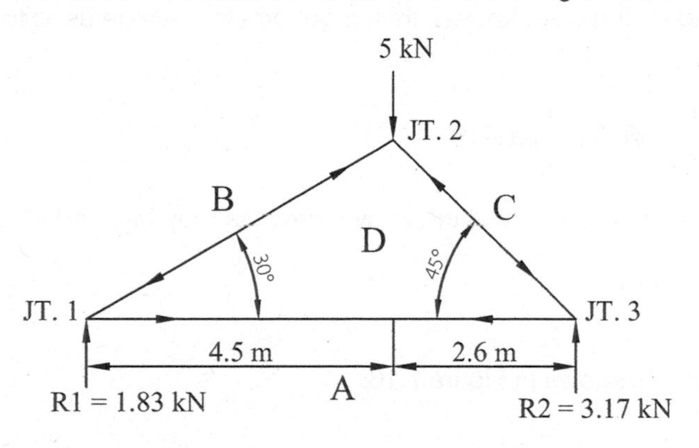

Figure 7.15

Solution:

F_{BD} represents the force in member BD, similarly F_{DA} represents the force in member DA.

At joint 1: Resolve forces in the members vertically, and put their algebraic sum equal to 0:

Vertical component of $F_{BD} - R_1 = 0$ (1)

Force in member BD and Reaction R_1 act in the opposite direction, therefore if 1 is considered positive, the other will be considered negative.

Vertical component of $F_{BD} = F_{BD} \times \sin\theta$ (explained earlier in section 7.6)

$= F_{BD} \times \sin 30° = 0.5\,F_{BD}$

Substituting values in equation (1), $0.5\,F_{BD} - 1.83 = 0$ or $\mathbf{F_{BD} = 3.66\ kN}$

Resolve forces in the members horizontally, and put their algebraic sum equal to 0:

Horizontal component of $F_{BD} - F_{DA} = 0$ (2)

(There is no horizontal component of R_1; the forces in members BD and DA act in the opposite direction, therefore if 1 is considered positive, the other will be considered negative.)

Horizontal component of $F_{BD} = F_{BD} \times \cos\theta$

$= 3.66 \times \cos 30° = 3.17$

From equation (2), $3.17 - F_{DA} = 0$, or $\mathbf{F_{DA} = 3.17\ kN}$

At joint 2: Resolve forces in the members vertically, and put their algebraic sum equal to 0:

Vertical component of F_{BD} + Vertical component of $F_{CD} = 5$ kN

$F_{BD} \times \sin 30° + F_{CD} \times \sin 45° = 5$

$3.66 \times 0.5 + F_{CD} \times 0.707 = 5$

$F_{CD} \times 0.707 = 3.17$, $\therefore\ \mathbf{F_{CD} = 4.48\ kN}$

(The forces in member CD could also be determined by resolving forces at joint 3.)

7.7 Area of triangles

If 2 sides and the included angle are known, then trigonometry may be used to determine the area of the triangle:

$$\text{Area} = \frac{1}{2}ab\sin C = \frac{1}{2}bc\sin A = \frac{1}{2}ca\sin B$$

Depending on the information given, 1 of the above formulae may be used for area calculation.

Example 7.11

Find the area of the triangle shown in Figure 7.16.

Solution:

In Figure 7.16, a = 100 cm; b = 130 cm;

Angle between sides a and b, i.e. $\angle C = 60°$

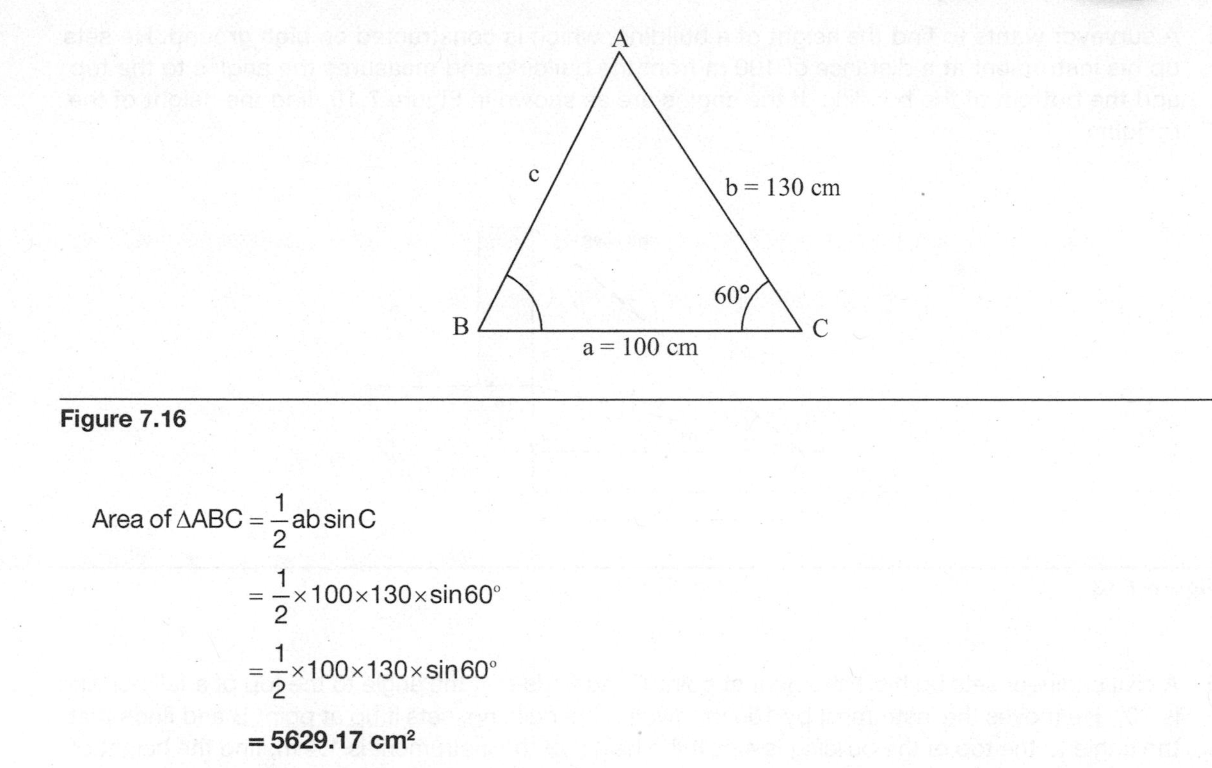

Figure 7.16

$$\text{Area of } \Delta ABC = \frac{1}{2} ab \sin C$$
$$= \frac{1}{2} \times 100 \times 130 \times \sin 60^\circ$$
$$= \frac{1}{2} \times 100 \times 130 \times \sin 60^\circ$$
$$\mathbf{= 5629.17\ cm^2}$$

Exercise 7.1

1. A land surveyor wants to find the distance between 2 buildings A and B. He sets up his instrument at building A and finds that the angle of elevation to the top of building B is 7°. If the height of building B is 8.0 m, find the distance between the 2 buildings. The height of the instrument is 1.35 m.
2. A 6 m long ladder is placed on horizontal ground and its top rests against a wall. How far should the foot of the ladder be from the wall so that the ladder makes an angle of 75° with the ground?
3. An engineer wants to find the width of a river and stands on one bank at point S, directly opposite a tree, as shown in Figure 7.17. She sets up another station at point R which is 60 m from point S, along the river bank. If angle TRS is 60°, find the width of the river.

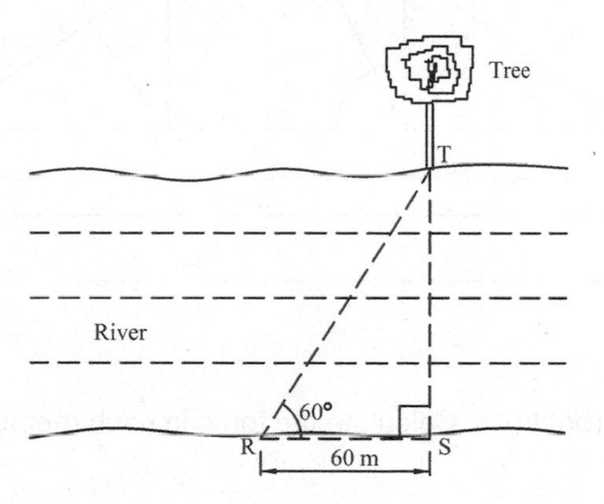

Figure 7.17

4. A surveyor wants to find the height of a building, which is constructed on high ground. He sets up his instrument at a distance of 100 m from the building and measures the angles to the top and the bottom of the building. If the angles are as shown in Figure 7.18, find the height of the building.

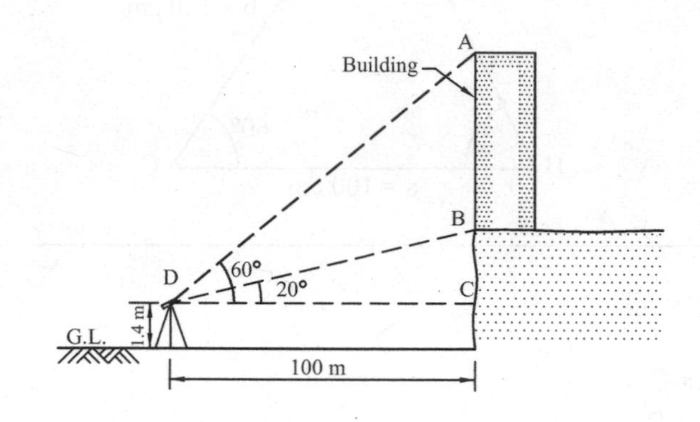

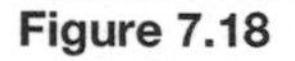
Figure 7.18

5. A civil engineer sets up his instrument at point A and finds that the angle to the top of a tall building is 20°. He moves the instrument by 100 m towards the building, sets it up at point B and finds that the angle to the top of the building is 42°. If the height of the instrument is 1.4 m, find the height of the building.
6. A pitched roof is 12 m long, 4 m high and has a span of 14 m. Calculate:
   a) Pitch of the roof.
   b) True lengths of the common rafters.
   c) Surface area of the roof.
7. A trussed rafter is shown in Figure 7.19. Find the length of members HD, HE and AH.

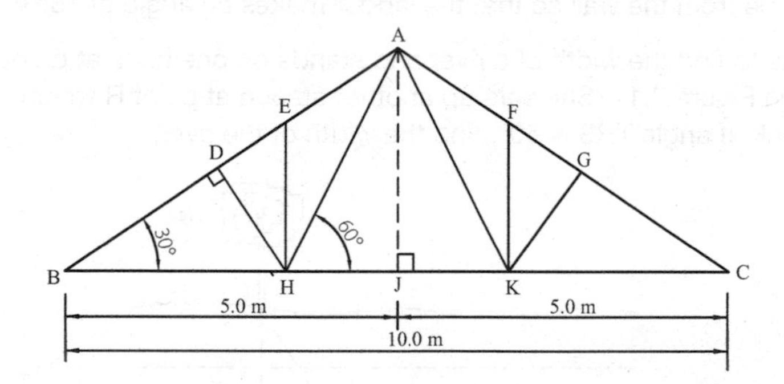

Figure 7.19 Trussed rafter

8. Figure 7.20 shows a fink roof truss. Calculate the force in each member of the truss.

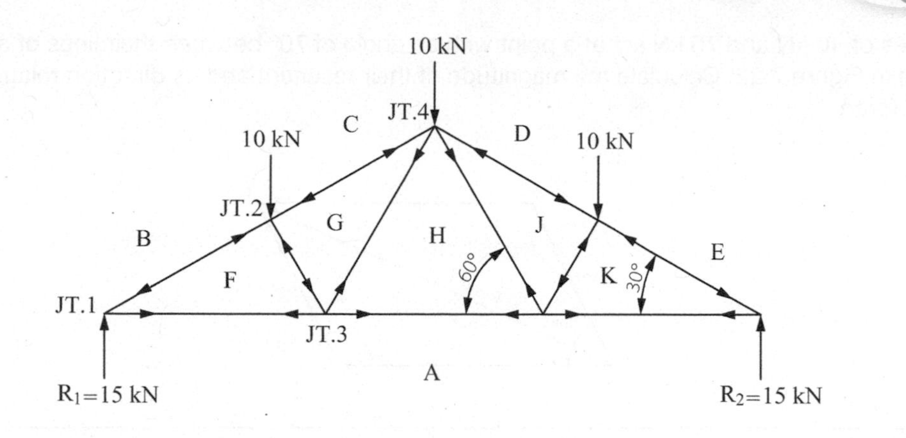

Figure 7.20

Exercise 7.2

1. In triangle PQR, ∠Q = 50°, ∠R = 70° and side PQ = 75 cm. Calculate ∠P and sides PR and QR.
2. In triangle ABC, sides a, b and c are 8 cm, 7 cm and 5 cm respectively. Use the cosine rule to find ∠B.
3. A surveyor wants to find the distance between 2 buildings A and B. However, the distance cannot be calculated directly as there is a small lake between the two buildings. The surveyor sets up a station at point C and measures angle ACB to be 120°. If distances AC and CB are 220 m and 260 m respectively, find the distance between the 2 buildings.
4. Figure 7.21 shows 2 members of a crane, 1 in tension (tie) and the other in compression (strut). While lifting a certain load, the angle between the strut and the tie was 55°. Find the length of the strut and the angle that it makes with the vertical wall.

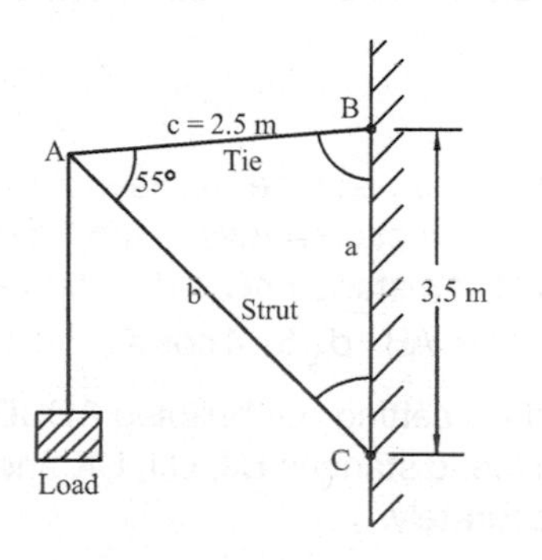

Figure 7.21

5. 2 forces of 40 kN and 70 kN act at a point with an angle of 70° between their lines of action, as shown in Figure 7.22. Calculate the magnitude of their resultant and its direction relative to the 70 kN force.

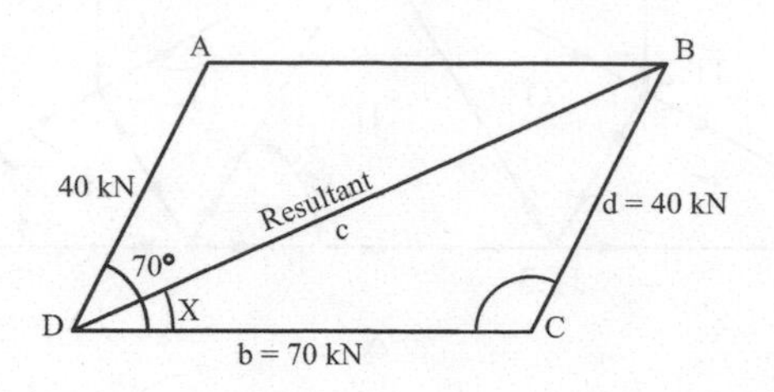

Figure 7.22

6. Figure 7.23 shows the dimensions of a building site. Find the area ABCD.

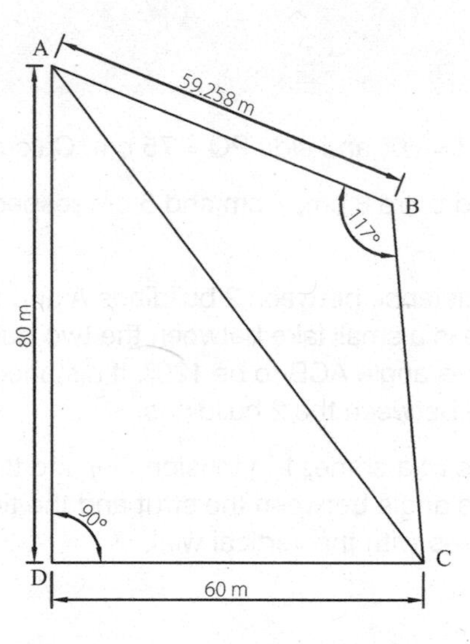

Figure 7.23

7. Bob Brickwalla wants to determine the distance between 2 points A and B. The distance cannot be measured directly as there is an obstruction between the 2 buildings. He sets up his instrument at point C, and measures angle ACB. If distance AC = d metres, ∠ACB = X° and distance CB is twice the distance AC, show that distance $AB = d\sqrt{5 - 4\cos X}$

8. Stations J and K are established for setting out building ABCD, as shown in Figure 7.24. Find the information, which is not shown (i.e. distances BJ, CJ, DK and AK), so that the setting out engineer is able to set out the building accurately.

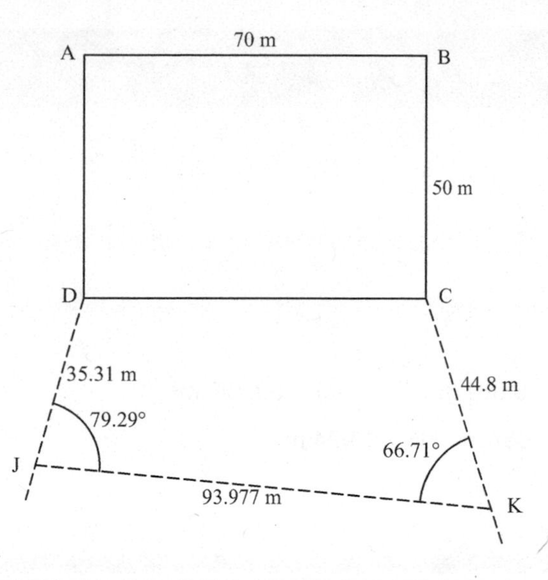

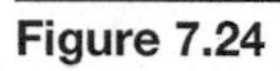
Figure 7.24

9. Station T is 65 m from corner A and 41 m from corner D of building ABCD (see Figure 7.25). If the whole circle bearings of TD and TA are 12.5° and 308° find the distance BT and CT, and angles ∠ATB and ∠CTD so that the building can be set out accurately.

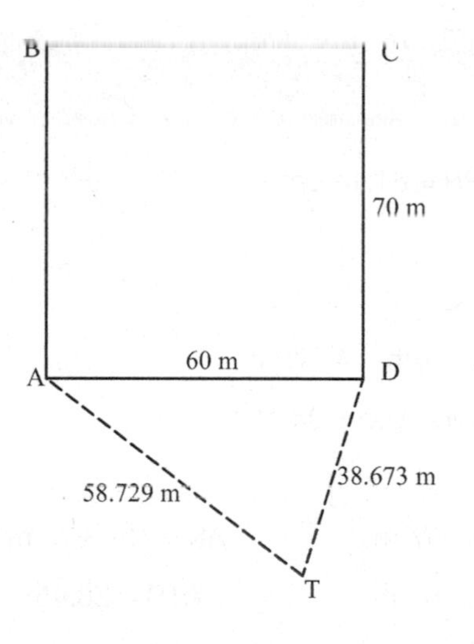

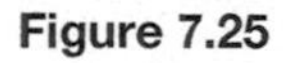
Figure 7.25

Answers – Exercise 7.1

1. 54.16 m
2. 1.553 m
3. 103.923 m
4. 136.808 m
5. 62.492 m
6. a) 29.745° b) 8.062 m c) 193.488 m^2
7. AH = 3.334 m HD = 1.667 m HE = 1.924 m
8.

Member	Compression/Tension	Force (kN)
BF, EK	Compression	30
FA, AK	Tension	26
CG, DJ	Compression	25
GF,KJ	Compression	8.7
GH, HJ	Tension	8.7
HA	Tension	17.3

Answers – Exercise 7.2

1. ∠P = 60° QR = 69.12 cm PR = 61.14 cm
2. 60°
3. 416.173 m
4. Angle = 35.8° Length = 4.272 m
5. Resultant = 91.735 kN Direction = 24.19°
6. 3931.765 m^2
8. DK = 94.049 m CJ = 86.657 m AK = 124.925 m BJ = 115.775 m
9. BT = 115.806 m CT = 107.04 m ∠ATB = 28.45° ∠CTD = 13.42°

CHAPTER 8

Trigonometry 2

Topics covered in this chapter:

- Trigonometric identities
- Trigonometric ratios of compound angles
- Trigonometric equations
- Trigonometric waveforms

8.1 Introduction

In the previous chapter the trigonometric ratios and their application in the construction problems were explained. This was followed by sine rule and cosine rule and their application in the field of land surveying. In this chapter, trigonometric identities, trigonometric equations and sinusoidal wave forms will be discussed.

8.2 Trigonometric identities

Identities may look similar to equations, but they are true for all values of the unknown variable. An equation on the other hand is only true for certain values of the unknown variable. Some of the trigonometric identities are dealt with in this section as they can be used to simplify trigonometric expressions/equations.

The most fundamental trigonometric identity is: $\sin^2\theta + \cos^2\theta = 1$, and to prove it consider a right-angled triangle ABC, shown in Figure 8.1. Using Pythagoras' theorem:

$b^2 + a^2 = c^2$

Divide both sides by c^2: $\dfrac{b^2}{c^2} + \dfrac{a^2}{c^2} = \dfrac{c^2}{c^2}$

$$\left(\frac{b}{c}\right)^2 + \left(\frac{a}{c}\right)^2 = 1$$

$$(\sin\theta)^2 + (\cos\theta)^2 = 1$$

$(\sin\theta)^2$ and $(\cos\theta)^2$ can be written as $\sin^2\theta$ and $\cos^2\theta$

or $\sin^2\theta + \cos^2\theta = 1$

From the above, after transposition $\sin^2\theta = 1 - \cos^2\theta$

and $\cos^2\theta = 1 - \sin^2\theta$

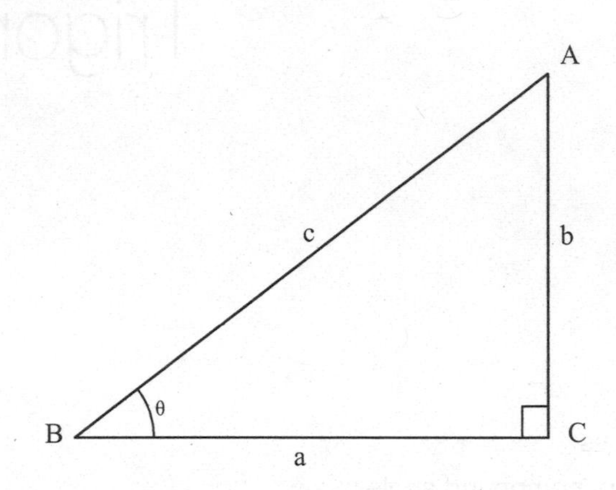

Figure 8.1

Example 8.1

Show that: $1 + \tan^2\theta = \sec^2\theta$

Solution:

Refer to Figure 8.1 that shows a right-angled triangle ABC. Using Pythagoras' theorem:

$a^2 + b^2 = c^2$

Divide both sides by a^2: $\frac{a^2}{a^2} + \frac{b^2}{a^2} = \frac{c^2}{a^2}$

$1 + \left(\frac{b}{a}\right)^2 = \left(\frac{c}{a}\right)^2$

$1 + (\tan\theta)^2 = (\sec\theta)^2$

or $1 + \tan^2\theta = \sec^2\theta$

Example 8.2

Prove the identity: $1 - 2\sin^2\theta = 2\cos^2\theta - 1$

Solution:

As shown earlier in example 8.1, $\sin^2\theta = 1 - \cos^2\theta$

Therefore, $1 - 2\sin^2\theta = 1 - 2(1 - \cos^2\theta)$

$= 1 - 2 + 2\cos^2\theta$

$= 2\cos^2\theta - 1$

Hence $1 - 2\sin^2\theta = 2\cos^2\theta - 1$

Example 8.3

Use a right-angled triangle to show that: $(1 - \sin^2\theta)(1 + \tan^2\theta) = 1$

Solution:

$(1 - \sin^2\theta)(1 + \tan^2\theta) = 1 - \sin^2\theta + \tan^2\theta - \sin^2\theta\tan^2\theta$

$$= 1 - \left(\frac{b}{c}\right)^2 + \left(\frac{b}{a}\right)^2 - \left(\frac{b}{c}\right)^2\left(\frac{b}{a}\right)^2 \qquad \text{(Refer to Figure 8.1)}$$

$$= 1 - \frac{b^2}{c^2} + \frac{b^2}{a^2} - \frac{b^2}{c^2} \times \frac{b^2}{a^2}$$

$$= 1 - \frac{c^2 - a^2}{c^2} + \frac{c^2 - a^2}{a^2} - \frac{c^2 - a^2}{c^2} \times \frac{c^2 - a^2}{a^2}$$

$$= 1 - 1 + \frac{a^2}{c^2} + \frac{c^2}{a^2} - 1 - \left(1 - \frac{a^2}{c^2}\right) \times \left(\frac{c^2}{a^2} - 1\right)$$

$$= \frac{a^2}{c^2} + \frac{c^2}{a^2} - 1 - \left(\frac{c^2}{a^2} - \frac{c^2}{a^2}\frac{a^2}{c^2} - 1 + \frac{a^2}{c^2}\right)$$

$$= \frac{a^2}{c^2} + \frac{c^2}{a^2} - 1 - \left(\frac{c^2}{a^2} - 1 - 1 + \frac{a^2}{c^2}\right)$$

$$= \frac{a^2}{c^2} + \frac{c^2}{a^2} - 1 - \frac{c^2}{a^2} + 2 - \frac{a^2}{c^2} = \mathbf{1}$$

Example 8.4

Prove that $\dfrac{\cot B \ \text{cosec} B}{1 + \cot^2 B} = \cos B$

Solution:

$$\frac{\cot B \ \text{cosec} B}{1 + \cot^2 B} = \frac{\cot B \ \text{cosec}\ B}{\text{cosec}^2 B} \qquad (1 + \cot^2 B = \text{cosec}^2 B)$$

$$= \frac{\cot B}{\text{cosec}\ B}$$

$$= \frac{\cos B}{\sin B} \times \frac{1}{\text{cosec} B} = \frac{\cos B}{\sin B} \times \sin B = \mathbf{\cos B}$$

8.3 Trigonometric ratio of compound angles

a) To show that $\sin(A+B) = \sin A \cos B + \cos A \sin B$, consider Figure 8.2; the right-angled triangles OMN and OPM contain angle A and angle B: $\angle PON = \angle A + \angle B$

From the geometry of the diagram, $\angle RPM = \angle A$

$$\sin(A + B) = \frac{TP}{OP} = \frac{TR + RP}{OP} = \frac{MN + RP}{OP} \qquad (RT = MN)$$

$$=\frac{MN}{OM}\times\frac{OM}{OP}+\frac{RP}{MP}\times\frac{MP}{OP}$$

$= \sin A \cos B + \cos A \sin B$

Hence $\sin(A + B) = \sin A \cos B + \cos A \sin B$

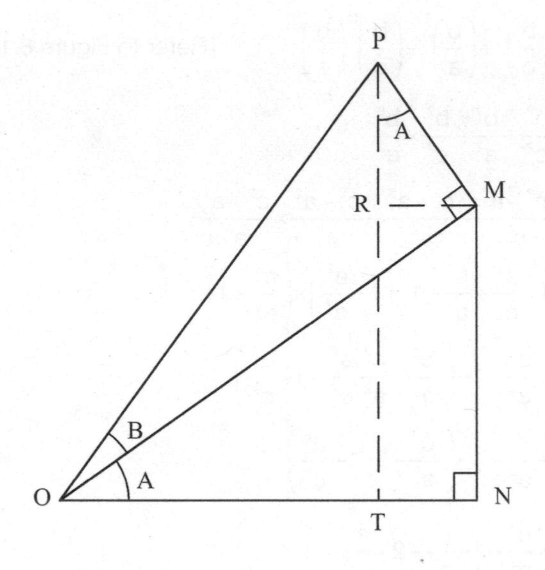

Figure 8.2

b) Show that: $\cos(A + B) = \cos A \cos B - \sin A \sin B$

Refer to ΔOPT in Figure 8.2,

$$\cos(A+B)=\frac{OT}{OP}=\frac{ON-TN}{OP}=\frac{ON-MR}{OP} \qquad (TN=MR)$$

$$=\frac{ON}{OP}-\frac{MR}{OP}$$

$$=\frac{ON}{OM}\times\frac{OM}{OP}-\frac{MR}{MP}\times\frac{MP}{OP}$$

$= \cos A \cos B - \sin A \sin B$

Hence $\cos(A + B) = \cos A \cos B - \sin A \sin B$

c) Similarly it can be shown that $\tan(A+B)=\dfrac{\tan A+\tan B}{1-\tan A\tan B}$

d) Replacing B with – B, the three equations become:

$\sin(A - B) = \sin A \cos B - \cos A \sin B$

$\cos(A - B) = \cos A \cos B + \sin A \sin B$

$$\tan(A-B)=\frac{\tan A-\tan B}{1+\tan A\tan B}$$

8.4 Double angle formulae

Double angle formulae can be established by putting A = B in the compound angle formulae:

a) $\sin(A + A) = \sin A \cos A + \sin A \cos A$

therefore, $\sin 2A = 2 \sin A \cos A$

b) $\cos(A + A) = \cos A \cos A - \sin A \sin A$

$\cos 2A = \cos^2 A - \sin^2 A$

$= 1 - \sin^2 A - \sin^2 A = 1 - 2 \sin^2 A$

Also, $\cos 2A = \cos^2 A - \sin^2 A$

$= \cos^2 A - (1 - \cos^2 A) = 2 \cos^2 A - 1$

c) Similarly it can be shown that $\tan 2A = \dfrac{2 \tan A}{1 - \tan^2 A}$

8.5 Trigonometric equations

There are many types of trigonometric equations, some of which are dealt with in this chapter. Any equation which contains a trigonometric ratio (sine, cosine etc.) is known as a trigonometric equation. As the trigonometric functions are periodic, trigonometric equations usually have an unlimited number of solutions. The solutions, however, may be restricted to those between 0° and 360°.

Figure 8.3 shows the 4 quadrants of a graph and also shows which trigonometric ratio is positive in a particular quadrant.

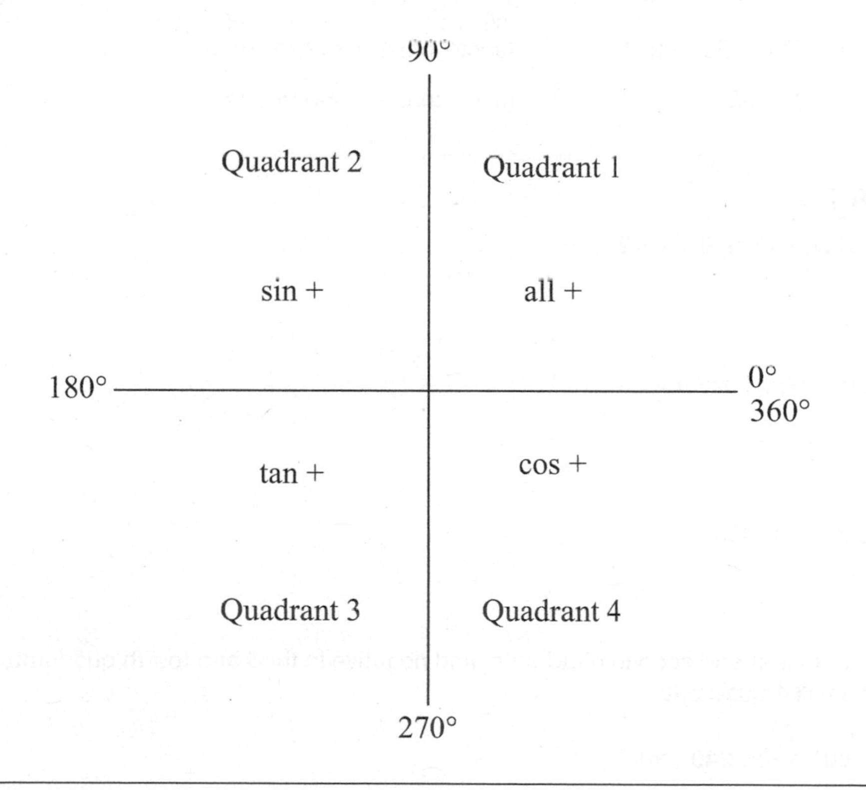

Figure 8.3

Example 8.5

Solve the equation: $\sin x = 0.629321$

Solution:

$\sin x = 0.629321$, therefore $\angle x = \sin^{-1} 0.629321$

or $\angle x = 39.0°$

Since the sine of the angle is positive, the solution must be in quadrants 1 and 2.

Hence **$x = 39°, 141°, 399°, 501°$** etc. $(141° = 180 - 39; 399° = 360 + 39)$

Example 8.6

Solve the equation $4 \sin \theta + 2 = 0$, for values of θ from 0° to 360°.

Solution:

$4 \sin \theta + 2 = 0$ can be written as $4 \sin \theta = -2$

From the above, $\sin \theta = -0.5$

Angle $\theta = -30°$

Since the value of $\sin \theta$ is negative, angle θ is in the third and fourth quadrants

Therefore, $\theta = 180° + 30° =$ **210°** (angle in the third quadrant)

or $\theta = 360° - 30° =$ **330°** (angle in the fourth quadrant)

Example 8.7

Solve the equation $4 \sin^2 \theta - 1 = 2$.

Solution:

After transposition $4 \sin^2 \theta = 2 + 1 = 3$

$$\sin^2 \theta = \frac{3}{4} = 0.75$$

$$\sin \theta = \sqrt{0.75} = \pm 0.866$$

or $\theta = \pm 60°$

$\sin \theta$ is positive in first and second quadrants, and negative in third and fourth quadrants, therefore, the solutions are in all 4 quadrants.

Hence θ **= 60°, 120°, 240°, 300°**

Example 8.8

Solve the equation $3\cos^2 x + \sin^2 x = 2$ for $0° \le x \le 180°$.

Solution:

$3\cos^2 x + \sin^2 x = 3(1 - \sin^2 x) + \sin^2 x = 2$

$3 - 3\sin^2 x + \sin^2 x = 2$

$3 - 2 = 3\sin^2 x - \sin^2 x$

$2\sin^2 x = 1$, therefore $\sin^2 x = 0.5$

$\sin x = \sqrt{0.5}$, or $\sin x = \pm 0.7071$

We are only interested in values of x between 0° and 180°. There are 2 answers:

$\angle x = \sin^{-1} 0.7071$, therefore $\angle x =$ **45°** (first quadrant)

and $\angle x = 180 - 45 =$ **135°** (second quadrant)

Example 8.9

Solve the equation: $2\,\text{cosec}^2 A + 6\cot A = 7$, for $0° \le x \le 360°$.

Solution:

$2(\cot^2 A + 1) + 6\cot A - 7 = 0 \quad (\text{cosec}^2 A - \cot^2 A + 1)$

After simplifying the above, $2\cot^2 A + 6\cot A - 5 = 0$

Solving the above equation by the quadratic formula:

$$\cot A = \frac{-6 \pm \sqrt{36 - 4(2)(-5)}}{2 \times 2}$$

$$= \frac{-6 \pm \sqrt{76}}{4} = \frac{-6 \pm 8.72}{4}$$

$$\cot A = \frac{-6 - 8.72}{4} = -3.68, \quad \text{and} \quad \cot A = \frac{-6 + 8.72}{4} = 0.68$$

$$\tan A = -\frac{1}{3.68} = -0.272 = -15.2°$$

tangent (and cotangent) is negative in the second and fourth quadrants.

$A = 180 - 15.2 =$ **164.8°**, and $A = 360 - 15.2 =$ **344.8°**

$$\tan A = \frac{1}{0.68} = 1.47 = 55.8°$$

tangent (and cotangent) is positive in the first and third quadrants.

$A =$ **55.8°**, and $A = 180 + 55.8 =$ **235.8°**

8.6 Trigonometric graphs

The graph of any trigonometric function can be drawn by obtaining their values at different angles, such as, 0°, 30°, 60°, 90°, 120° etc.; Table 8.1 shows the values of sin x and cos x at different angles. The angles are marked on the x-axis, their values on the y-axis and the graph is produced by drawing a smooth curve through the points, as shown in Figure 8.4.

Table 8.1

x	0°	30°	60°	90°	120°	150°	180°	210°	240°	270°	300°	330°	360°
sin x	0	0.5	0.87	1	0.87	0.5	0	–0.5	–0.87	–1	–0.87	–0.5	0
cos x	1	0.87	0.5	0	–0.5	–0.87	–1	–0.87	–0.5	0	0.5	0.87	1

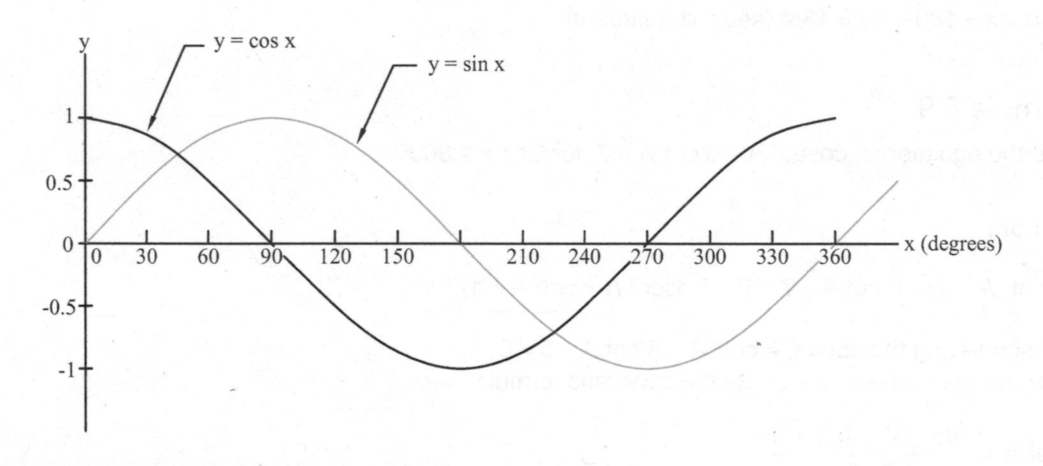

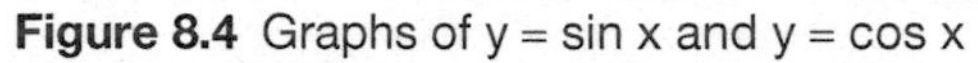
Figure 8.4 Graphs of y = sin x and y = cos x

Some of the important features of the 2 graphs are:

1) The graph of cos x has the same shape as the graph of sin x, but it is ahead of sin x by 90°. The cosine function can be written as:

   y = sin (x + 90°), 90° is the **phase angle**.

2) Each function has a maximum value of +1 and a minimum value of –1; these values are known as the **amplitude**.

3) The graphs of sin x and cos x are periodic, i.e. they repeat themselves after every 360°, therefore, the **period** of sin x and cos x is 360°.

   The graph of sin x is also known as the sine wave and another name for this wave is **sinusoidal waveform.**

The sin x graph may also be produced by considering the rotating radius of a circle. In Figure 8.5 OM is the rotating radius or rotating vector that is free to rotate anticlockwise at a velocity of ω radians per second (rad/s). A rotating vector is known as a **phasor.** The length of the phasor is equal to the radius of the circle, i.e. R.

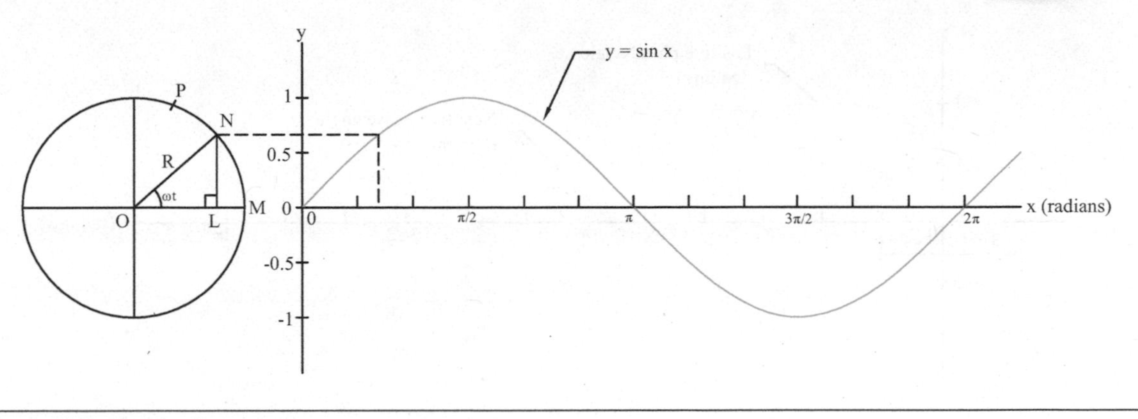

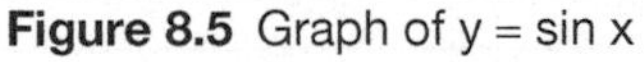
Figure 8.5 Graph of y = sin x

After time 't' seconds phasor OM moves to ON and the angle through which it turns (∠NOM) is ωt. If NL is perpendicular to OM, then $\frac{NL}{ON} = \sin\omega t$ or NL = ON sin ωt. As the phasor moves to P and other positions, and their vertical components are projected onto a graph, the result is the sine wave or the graph of **y = R sin ωt**. The graph shows that the peak value of the sine wave is R, i.e. the length of the rotating vector. When the phasor returns to position OM, the sine wave completes 1 cycle.

Period: the time taken for the waveform to complete 1 cycle is known as the period. We can also say that it is the time taken by phasor OM to complete 1 revolution or 2π radians.

$$\text{Period} = \frac{2\pi}{\omega}\ \text{seconds}$$

Frequency: the number of cycles per second is known as the frequency of the wave.

$\frac{2\pi}{\omega}$ seconds are taken to complete 1 cycle

1 second is taken to complete $\frac{1}{\frac{2\pi}{\omega}}$ or $\frac{\omega}{2\pi}$ cycles

Therefore frequency, $f = \frac{\omega}{2\pi}\ \text{Hz}$ (Hz or hertz is the unit of frequency)

Also, $\text{Frequency} = \frac{1}{\text{Period}}$

Phase angle: it is not necessary that all sinusoidal waveform will pass through the zero point on the x-axis at the same time, but may be shifted to the left or to the right as compared to another sinusoidal waveform. This is known as the phase shift or phase difference and is defined as the angle α in degrees or radians that the waveform has shifted from a reference point along the horizontal axis. The equation for the sine function needs to be modified to include the phase angle and is given by:

$$y = R\sin(\omega t \pm \alpha)$$

If α is positive, the sine wave is said to lead R sin ωt. If α is negative the sine wave is said to lag R sin ωt. Figure 8.6 illustrates these 2 cases.

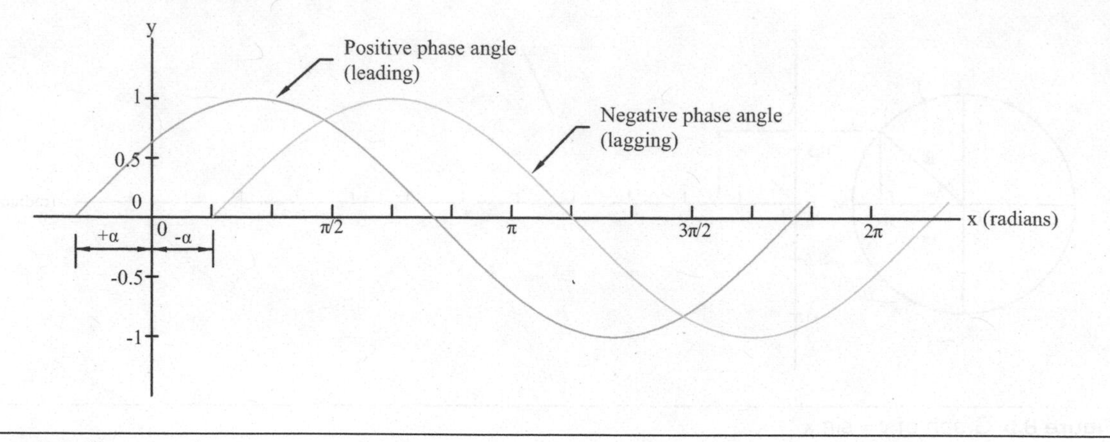

Figure 8.6 Phase angles

Example 8.10

A power tool on a building site can be used with a maximum voltage of 155 volts. If the voltage is sinusoidal and its frequency is 60 Hz, express the voltage (v) in the form, $v = A \sin(\omega t \pm \alpha)$; at time $t = 0$, voltage = 60 V.

Solution:

A = amplitude = maximum voltage = 155 volts

Angular velocity, $\omega = 2\pi f = 2\pi \times 60 = 120\pi$ rad/s

Therefore, $v = 155 \sin(120\pi t + \alpha)$

$60 = 155 \sin(120\pi \times 0 + \alpha)$ (v = 60 V at t = 0)

$60 = 155 \sin \alpha$ or $\sin\alpha = \frac{60}{155} = 0.3871$

$\alpha = \sin^{-1} 0.3871 = 0.397$

Therefore, $\mathbf{v = 155 \sin(120\pi t + 0.397)}$.

Example 8.11

A sinusoidal ocean wave has a frequency of 0.25 Hz, and at time t = 0.2 s the height of the crest of the wave, from the horizontal base line, is 0.618 m (see Figure 8.7).

a) Find the maximum height of the wave's crest from the base line.
b) Find maximum height of the wave (height of the wave is from trough to crest).
c) Express the height of the wave in the form: height = $A \sin \omega t$.

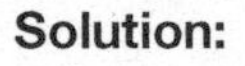

Solution:

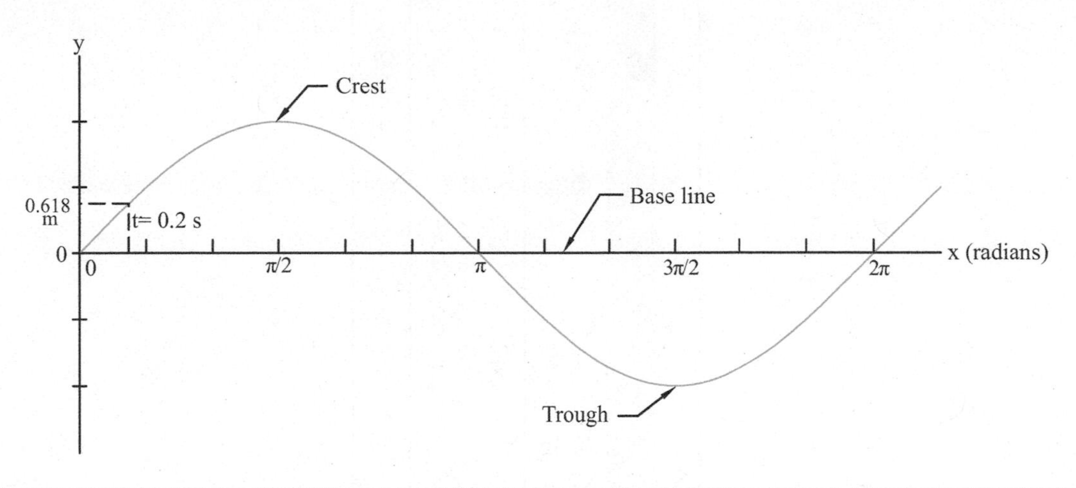

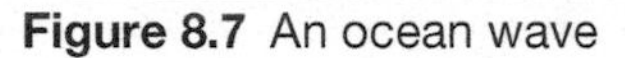

Figure 8.7 An ocean wave

a) Angular velocity, $\omega = 2\pi f = 2\pi \times 0.25 = 0.5\pi$ rad/s

Height of the crest from the base line = $A \sin \omega t = A \sin(0.5\pi t)$

In the above equation, A = amplitude = the height of the crest of the wave, from the horizontal base line.

$0.618 = A \sin(0.5\pi \times 0.2) = A \sin(0.1\pi)$

$0.618 = A \times 0.309$

Therefore, $A = \frac{0.618}{0.309} = \mathbf{2.0\,m}$

b) The height of the wave is from the trough to the crest. Assuming that the distance between the trough and the base line is same as the distance between the crest and the base line, the height of the wave is:

Height of the wave = $2 \times 2.0 =$ **4.0 m**

c) Height of the crest of the wave, above the base line = **$2 \sin(0.5\pi t)$** m

8.7 Addition of sine waves

In mechanical mechanisms and other processes waves of same frequency or different frequencies are produced, and sometimes 2 waves need to be added to see their combined effect. If 2 sine waves have same frequency then their combination will produce a sine wave of the same frequency. If 2 sine waves have different frequencies then their combination will produce a complicated waveform, which will not be in the form of a sine wave. Sine waves may be added graphically, as explained in example 8.12.

Table 8.2

θ°	0	30	60	90	120	150	180	210	240	270	300	330	360
sin θ	0	0.5	0.866	1.0	0.866	0.5	0	–0.5	–0.866	–1.0	–0.866	–0.5	0
10 sin θ	0	5	8.66	10	8.66	5	0	–5	–8.66	–10	–8.66	–5	0
(θ – 30°)	– 30	0	30	60	90	120	150	180	210	240	270	300	330
sin (θ – 30)	–0.5	0	0.5	0.866	1	0.866	0.5	0	–0.5	–0.866	–1	–0.866	–0.5
15 sin (θ – 30)	–7.5	0	7.5	12.99	15	12.99	7.5	0	–7.5	–12.99	–15	–12.99	–7.5
y1 + y2	–7.5	5	16.16	22.99	23.66	17.99	7.5	–5	–16.16	–22.99	–23.66	–17.99	–7.5

Example 8.12

A construction process at a site produces movements given by $y_1 = 10 \sin \theta$ and $y_2 = 15 \sin(\theta - 30°)$. If the movements are measured in mm, draw graphs of the 2 sine waves and hence calculate the resultant movement by adding the 2 waves together.

Solution:

The graphs of y_1 and y_2 are shown in Figure 8.8; the data have been plotted for 1 cycle of 360°. The ordinates of the combined graph can be determined either from the graph or from Table 8.2. Because the 2 sine waves have the same frequency, the combined graph will be a sine wave with the same frequency. The phase angle of the combined graph is 18° and its peak value is 24 mm. Therefore the formula of the combined wave is: $y_3 = 24 \sin(\theta - 18°)$.

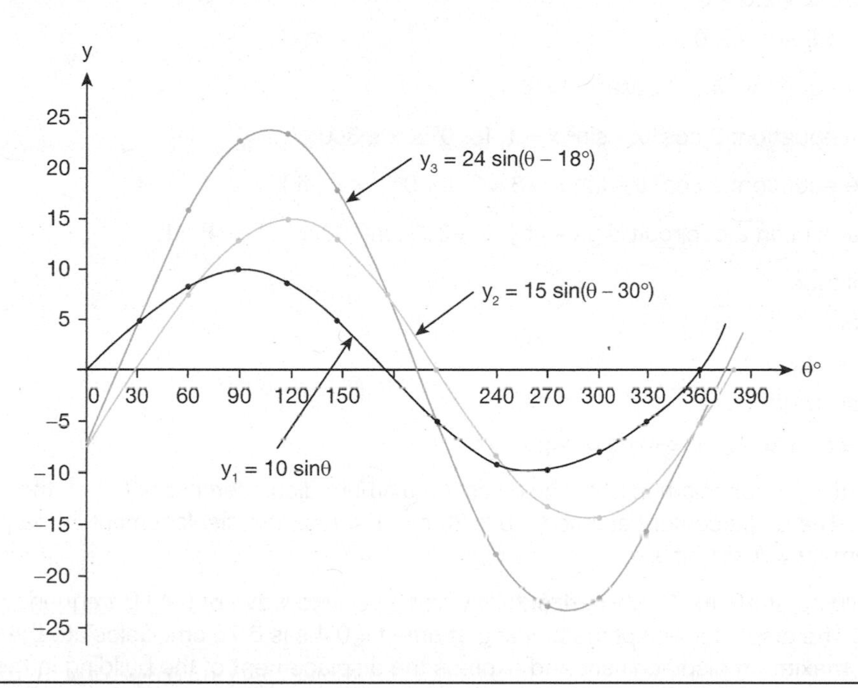

Figure 8.8 Addition of two sine waves of same frequency

Exercise 8.1

1. Simplify: a) $\dfrac{\cos A \tan A}{\sin^2 A}$ b) $\sin A \cos A \cot A$.
2. Use a right-angled triangle to prove that $1 + \cot^2 \theta = \operatorname{cosec}^2 \theta$.
3. Use a right-angled triangle to prove that $\dfrac{\sin\theta + \tan\theta}{\cot\theta + \operatorname{cosec}\theta} = \sin\theta \tan\theta$.
4. Prove that:

   a) $\tan A + \cot A = \operatorname{cosec} A \sec A$

   b) $\dfrac{\cos A}{1 + \sin A} = \dfrac{1 - \sin A}{\cos A}$

c) $\frac{\sec A}{\tan A + \cot A} = \sin A$

5. Prove that: $\frac{1+\cos\theta}{\sin\theta} + \frac{\sin\theta}{1+\cos\theta} = \frac{2}{\sin\theta}$

6. Show that: $\tan(A+B) = \frac{\tan A + \tan B}{1 - \tan A \tan B}$

7. Solve the equations:
   a) $\cos 2x = 0.5$, for x in the range 0° and 360°
   b) $2.5 \tan x - 2.5 = 0$
   c) $5.2 - 4.5 \sec x = 0$

8. Solve the equation: $4 + 2 \operatorname{cosec}^2 B = 6$

9. Solve the equation: $2 \cos^2 x + \sin^2 x = 1$, for $0° \leq x \leq 360°$

10. Solve the equation: $2 \sec^2 \theta + \tan \theta - 3 = 0$, for $0° \leq x \leq 360°$

11. The voltage in an a.c. circuit is given by, $v = 330 \sin(100\pi t - 0.5)$. Find:
    a) amplitude
    b) periodic time
    c) frequency
    d) phase angle
    e) is the phase angle leading or lagging?

12. An oscillating mechanism in a machine has a maximum displacement of 0.5 m and a frequency of 40 Hz. The displacement at time t = 0 is 20 cm. Express the displacement in the general form: displacement = $A \sin(\omega t \pm \alpha)$.

13. A tall building starts to vibrate horizontally when a seismic wave of 0.4 Hz frequency hits the building. The displacement of the building at time t = 0.4 s is 6.75 cm. Calculate the angular velocity, maximum displacement and express the displacement of the building in the form: Displacement = $A \sin \omega t$.

14. A construction process at a site produces 2 movements given by $y_1 = 5 \sin \theta$ and $y_2 = 10 \sin(\theta + 20°)$. If the movements are measured in mm, draw graphs of the 2 sine waves and hence calculate the resultant movement by adding the 2 waves together.

Answers – Exercise 8.1

1. a) cosec A
   b) $\cos^2 A$
2. See solutions in Appendix 3.
3. See solutions in Appendix 3.
4. See solutions in Appendix 3.

5. See solutions in Appendix 3.
6. See solutions in Appendix 3.
7. a) $\theta = 30°$ and $\theta = 330°$
   b) $x = 45°$, and $x = 225°$
   c) $x = 30.07°$, and $x = 329.93°$
8. $B = 90°$, and $B = 270°$
9. $x = 90°$, and $x = 270°$
10. $\theta = 135°$, $\theta = 315°$
    $\theta = 26.57°$, $\theta = 206.57°$
11. a) 330 V b) 0.02 s c) 50 Hz d) 0.5 rad. e) Lagging
12. Displacement $= 0.5 \sin(80\pi t + 0.412)$ m
13. $\omega = 0.8\pi$ rad/s; $A = 0.08$ m; Displacement $= 0.08 \sin(0.8\pi t)$ m
14. $y_3 = 14.75 \sin(\theta + 13.3°)$.

CHAPTER 9

Logarithmic, exponential and hyperbolic functions

Topics covered in this chapter:

- Logarithms and antilogarithms
- Exponential functions and their application
- Hyperbolic functions

9.1 Logarithmic function

Before the invention of scientific calculators, logarithms were used in the form of logarithmic tables (log tables) to perform a range of calculations. The credit for inventing logarithms goes to John Napier although many scientists and mathematicians contributed to the final form that we use now.

If a positive number x is expressed in the form: $x = a^y$, then y is called the logarithm of x to the base a, which is written as:

$$y = \log_a x$$

Consider the number 100; this can be expressed as:

$100 = 10^2$ (Base (a) = 10, Index (y) = 2, x = 100)

We can say that base 10 must be raised to the power 2 to get 100, or $\log_{10} 100 = 2$. Similarly, $2^3 = 8$ (Base (a) = 2, Index (y) = 3)

Therefore, $\log_2 8 = 3$

Although we can use any base, the most commonly used are base 10 and base 'e'. Logarithm having a base of 10 is known as the common logarithm; it is usually shown as $\log_{10}$ or just log. When 'e' is used as the base, the logarithms are known as natural, Napierian or hyperbolic logarithms; 'e' is an irrational number, and its approximate value is 2.71828 (5 d.p.). Logarithms to the base 'e' are shown as $\log_e$ or just ln.

On most scientific calculators there are 2 log keys whereas on others there might be 3.

The key [log] is used for calculations involving 10 as the base.

The key [ln] is used when the base is 'e'.

On some calculators there is a key to select any number as the base.

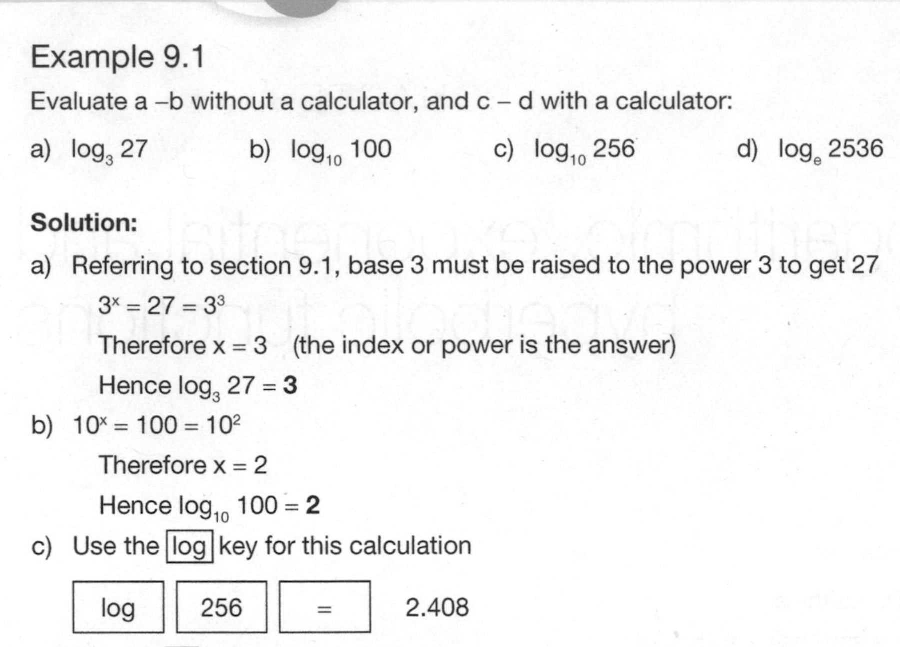

Example 9.1

Evaluate a –b without a calculator, and c – d with a calculator:

a) $\log_3 27$ b) $\log_{10} 100$ c) $\log_{10} 256$ d) $\log_e 2536$

Solution:

a) Referring to section 9.1, base 3 must be raised to the power 3 to get 27

$3^x = 27 = 3^3$

Therefore $x = 3$ (the index or power is the answer)

Hence $\log_3 27 = \mathbf{3}$

b) $10^x = 100 = 10^2$

Therefore $x = 2$

Hence $\log_{10} 100 = \mathbf{2}$

c) Use the [log] key for this calculation

[log] [256] [=] 2.408

d) Use the [ln] key for this calculation.

[ln] [2536] [=] 7.838

9.1.1 Antilogarithm (antilog)

Antilogarithm is the reverse of logarithm. If we know the logarithm of a number, then we can use the antilogarithm key ([10^x]) of a scientific calculator to determine the original number. Antilogarithms can be used to simplify equations that involve logarithms, as they cancel each other out; this is illustrated in example 9.5b.

Example 9.2

Calculate the antilogarithms of:

a) 3.7 (base = 10)
b) 4.8 (base = e)

Solution:

Press the following keys of a scientific calculator

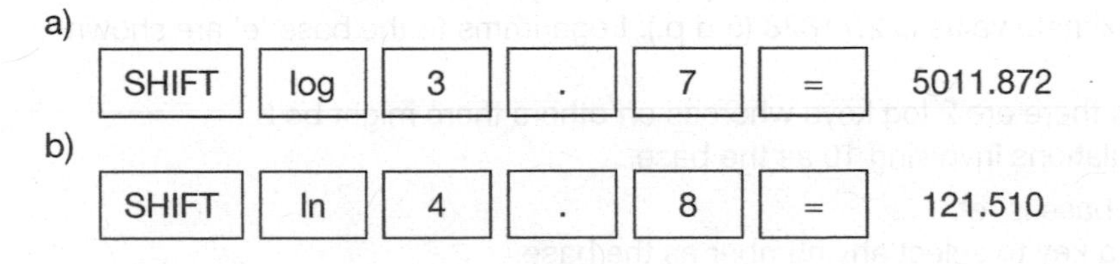

a)

[SHIFT] [log] [3] [.] [7] [=] 5011.872

b)

[SHIFT] [ln] [4] [.] [8] [=] 121.510

9.1.2 Laws of logarithms

In each of the following laws, every logarithm must have the same base:

a) When two numbers are multiplied: $\log(a \times b) = \log a + \log b$.

b) When a number is divided by another number: $\log\left(\frac{a}{b}\right) = \log a - \log b$.

c) When a number is raised to a power: $\log a^n = n \log a$.

Example 9.3

Express as the sum or difference of the simplest possible logarithms:

a) $\log \frac{3x}{x-1}$

b) $\ln 4(x+2)$

c) $\ln(x^2 - 1)$

Solution:

a) Use the laws explained in section 9.1.2.

$$\log \frac{3x}{x-1} = \log 3x - \log(x-1)$$

$$= \mathbf{\log 3 + \log x - \log(x-1)}$$

b) $\ln 4(x+2) = \mathbf{\ln 4 + \ln(x+2)}$

c) $\ln(x^2 - 1) = \ln(x+1)(x-1) \quad [(x^2 - 1) = (x+1)(x-1)]$

$$= \mathbf{\ln(x+1) + \ln(x-1)}$$

Example 9.4

Simplify the expression: a) $\log 9 + \log 27$
b) $\log 8 + \log 16 - \log 4$

Solution:

a) $\log 9 + \log 27 = \log(3 \times 3) + \log(3 \times 3 \times 3)$

$$= \log 3 + \log 3 + \log 3 + \log 3 + \log 3 = \mathbf{5 \log 3}$$

b) $\log 8 + \log 16 - \log 4 = \log(2 \times 2 \times 2) + \log(2 \times 2 \times 2 \times 2) - \log(2 \times 2)$

$$= \log 2 + \log 2 + \log 2 + \log 2 + \log 2 + \log 2 + \log 2 - \log 2 - \log 2$$

$$= \mathbf{5 \log 2}$$

Example 9.5

Solve the following equations:

a) $\log_4 x = 2$
b) $x^{1.7} = 3.259$
c) $2.55^x = 5.955$

Solution:

a) Refer to section 9.1; if $\log_a x = y$ then $a^y = x$
Comparing this to the equation, $\log_4 x = 2$, we have $a = 4$ and $y = 2$
Therefore, $4^2 = x$, Hence $\mathbf{x = 16}$

b) $x^{1.7} = 3.259$
Take logarithms (base10) of both sides, $\log x^{1.7} = \log 3.259$
$1.7 \log x = 0.5131$

$$\log x = \frac{0.5131}{1.7} = 0.3018$$

Take antilogarithms of both sides, antilog (log x) = antilog 0.3018
Antilog and log, being opposite are cancelled out, therefore, x = antilog 0.3018
Hence $\mathbf{x = 2.004}$

c) $2.55^x = 5.955$
Take logarithms (base10) of both sides, $\log 2.55^x = \log 5.955$
$x \log 2.55 = 0.77488$
$x \times 0.40654 = 0.77488$

$$\text{or } x = \frac{0.77488}{0.40654} = 1.906$$

Example 9.6

Solve: $\log(2x + 5) = \log(x^2 - 3)$

Solution:

$\log(2x + 5) = \log(x^2 - 3)$

Take antilogarithms of both sides:

antilog $[\log(2x + 5)]$ = antilog $[\log(x^2 - 3)]$

Antilog and log, being opposite, cancel each other out, therefore:

$2x + 5 = x^2 - 3$, which after simplification, becomes, $x^2 - 2x - 8 = 0$

This is a quadratic equation and can be solved by using the quadratic formula, factorisation or other methods. The first method is used here:

$$x = \frac{-(-2) \pm \sqrt{(-2)^2 - 4(1)(-8)}}{2 \times 1} \qquad (a = 1; b = -2; c = -8)$$

$$x = \frac{2 \pm \sqrt{4 + 32}}{2} = \frac{2 \pm \sqrt{36}}{2}$$

Now $x = \frac{2+6}{2}$ or $x = \frac{2-6}{2}$

Hence, either $x = 4$, or $x = -2$

9.2 Exponential function

An exponential function is a mathematical function of the form: $f(x) = a^x$, where the value of 'a' is: $a > 0$ and $a \neq 1$. 'a' is called the base and 'x', known as the exponent, is a variable. Exponential functions can be used in many applications, for example, calculation of compound interest on savings, population growth, thermal expansion of construction materials, etc. The function is usually modified by adding constants before it can be used in practical applications.

In applied mathematics, the most commonly encountered exponential function base is 'e' whose value is 2.71828 (approx.); thus the expression for the exponential function becomes $f(x) = e^x$. Figure 9.1 shows the graphs of exponential growth (e^x) and exponential decay (e^{-x}).

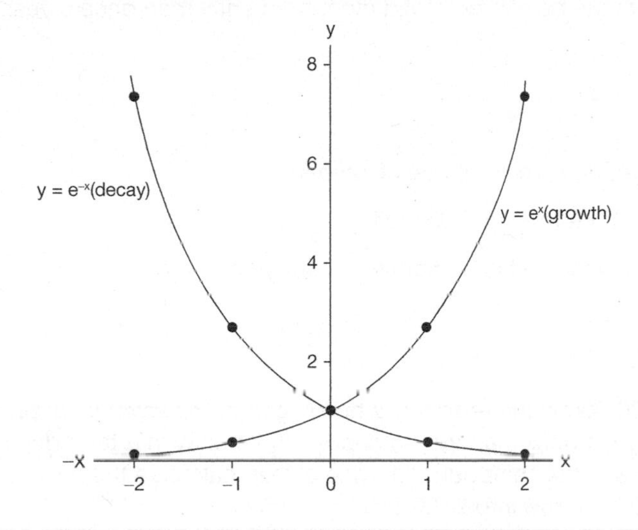

Figure 9.1

Example 9.7

Solve $25(1 - e^{-x/2}) = 20$

Solution:

$$1 - e^{-\frac{x}{2}} = \frac{20}{25} = 0.8$$

$$e^{-\frac{x}{2}} = 1 - 0.8 = 0.2$$

$$\frac{1}{e^{-\frac{x}{2}}} = \frac{1}{0.2}, \text{ therefore } e^{\frac{x}{2}} = 5$$

Take logarithms of both sides, $\ln\left(e^{\frac{x}{2}}\right) = \ln(5)$

(ln and e cancel each other out)

$\frac{x}{2} = 1.6094$, therefore, $x = 3.219$

9.2.1 Compound interest on savings

The banks normally show the interest earned on savings as an annual rate, e.g. 2%, 3%, etc. There are basically 2 types of interest, i.e. simple interest and compound interest. In simple interest the interest is paid on the principal amount, whereas in compound interest the interest is also paid on the interest already earned. The amount of compound interest after a number of years can be calculated from the following formula:

$$Y = P(1 + r)^t$$

However, the interest may be compounded more frequently than once a year, therefore this formula is modified to:

$$Y = P\left(1 + \frac{r}{n}\right)^{nt}$$

Y = amount after 't' years; r = annual rate of interest

P = principal amount; t = term of the deposit

n = number of times the interest is compounded per year

Example 9.8

John Tahli requires £100 000 in future to buy a building plot. His savings, at present, are £75 000, and he has been advised by his bank manager to deposit the money in a term deposit yielding an annual interest of 6%. If the interest is compounded twice a year, calculate the time required (in years) for his initial amount of £75 000 to grow into £100 000.

Solution:

Y = the value of the matured account = £100,000

P = principal (original) amount = £75,000

r = 6%; n = 2; t = term of the deposit in years

$$Y = P\left(1 + \frac{r}{n}\right)^{nt}$$

$$100{,}000 = 75{,}000\left(1 + \frac{6}{100 \times 2}\right)^{2t} \left(r = 6\% = \frac{6}{100}\right)$$

$$\frac{100000}{75000} = (1.03)^{2t}$$

$$1.33333 = (1.03)^{2t}$$

Take logarithms of both sides, $\ln(1.33333)=\ln(1.03)^{2t}$

$\ln(1.33333) = 2t (\ln 1.03)$

$0.287682 = 2t \times 0.0295588$

$0.287682 = t \times 0.0591176$

$t = \frac{0.287682}{0.0591176} = 4.87$ years

The growth of £75,000 into £100,000 requires 4.87, say 5 years.

9.2.2 Newton's law of cooling

According to Newton's law of cooling, the rate of temperature change of an object is proportional to its own temperature and the temperature of its surroundings (ambient temperature). If an object is not at the temperature of its surrounding environment, then its temperature continues to change until it reaches the ambient temperature.

$T(t) = T_s + (T_0 - T_s)e^{-kt}$

$T(t)$ = temperature of an object at time 't'

T_s = temperature of the surrounding environment (or ambient temperature)

T_0 = the initial temperature of the object

k = decay constant

The assumption in this law is that the ambient temperature remains constant.

Example 9.9

In order to assess the effectiveness of a cavity wall insulation material an experiment was performed which involved taking a container full of hot water at 85 °C that was surrounded by the insulation material. The hot water temperature reduced to 70 °C after 5 minutes, while the ambient temperature remained constant at 20 °C. Calculate:

a) the exponential function to represent this situation.

b) how long will it take for the water to reach 60 °C.

Solution:

a) $T(t) = T_s + (T_0 - T_s)e^{-kt}$

$T(t) = 70$ °C; $T_s = 20$ °C; $T_0 = 85$ °C; $t = 5$ minutes

$70 = 20 + (85 - 20)e^{-k\times5}$

$50 = 65e^{-5k}$, which gives $\frac{50}{65} = e^{-5k}$

Take logarithm of both sides, $\ln\left(\frac{50}{65}\right)=\ln(e^{-5k})$

$-0.26236 = -5k$, therefore, $k = 0.05247$

b) $T(t) = 60$ °C

$60 = 20 + (85 - 20)e^{-0.05247\,t}$

$40 = 65e^{-0.05247\,t}$, which gives $\frac{40}{65} = e^{-0.05247t}$

Take natural logarithms of both sides, $\ln\left(\frac{40}{65}\right) = \ln e^{-0.05247\,t}$

$-0.48551 = -0.05247$ t, Therefore, t = **9.25 minutes**

9.2.3 Thermal movement of building components

All components in the superstructure of a building expand/contract due to temperature variations in the surrounding environment. Assuming that the co-efficient of thermal expansion α is constant, the linear thermal expansion of a material/component is given by:

$L = L_0\, e^{\alpha(t - t_0)}$

where, L_0 and L are the original and the length after expansion, respectively

t_0 and t are the initial and final temperatures, respectively.

Example 9.10

Calculate the thermal expansion of 10 m long PVC guttering if the air temperature increases from 15 °C to 27 °C. The co-efficient of linear expansion of PVC is

7×10^{-5}/°C.

Solution:

$L = L_0\, e^{\alpha(t - t_0)}$

$L_0 = 10.0$ m; $\alpha = 7 \times 10^{-5}$/°C; $t_0 = 15$ °C; t = 27 °C

$L = 10.0(e^{7\times10^{-5}(27-15)}) = 10.0(e^{84\times10^{-5}})$

$L = 10 \times 1.00084 = 10.0084$ m

The linear expansion of the PVC guttering = $L - L_0 = 10.0084 - 10.0$

= 0.0084 m or 8.4 mm

9.2.4 Laws of growth and decay

The laws of exponential growth and decay are of the form, $y = A\, e^{kx}$ and $y = A\, e^{-kx}$, where A and k are constants. The negative value of k is used in problems involving reduction or decay.

Example 9.11

The municipal council of Alpha-city needs to predict the future population of the city so that the road network could be planned. If the population increases at a rate of 1.8% each year, what will be

Alpha-city's population after 20 years. The population now is 100 000 and assume that it increases continuously.

Solution:

Population after growth, $P = P_0 e^{rt}$

$P_0 = 100{,}000$; $r = 1.8\%$ or 0.018; $t = 20$ years

$P = 100{,}000 \times e^{0.018 \times 20} = 100{,}000 \times e^{0.36}$

$= 100{,}000 \times 1.4333294 =$ **143,333**

9.2.5 Decay of sound energy

Reverberation is the continued presence of sound in a room/hall due to repeated reflections of sound waves from the various surfaces. The reverberant sound decays exponentially with time as the sound energy is absorbed by the room surfaces. The sound energy, S, after time, t, is given by:

$$S = S_0 e^{-\frac{t}{c}}$$

where S_0 is the initial sound level

c is the time constant of the exponential.

Example 9.12

The sound in a hall decays exponentially with time, t, as shown in Figure 9.2. From the given information on the 2 points A and B, find:

a) an exponential equation of the form $S = S_0 e^{\frac{t}{c}}$.

b) the sound level at time, $t = 0$.

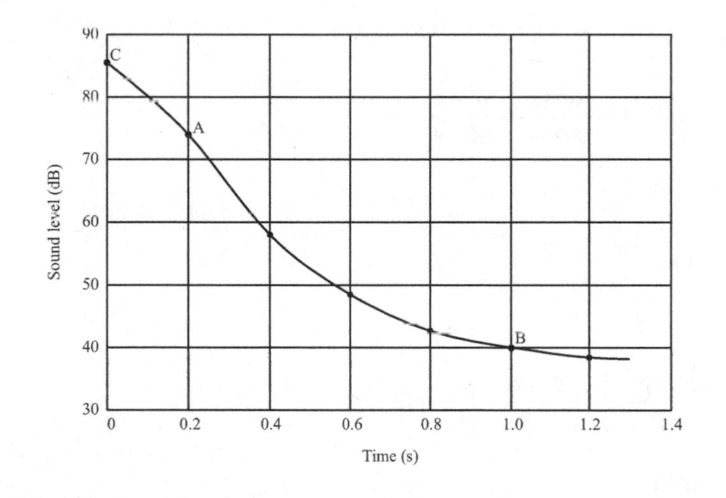

Figure 9.2 Sound level decay

Solution:

a) Time, t, between points A and B = 1.0 – 0.2 = 0.8 s

S (at point B) = 40 dB; S_0 (at point A) = 73 dB;

Put these values in the equation, $40 = 73e^{-\frac{0.8}{c}}$

$$\frac{40}{73} = e^{-\frac{0.8}{c}}$$

Take logarithms of both sides, $\ln(0.548) = \frac{-0.8}{c}$

$$c = \frac{-0.8}{\ln(0.548)} = 1.33$$

Therefore, $S = S_0 e^{-\frac{t}{1.33}}$

b) S = 73 dB at 0.2 s ; S_0 = sound level at time zero seconds (point C)

Time t between points A and C = 0.2 – 0 = 0.2 s

$$S = S_0 e^{-\frac{t}{1.33}}$$

$$73 = S_0 e^{-\frac{0.2}{1.33}}$$

$73 = S_0 \times 0.8603844;$ therefore, S_0 = **84.8 dB**

9.3 Hyperbolic Function

The hyperbolic functions are defined using the exponential function e^x. sinh x (pronounced as 'shine x') and cosh x (pronounced as 'kosh x') are defined as:

$$\sinh x = \frac{e^x - e^{-x}}{2}$$

$$\cosh x = \frac{e^x + e^{-x}}{2}$$

As in trigonometry, $\tanh x = \frac{\sinh x}{\cosh x} = \frac{e^x - e^{-x}}{e^x + e^{-x}}$

Example 9.13

Simplify cosh x + sinh x.

Solution:

$$\cosh x + \sinh x = \left(\frac{e^x + e^{-x}}{2}\right) + \left(\frac{e^x - e^{-x}}{2}\right)$$

$$= \frac{1}{2}\left(e^x + e^{-x} + e^x - e^{-x}\right)$$

$$= \frac{1}{2}\left(2e^x\right) = \mathbf{e^x}$$

Example 9.14

Show that: $\cosh^2 x - \sinh^2 x = 1$.

Solution:

$$\cosh^2 x - \sinh^2 x = \left(\frac{e^x + e^{-x}}{2}\right)^2 - \left(\frac{e^x - e^{-x}}{2}\right)^2$$

$$= \frac{e^{2x} + 2 + e^{-2x}}{4} - \frac{e^{2x} - 2 + e^{-2x}}{4}$$

$$= \frac{1}{4}\left(e^{2x} + 2 + e^{-2x} - e^{2x} + 2 - e^{-2x}\right)$$

$$= \frac{1}{4}(4) = \mathbf{1}$$

Example 9.15

Prove that: $\sinh 2x = 2 \sinh x \cosh x$.

Solution:

$$\sinh 2x = \frac{e^{2x} - e^{-2x}}{2}$$

$$2 \sinh x \cosh x = 2\left(\frac{e^x - e^{-x}}{2}\right)\left(\frac{e^x + e^{-x}}{2}\right)$$

$$= 2\left(\frac{e^{2x} - e^0 + e^0 - e^{-2x}}{4}\right)$$

$$= \frac{e^{2x} - e^{-2x}}{2} = \sinh 2x$$

Hence, $\sinh 2x = 2 \sinh x \cosh x$

Example 9.16

Prove that: $\operatorname{sech}^2 x + \tanh^2 x = 1$.

Solution:

$$\operatorname{sech}^2 x + \tanh^2 x = \left(\frac{2}{e^x + e^{-x}}\right)^2 + \left(\frac{e^x - e^{-x}}{e^x + e^{-x}}\right)^2$$

$$= \frac{4}{e^{2x} + 2 + e^{-2x}} + \frac{e^{2x} - 2 + e^{-2x}}{e^{2x} + 2 + e^{-2x}}$$

$$= \frac{4 + e^{2x} - 2 + e^{-2x}}{e^{2x} + 2 + e^{-2x}} = \frac{e^{2x} + 2 + e^{-2x}}{e^{2x} + 2 + e^{-2x}} = \mathbf{1}$$

Example 9.17

If 2 cosh 2x + 8 sinh 2x = 3, find the value of x.

Solution:

$$2\cosh 2x + 8\sinh 2x = 2\left(\frac{e^{2x}+e^{-2x}}{2}\right)+8\left(\frac{e^{2x}-e^{-2x}}{2}\right)$$

Therefore, $2\left(\frac{e^{2x}+e^{-2x}}{2}\right)+8\left(\frac{e^{2x}-e^{-2x}}{2}\right)=3$

$e^{2x}+e^{-2x}+4e^{2x}-4e^{-2x}-3=0$

Multiply by e^{2x} : $(e^{2x})^2+(e^{-2x})(e^{2x})+4(e^{2x})^2-4(e^{-2x})(e^{2x})-3(e^{2x})=0$

$(e^{2x})^2+e^0+4(e^{2x})^2-4(e^0)-3(e^{2x})=0$

$5(e^{2x})^2+1-4-3(e^{2x})=0 \quad (e^0=1)$

$5(e^{2x})^2-3(e^{2x})-3=0$

Solve for e^{2x} using the quadratic formula:

$$e^{2x}=\frac{-(-3)\pm\sqrt{(-3)^2-4(5)(-3)}}{2\times 5}$$

$$=\frac{3\pm\sqrt{69}}{10}=\frac{3\pm 8.3066}{10}$$

$e^{2x}=1.13066$ or -0.5307

e^{2x} is always positive, therefore $e^{2x}=1.13066$

Take logarithms of both sides, $\ln(e^{2x})=\ln(1.13066)$

$2x=0.1228$ (logarithm and e^x cancel out to leave behind x)

Therefore, $x=\mathbf{0.0614}$

Exercise 9.1

1. Evaluate (without using a calculator): a) $\log_2 32$
   b) $\log_{10} 10000$
2. Use a scientific calculator to find the logarithm (base 10) of: a) 25, b) 150
3. Use a scientific calculator to find the natural logarithm of: a) 25, b) 150
4. Find the antilogarithm of a) 2.5, b) 0.0014.
5. Express as the sum or difference of the simplest possible logarithms:

   a) $\log\frac{x}{x+1}$

   b) $\log\sqrt{2x+1}$

   c) $\ln x(x+4)$

6. Simplify the expression: a) $\log 81 - \log 27$
   b) $\log 81 - \log 9 - \log 3$
7. Solve: a) $\log_{10} x = 3$
   b) $x^{2.5} = 15.598$
   c) $3^{x-1} = 2^{x+3}$
8. Solve: a) $\log(x^2 - 1) = \log(3x - 3)$
   b) $\log(x + 3) + \log(x - 4) = 2\log(x - 1)$

Exercise 9.2

1. Solve: a) $3e^{2t} = 1.2$
   b) $5.1 = 2.2e^{-1.5x}$
   c) $10 = 14(1 - e^{-\frac{x}{2}})$
2. Rob Kikkar requires £100 000 in future to buy a building plot. His savings, at present, are £80 000, and he has been advised by his bank manager to deposit the money in a term deposit yielding an annual interest of 5.5%. If the interest is compounded 4 times a year, calculate the time required (in years) for his initial amount of £80 000 to grow into £100 000.
3. In order to assess the effectiveness of an insulation material, a container full of hot water at 86 °C was surrounded by the insulation material. The hot water temperature dropped to 70 °C after 6 minutes, while the ambient temperature remained constant at 20 °C. Calculate:
   a) the exponential function to represent this situation.
   b) how long it will take for the water to reach 60 °C.
4. Calculate the thermal expansion of 8 m long PVC guttering if the air temperature rises from 15 °C to 30 °C. The co-efficient of linear expansion of PVC is 7×10^{-5}/°C.

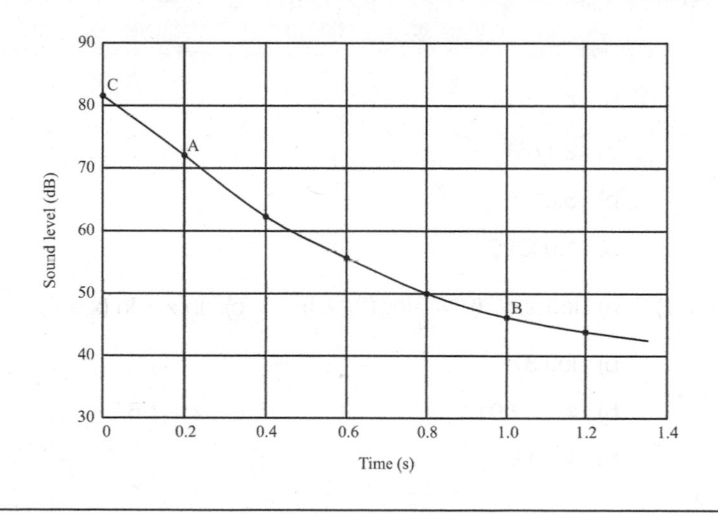

Figure 9.3

5. The city council of Blizton needs to predict the future population of the city so that the road network and other facilities could be planned. If the population increases at a rate of 1.6% each year, what will be Blizton's population after 25 years? The current population is 100 000 and assume that it increases continuously.

6. The population of a country is decreasing at a rate of 1.3% per year. If the population at present is 20 million, what will the population be after 20 years? Assume that the population decreases continuously.

7. The sound in a hall decays exponentially with time, t, as shown in Figure 9.3. From the given information on the two points A and B, find:

   a) an exponential equation of the form $S = S_0 e^{-\frac{t}{c}}$.

   b) the sound level at time, t = 0.

   c) the sound level after 0.6 seconds.

Exercise 9.3

1. Simplify: cosh x – sinh x.
2. Prove that $\cosh 2x = (\cosh x)^2 + (\sinh x)^2$.
3. Prove that $\coth^2 x - \text{cosech}^2 x = 1$.
4. If $\sinh x = \frac{3}{5}$, find the value of x.
5. If 2 cosh x = 3.4, find the value of x.
6. Solve the following equations:

   a) 2 sech x – 1 = 0.

   b) 2.4 cosh x + 5 sinh x = 7.5.

Answers – Exercise 9.1

1. a) 5 b) 4
2. a) 1.3979 b) 2.1761
3. a) 3.2189 b) 5.0106
4. a) 316.2278 b) 1.00323
5. a) log x – log (x + 1) b) $\log(2x+1)^{\frac{1}{2}} = \frac{1}{2}\log(2x+1)$ c) ln x + ln (x + 4)
6. a) log 3 b) log 3
7. a) x = 1000 b) x = 3.001 c) x = 7.84
8. a) x = 2, or x = 1 b) x = 13

Answers – Exercise 9.2

1. a) – 0.458 b) – 0.5605 c) 2.506
2. 4.08 years
3. a) 0.04627 b) 10.82 min
4. 8.4 mm
5. 149 182
6. 15 421 032
7. a) $S = S_0 e^{-\frac{t}{1.7856}}$ b) 80.5 dB c) 57.5 dB

Answers – Exercise 9.3

1. e^{-x}
2. See solutions in Appendix 3.
3. See solutions in Appendix 3.
4. 0.5687
5. 1.3008
6. a) 2.01, – 0.624 b) 0.7828

CHAPTER 10

Differentiation

Topics covered:

- Differentiation from first principles
- Differentiation using the product rule, quotient rule, function of a function
- Differentiation of trigonometric, logarithmic and exponential functions

10.1 Introduction

If we want to find the gradient of a curve at a specified point, e.g. point P in Figure 10.1, then we must produce the graph of the curve and draw tangent MN to the curve at that point. The gradient of the tangent is the gradient of the curve at point P. However, if we know the equation of the curve, an alternative method known as differentiation may be used, to produce accurate results.

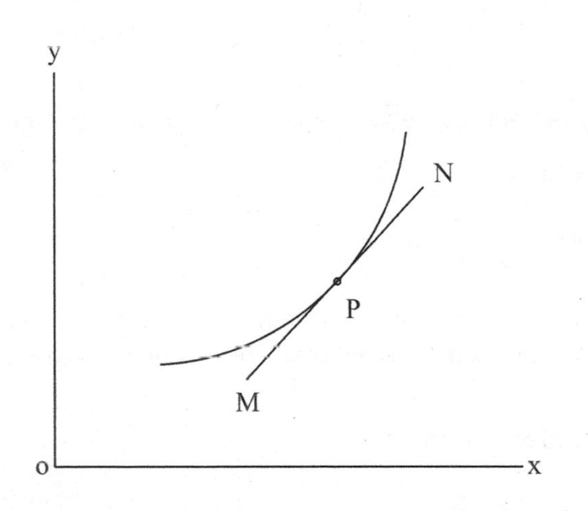

Figure 10.1

10.2 Differentiation from first principles

Consider the curve $y = x^2$, shown in Figure 10.2. Let P be the point on this curve at which $x = 1$ and $y = 1$. Let Q be another point on the curve whose x and y co-ordinates are 3 and 9 respectively.

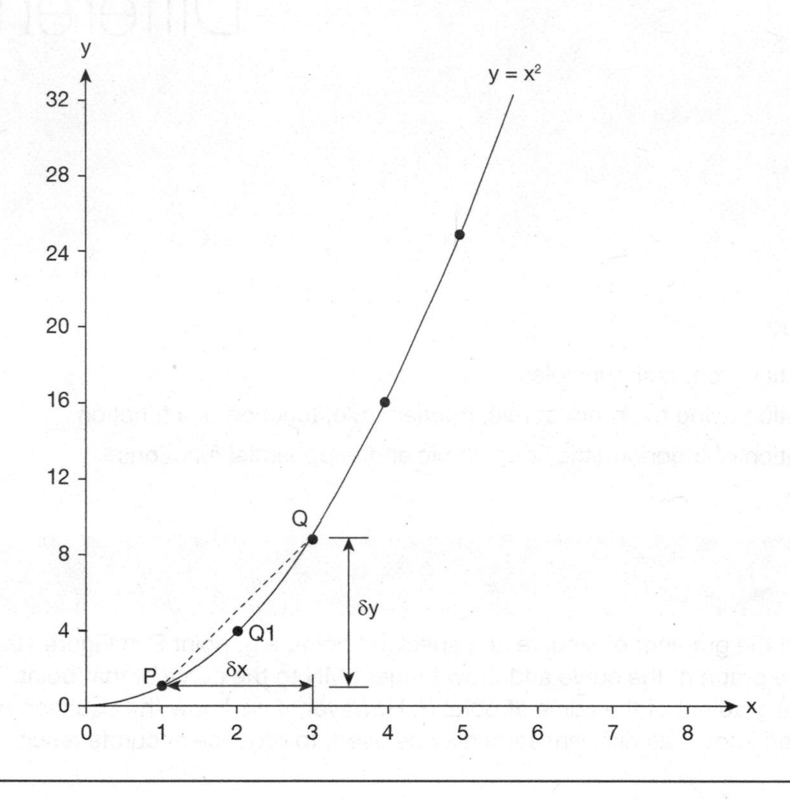

Figure 10.2

Let δx and δy represent increments of x and y respectively, as shown in Figure 10.2.

$$\delta x = 3 - 1 = 2, \quad \text{and } \delta y = 9 - 1 = 8$$

The gradient of chord $PQ = \dfrac{\delta y}{\delta x} = \dfrac{8}{2} = 4$

As point Q moves nearer to point P, the gradient of the curve changes. This change is due to the shape of the curve (only a straight line has a constant gradient). When Q moves to position Q_1, the new co-ordinates are 2, 4.

$$\delta x = 2 - 1 = 1, \quad \text{and } \delta y = 4 - 1 = 3$$

The gradient of chord $PQ = \dfrac{\delta y}{\delta x} = \dfrac{3}{1} = 3$

When Q is very close to P, the gradient of chord PQ may be considered as the average gradient of the portion of the curve PQ. Let x and y co-ordinates for Q be $x + \delta x$ and $y + \delta y$ respectively.

Since the equation of the curve is: $y = x^2$

Therefore $y + \delta y = (x + \delta x)^2$

or $y + \delta y = x^2 + 2x\delta x + (\delta x)^2$

Since $y = x^2$, $x^2 + \delta y = x^2 + 2x\delta x + (\delta x)^2$

or $\delta y = 2x\delta x + (\delta x)^2$

Divide both sides by δx to obtain the gradient of the chord

$$\frac{\delta y}{\delta x} = \frac{2x\delta x}{\delta x} + \frac{\delta x^2}{\delta x}$$

$$= 2x + \delta x \qquad \text{(equation 1)}$$

When Q approaches P, δx becomes smaller, and finally when P and Q coincide, δx becomes zero. As δx approaches zero ($\delta x \to 0$), the gradient of the chord becomes the gradient of the tangent. The limiting value of $\frac{\delta y}{\delta x}$ is written as $\frac{dy}{dx}$

$$\text{Lim } \delta x \to 0 \left(\frac{\delta y}{\delta x}\right) = \frac{dy}{dx}$$

$$\text{Lim } \delta x \to 0 \left(\frac{\delta y}{\delta x}\right) = \frac{dy}{dx} = 2x + 0 = 2x \qquad (\delta x \text{ in equation 1 becomes zero})$$

At $x = 1$, $\frac{dy}{dx}$ = gradient of tangent = 2

At $x = 2$, $\frac{dy}{dx}$ = gradient of tangent = 4

Similarly gradient at other points may be determined, $\frac{dy}{dx}$ is known as 'the differential coefficient of y with respect to x' and the process of determining it is known as **differentiation.**

Example 10.1

Differentiate from first principles: i) $y = x^3$

ii) $y = \frac{1}{x}$

iii) $y = 2x^2$

Solution:

i) Let x increase by a small amount δx and the corresponding increase in y be δy

$y + \delta y = (x + \delta x)^3$

or $y + \delta y = x^3 + 3x^2\,\delta x + 3x\,(\delta x)^2 + (\delta x)^3$ (equation 2)

As $y = x^3$, equation 2 becomes:

$x^3 + \delta y = x^3 + 3x^2\,\delta x + 3x\,(\delta x)^2 + (\delta x)^3$

or $\delta y = 3x^2\,\delta x + 3x\,(\delta x)^2 + (\delta x)^3$

Divide both sides by δx

$$\frac{\delta y}{\delta x} = \frac{3x^2\,\delta x}{\delta x} + \frac{3x\,\delta x^2}{\delta x} + \frac{\delta x^3}{\delta x}$$

$$= 3x^2 + 3x\,\delta x + (\delta x)^2$$

$$\text{Limit } \delta x \to 0\left(\frac{\delta y}{\delta x}\right) = \frac{dy}{dx} = \mathbf{3x^2} \quad (\delta x \text{ and } (\delta x)^2 \text{ are zero})$$

ii) Let x increase by a small amount δx and the corresponding increase in y be δy

$$y + \delta y = \frac{1}{x + \delta x} \qquad \text{(equation 3)}$$

As $y = \frac{1}{x}$, equation 3 becomes:

$$\frac{1}{x} + \delta y = \frac{1}{x + \delta x}$$

By transposition, $\delta y = \frac{1}{x+\delta x} - \frac{1}{x}$

$$\delta y = \frac{x - x - \delta x}{x(x+\delta x)} = \frac{-\delta x}{x(x+\delta x)}$$

Divide both sides by δx

$$\frac{\delta y}{\delta x} = \frac{-\delta x}{x(x+\delta x)\delta x}$$

As δx is zero, Limit $\delta x \to 0\left(\frac{\delta y}{\delta x}\right) = \frac{dy}{dx} = \frac{-1}{x(x+0)}$

or $\frac{dy}{dx} = \frac{-1}{x^2}$

iii) Let x increase by a small amount δx and the corresponding increase in y be δy

$y + \delta y = 2(x + \delta x)^2$

or $y + \delta y = 2(x^2 + 2x\,\delta x + (\delta x)^2)$

$y + \delta y = 2x^2 + 4x\,\delta x + 2(\delta x)^2)$ (equation 4)

As $y = 2x^2$, equation 4 becomes:

$2x^2 + \delta y = 2x^2 + 4x\,\delta x + 2(\delta x)^2$

or $\delta y = 4x\,\delta x + 2(\delta x)^2$

Divide both sides by δx

$$\frac{\delta y}{\delta x} = \frac{4x\,\delta x}{\delta x} + \frac{2(\delta x)^2}{\delta x}$$

$$= 4x + 2\delta x$$

$$\text{Limit } \delta x \to 0\left(\frac{\delta y}{\delta x}\right) = \frac{dy}{dx} = 4x + 0 = \mathbf{4x} \,(\delta x \text{ is zero})$$

10.2.1 Differentiation of $y = x^n$

It has been shown in section 10.2 that if $y = x^2$, then $\frac{dy}{dx} = 2x$

Similarly in example 10.1 it has been shown that if $y = x^3$, then $\frac{dy}{dx} = 3x^2$

From these 2 results a rule may be derived for differentiating functions such as $y = x^n$, where n could be any value:

If $y = x^n$, then $\frac{dy}{dx} = nx^{n-1}$

[multiply by the original power and reduce the power by 1]

This rule enables us to differentiate functions without using the first principles, which are more time consuming.

Also, we can conclude from the solution of example 10.1 (iii) that a constant (i.e. 2) is not involved in the differentiation process. A number, which is on its own, will result into 0 after a function has been differentiated.

The rule for differentiating $y = ax^n$, where a is a constant, is:

$$\frac{dy}{dx} = nax^{n-1}$$

These results are summarised in Table 10.1.

Table 10.1

y	$\frac{dy}{dx}$
x^n	nx^{n-1}
ax^n	nax^{n-1}

Example 10.2

Differentiate i) $y = x$

ii) $y = 3x$

iii) $y = 2x^3 + 4x^2 + 5x$

Solution:

i) $y = x$ may be written as $y = x^1$

Using the rule: $\frac{dy}{dx} = nx^{n-1}$

$\frac{dy}{dx} = 1\,x^{1-1}$, as $n = 1$

$\frac{dy}{dx} = x^0 = \mathbf{1}$ (according to law of zero index, $x^0 = 1$)

ii) $y = 3x$ may be written as $y = 3x^1$

Using the rule $\frac{dy}{dx} = nax^{n-1}$

$$\frac{dy}{dx} = 1 \times 3x^{1-1} \qquad (n = 1 \text{ and } a = 3)$$

$$\text{or } \frac{dy}{dx} = 3x^0 = 3 \times 1 = 3$$

iii) $y = 2x^3 + 4x^2 + 5x$

To find the solution, each term is differentiated separately

$$\frac{dy}{dx} = 3 \times 2x^{3-1} + 2 \times 4x^{2-1} + 1 \times 5x^{1-1}$$

$$= 6x^2 + 8x^1 + 5x^0 = \mathbf{6x^2 + 8x + 5}$$

10.3 Trigonometric functions

Figure 10.3a shows a graph of y = sin x. If tangents to the curve are drawn at points A, B, C, D and E and their gradients determined, then it will be found that the gradients are 1.0, 0, –1.0, 0 and 1.0 respectively. If these gradients and the gradients at other points on the curve are plotted, then we get a curve as shown in Figure 10.3b. The shape of this curve is same as the shape of the y = cos x graph. Hence the gradient of sin x curve at any value of angle x is the same as the value of cos x. Mathematically, this is written as:

If $y = \sin x$ then $\frac{dy}{dx} = \cos x$

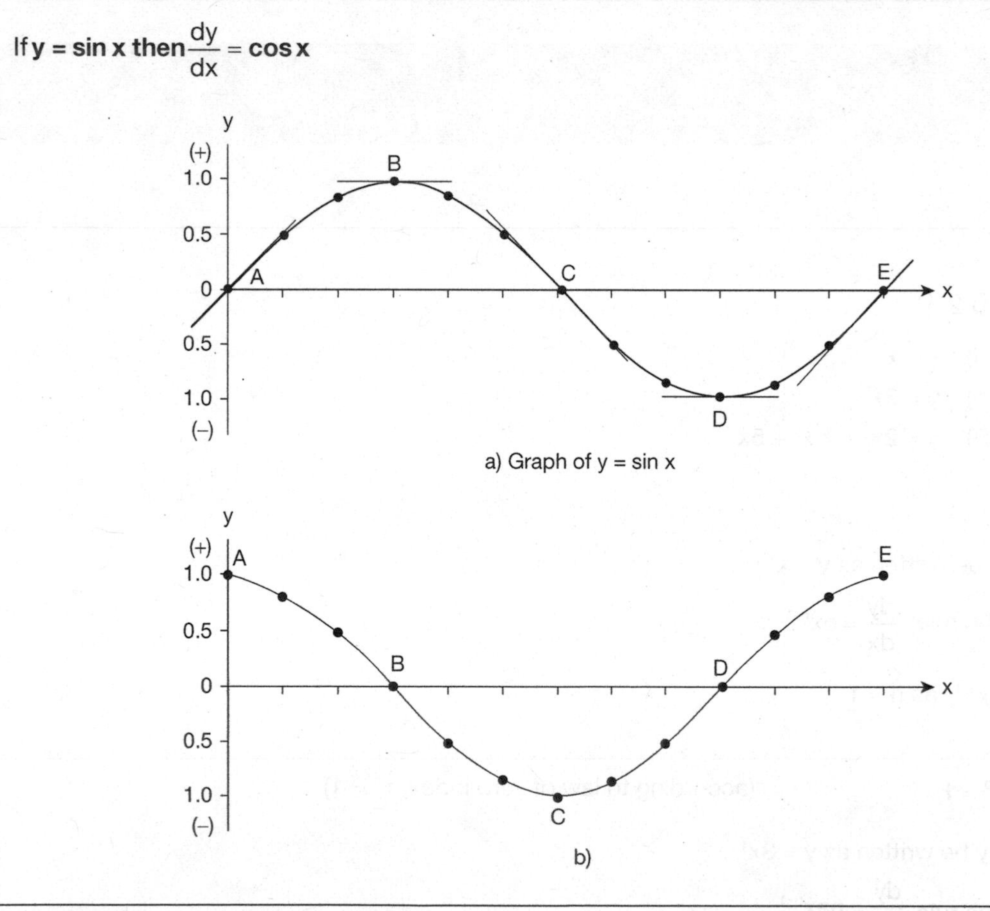

Figure 10.3

Similarly it can be shown that if **$y = \cos x$ then $\frac{dy}{dx} = -\sin x$**

These results and their extensions are shown in Table 10.2.

Table 10.2

y	$\frac{dy}{dx}$
sin x	cos x
cos x	–sin x
sin ax	a cos ax
cos ax	–a sin ax
tan x	$\sec^2 x$

Example 10.3

Differentiate i) $y = \sin 4x$ ii) $y = \cos 3x$

Solution:

i) $y = \sin 4x$, $a = 4$ in this question

From Table 10.2, $\frac{d}{dx}(\sin ax) = a \cos ax$

Therefore, $\frac{d}{dx}(\sin 4x) = \mathbf{4 \cos 4x}$

ii) $y = \cos 3x$, $a = 3$ in this question

From Table 10.2, $\frac{d}{dx}(\cos ax) = -a \sin ax$

Therefore, $\frac{d}{dx}(\cos 3x) = \mathbf{-3 \sin 3x}$

10.4 Differentiation of function of a function

It may be either inconvenient or impossible to expand functions such as, $y = (x^2 + 3x - 4)^9$ into polynomials so that the method explained in section 10.2 cannot be used. The function of a function method involves the substitution of $x^2 + 3x - 4$ with u so that:

$u = x^2 + 3x - 4$; therefore, $y = u^9$

$\frac{dy}{du}$ and $\frac{du}{dx}$ are determined, and finally:

$\frac{dy}{dx} = \frac{dy}{du} \times \frac{du}{dx}$; this is also known as the chain rule

Example 10.4

Differentiate $y = (x^2 + 3x - 4)^9$.

Solution:

Let $u = x^2 + 3x - 4$, so that $y = (u)^9$

Use the method explained in section 10.2.1

$$\frac{du}{dx} = 2x + 3, \qquad \text{and} \qquad \frac{dy}{du} = 9u^{9-1} = 9u^8$$

$$\frac{dy}{dx} = \frac{dy}{du} \times \frac{du}{dx}$$

$$= 9u^8 \times (2x + 3)$$

Now replace u with $x^2 + 3x - 4$

$$\frac{dy}{dx} = 9\,(x^2 + 3x - 4)^8 \times (2x + 3)$$

$$= \mathbf{9(2x + 3)(x^2 + 3x - 4)^8}$$

Example 10.5

Differentiate i) $y = \sin^2 \theta$ ii) $y = \cos^3 \theta$ iii) $y = \cos^4 2x$

Solution:

i) $y = \sin^2 \theta$ may also be written as $y = (\sin \theta)^2$

Let $u = \sin \theta$, so that $y = u^2$

Use the methods explained in sections 10.3 and 10.2.1

$$\frac{du}{d\theta} = \cos \theta, \qquad \text{and} \qquad \frac{dy}{du} = 2u^{2-1} = 2u$$

$$\frac{dy}{d\theta} = \frac{dy}{du} \times \frac{du}{d\theta}$$

$$= 2u \times \cos \theta = 2u \cos \theta$$

$$\frac{dy}{d\theta} = \mathbf{2 \sin\theta \cos\theta}$$

ii) $y = \cos^3 \theta$, or $y = (\cos \theta)^3$

Let $u = \cos \theta$, so that $y = u^3$

Use the methods explained in sections 10.3 and 10.2.1

$$\frac{du}{d\theta} = -\sin\theta, \qquad \text{and} \qquad \frac{dy}{du} = 3u^{3-1} = 3u^2$$

$$\frac{dy}{d\theta} = \frac{dy}{du} \times \frac{du}{d\theta}$$

$$= 3u^2 \times (-\sin\theta) = \mathbf{-\,3u^2 \sin\theta}$$

$$\frac{dy}{d\theta} = \mathbf{-\,3\cos^2\theta \sin\theta}$$

iii) $y = \cos^4 2x$, or $y = (\cos 2x)^4$

Let $u = \cos 2x$, so that $y = u^4$

Use the methods explained in sections 10.3 and 10.2.1

$$\frac{du}{dx} = -2\sin 2x, \qquad \text{and } \frac{dy}{du} = 4u^{4-1} = 4u^3$$

$$\frac{dy}{dx} = \frac{dy}{du} \times \frac{du}{dx}$$

$$= 4u^3 \times (-\,2\sin 2x) = -\,8u^3 \sin 2x$$

$$\frac{dy}{dx} = -8(\cos 2x)^3 \sin 2x = \mathbf{-\,8\cos^3 2x \sin 2x}$$

10.5 The chain rule by recognition

It is possible to differentiate composite functions without resorting to substitution, as discussed in section 10.4. For example, if we want to differentiate $y = (x^2 + 3x - 4)^9$; first deal with the outside function (i.e. the power) and then differentiate the inside function (i.e. $(x^2 + 3x - 4)$). The process is shown in example 10.6.

Example 10.6

Differentiate $y = (x^2 + 3x - 4)^9$.

Solution:

$$y = (x^2 + 3x - 4)^9$$

$$\frac{dy}{dx} = 9(x^2 + 3x - 4)^{9-1} \frac{d}{dx}(x^2 + 3x - 4)$$

$$= 9(x^2 + 3x - 4)^8 (2x + 3) = \mathbf{9(2x + 3)(x^2 + 3x - 4)^8}$$

10.6 Differentiation of exponential and logarithm functions

The exponential function e^x is given by:

$$e^x = 1 + x + \frac{x^2}{2!} + \frac{x^3}{3!} + \ldots\ldots\ldots\ldots$$

$$= 1 + x + \frac{x^2}{2 \times 1} + \frac{x^3}{3 \times 2 \times 1} + \ldots\ldots\ldots\ldots$$

$$\frac{d(e^x)}{dx} = 0 + 1 + \frac{2x}{2 \times 1} + \frac{3x^2}{3 \times 2 \times 1} + \ldots\ldots\ldots\ldots$$

$$= 1 + x + \frac{x^2}{2 \times 1} + \frac{x^3}{3 \times 2 \times 1} + \ldots\ldots\ldots\ldots \text{which is the original function}$$

Therefore, $\frac{d(e^x)}{dx} = e^x$

It can also be shown that a simple logarithm function after differentiation results in:

$$\frac{d(\log_e x)}{dx} = \frac{1}{x}$$

These results and their extensions are shown in Table 10.3.

Table 10.3

y	$\frac{dy}{dx}$
e^x	e^x
e^{ax}	ae^{ax}
$\log_e x$	$\frac{1}{x}$
$\log_e ax$	$\frac{1}{x}$

Example 10.7

Differentiate with respect to x.

a) e^{5x+1} b) $\log_e 8x$

Solution:

a) $y = e^{5x+1}$

$$\frac{dy}{dx} = e^{5x+1} \times \frac{d}{dx}(5x+1)$$

$$= e^{5x+1} \times 5 = 5\,e^{5x+1}$$

b) $y = \log_e 8x$

$$\frac{dy}{dx} = \frac{1}{8x} \times \frac{d}{dx}(8x)$$

$$= \frac{1}{8x} \times 8 = \frac{1}{x}$$

10.7 Differentiation of a product

If $y = u \times v$, where u and v are two different functions of x, then:

$$\frac{dy}{dx} = \mathbf{v}\frac{du}{dx} + \mathbf{u}\frac{dv}{dx}$$

Example 10.8

Differentiate i) $y = 2x^3 \log_e x$

ii) $y = \frac{1}{3}x^3 \sin 3x$

Solution:

i) $y = 2x^3 \log_e x$

Let $u = 2x^3$ and $v = \log_e x$

$$\frac{du}{dx} = 6x^2 \text{ and } \frac{dv}{dx} = \frac{1}{x}$$

$$\text{As } \frac{dy}{dx} = v\frac{du}{dx} + u\frac{dv}{dx}$$

$$= \log_e x \times 6x^2 + 2x^3 \times \frac{1}{x}$$

$$= 6x^2 \log_e x + 2x^2$$

or $\frac{dy}{dx} = \mathbf{2x^2(3\log_e x + 1)}$ ($2x^2$ is the common factor)

ii) $y = \frac{1}{3}x^3 \sin 3x$

Let $u = \frac{1}{3}x^3$ and $v = \sin 3x$

$\frac{du}{dx} = \frac{1}{3} \times 3x^2 = x^2$ and $\frac{dv}{dx} = \cos 3x \times 3 = 3 \cos 3x$

$$\text{As } \frac{dy}{dx} = v\frac{du}{dx} + u\frac{dv}{dx}$$

$$= \sin 3x \times x^2 + \frac{1}{3}x^3 \times 3\cos 3x$$

$$= x^2 \sin 3x + x^3 \cos 3x$$

or $\frac{dy}{dx} = \mathbf{x^2(\sin 3x + x \cos 3x)}$ (x^2 is the common factor)

10.8 Differentiation of a quotient

If u and v are functions of x and $y = \frac{u}{v}$, then:

$$\frac{dy}{dx} = \frac{v\frac{du}{dx} - u\frac{dv}{dx}}{v^2}$$

Example 10.9

Find $\frac{dy}{dx}$ if: i) $y = \frac{e^x}{x^2}$

ii) $y = \frac{\log_e 2x}{\sin 2x}$

Solution:

i) $y = \frac{e^x}{x^2}$

Let $u = e^x$, $v = x^2$

$v^2 = (x^2)^2 = x^4$

$\frac{du}{dx} = e^x$, $\frac{dv}{dx} = 2x$

$$\frac{dy}{dx} = \frac{v\frac{du}{dx} - u\frac{dv}{dx}}{v^2}$$

$$= \frac{x^2 \times e^x - e^x \times 2x}{x^4}$$

$$= \frac{x\,e^x\,(x-2)}{x^4}$$

Therefore $\frac{dy}{dx} = \frac{e^x\,(x-2)}{x^3}$

ii) $y = \frac{\log_e 2x}{\sin 2x}$

Let $u = \log_e 2x$, $v = \sin 2x$

$v^2 = (\sin 2x)^2 = \sin^2 2x$

$\frac{du}{dx} = \frac{1}{2x} \times 2 = \frac{1}{x}$, $\frac{dv}{dx} = 2 \cos 2x$

$$\frac{dy}{dx} = \frac{\sin 2x \times \frac{1}{x} - \log_e 2x \times 2\cos 2x}{\sin^2 2x}$$

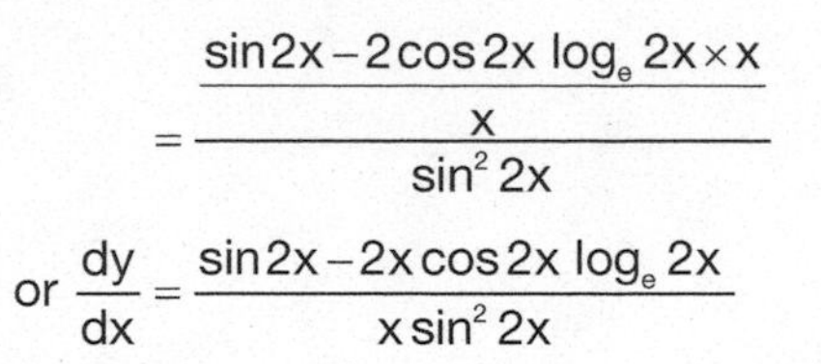

$$= \frac{\frac{\sin 2x - 2\cos 2x \log_e 2x \times x}{x}}{\sin^2 2x}$$

$$\text{or } \frac{dy}{dx} = \frac{\sin 2x - 2x\cos 2x \log_e 2x}{x \sin^2 2x}$$

10.9 Numerical values of differential coefficients

Example 10.10

Find the value of $\frac{dy}{dx}$ for the following curves:

i) $y = 2x^3 - 4x^2 - x + 5$, at the point where $x = 2$.

ii) $y = \cos 2x - 2 \sin x$, at $x = \frac{\pi}{4}$

Solution:

i) $y = 2x^3 - 4x^2 - x + 5$

$$\frac{dy}{dx} = 6x^2 - 8x - 1$$

$$\text{At } x = 2,\ \frac{dy}{dx} = 6(2)^2 - 8 \times 2 - 1$$

$$\text{Therefore } \frac{dy}{dx} = 24 - 16 - 1 = \mathbf{7}$$

ii) $y = \cos 2x - 2 \sin x$

The angle is in radians here, so the mode of your calculator must be set to radians

$$\frac{dy}{dx} = -2\sin 2x - 2\cos x$$

$$\text{At } x = \frac{\pi}{4},\ \frac{dy}{dx} = -2 \times \sin(2 \times \frac{\pi}{4}) - 2 \times \cos\frac{\pi}{4}$$

$$\text{or } \frac{dy}{dx} = -2 \times 1 - 2 \times 0.707 = \mathbf{-3.414}$$

Example 10.11

If $y = e^{2x} + e^{0.4x}$, find the value of $\frac{dy}{dx}$ where $x = 0.5$

Solution:

$$y = e^{2x} + e^{0.4x}$$

$$\frac{dy}{dx} = 2e^{2x} + 0.4e^{0.4x}$$

At $x = 0.5$, $\frac{dy}{dx} = 2e^{2\times0.5} + 0.4e^{0.4\times0.5}$

$$= 2\ e^{1} + 0.4\ e^{0.2}$$

or $\frac{dy}{dx} = 2\times2.7183 + 0.4\times1.2214 = \mathbf{5.925}$

Exercise 10.1

Differentiate the following from first principles:

1. $y = 2x$
2. $y = 3x + 5$

For questions 3 to 6, find $\frac{dy}{dx}$ if:

3. $y = 3x^2 + 2$
4. $y = \frac{-6}{\sqrt{x}}$
5. $y = 5x^2 + 2x + \frac{2}{x} + 5$

Differentiate the functions, given in questions 6 to 11, with respect to the variable:

6. $2 \sin 3x$
7. $\cos 2\theta - 3 \sin 4\theta$
8. $4z^2 - 2 \cos 2z$
9. $\frac{1}{2}\log_e 2x$
10. e^{7x}
11. $\frac{1}{e^{2x}}$

Exercise 10.2

Use the chain rule to differentiate the following with respect to the variable:

1. $(x^2 + x)^5$
2. $\sqrt{(2x^2 - 3x + 1)}$
3. $\frac{1}{(2x+5)^3}$
4. $\tan (2x + 4)$
5. $\cos 4x$
6. $\sin^5 x$

7. $\cos^3 3x$
8. $\sin(3x + 8)$
9. $\frac{1}{\sin^2 x}$
10. $\log_e (7 - 3x)$
11. $4e^{2x-1}$
12. $\frac{1}{e^{2x+4}}$

Exercise 10.3

Use the product/quotient rule to differentiate the following:

1. $y = x \cos x$
2. $y = x^2 \log_e x$
3. $y = e^{2x} \sin 3x$
4. $y = (2x + 1) \tan x$
5. $y = 3(x^2 + 3) \sin x$
6. $y = 4 \sin\theta \cos\theta$
7. $y = (z^2 + 2z - 2) \sin z$
8. $y = \frac{\cos x}{x}$
9. $y = \frac{2x}{x+4}$
10. $y = \frac{x+3}{\sin x}$
11. $y = \sec x$
12. $y = \text{cosec } x$
13. $y = \frac{e^{2x}}{x^2}$
14. $y = \frac{2\log_e x}{\sin 2x}$

Exercise 10.4

The following questions involve the calculation of numerical values of differential coefficients; these have been taken from Exercises 10.1 and 10.2:

1. Find the value of $\frac{dy}{dx}$ for the following functions when $x = 1.5$

   a) $y = \frac{-6}{\sqrt{x}}$

   b) $y = 5x^2 + 2x + \frac{2}{x} + 5$

2. If $y = 4z^2 - 2\cos 2z$, find the value of $\frac{dy}{dz}$ when $z = 0.3$
3. Find the gradient of the curve $\cos 2\theta - 3\sin 4\theta$, at the point where $\theta = \frac{\pi}{5}$

Find the value of $\frac{dy}{dx}$ for the following functions where $x = 0.4$

4. $\log_e(7 - 3x)$
5. $4e^{2x-1}$
6. $\frac{1}{e^{2x+4}}$

Answers – Exercise 10.1

1. 2
2. 3
3. $6x$
4. $\frac{3}{x^{3/2}}$
5. $10x + 2 - \frac{2}{x^2}$
6. $6\cos 3x$
7. $-2\sin 2\theta - 12\cos 4\theta$
8. $8z + 4\sin 2z$
9. $\frac{1}{2x}$
10. $7e^{7x}$
11. $-2e^{-2x}$

Answers – Exercise 10.2

1. $5(x^2 + x)^4(2x + 1)$
2. $\frac{1}{2}(2x^2 - 3x+1)^{-1/2}(4x - 3)$
3. $-6(2x + 5)^{-4}$
4. $2\sec^2(2x + 4)$
5. $-4\sin 4x$
6. $5\cos x \sin^4 x$
7. $-9\cos^2 3x \sin 3x$
8. $3\cos(3x + 8)$
9. $\frac{-2\cos x}{\sin^3 x}$

10. $\frac{-3}{7-3x}$

11. $8e^{2x-1}$

12. $\frac{-2}{e^{2x+4}}$

Answers – Exercise 10.3

1. $\cos x - x \sin x$
2. $2x \log_e x + x$
3. $e^{2x} (2 \sin 3x + 3 \cos 3x)$
4. $2 \tan x + (2x + 1) \sec^2 x$
5. $6x \sin x + 3(x^2 + 3) \cos x$
6. $4(\cos^2 \theta - \sin^2 \theta)$
7. $(2z + 2) \sin z + (z^2 + 2z - 2) \cos z$
8. $\frac{-x \sin x - \cos x}{x^2}$
9. $\frac{8}{(x + 4)^2}$
10. $\frac{\sin x - (x + 3)\cos x}{\sin^2 x}$
11. $\tan x \sec x$
12. $-\cot x \operatorname{cosec} x$
13. $\frac{2x e^{2x}(x-1)}{x^4}$
14. $\frac{2\sin 2x - 4x \log_e x \cos 2x}{x \sin^2 2x}$

Answers – Exercise 10.4

1. a) 1.633
   b) 16.111
2. 4.659
3. 7.806
4. – 0.5172
5. 6.5496
6. –0.01646

CHAPTER 11

Applications of differentiation

Topics covered:

- Application of differentiation to solve civil engineering problems
- Second derivatives and their application
- Maximum and minimum, and their application to solve construction related problems

11.1 Application in structural mechanics

In the structural design of beams and slabs the engineers have to determine, as the first step, the maximum bending moment (BM) and the position where it occurs. If a beam is subjected to uniformly distributed load or uniformly increasing load, then differentiation may be used to determine the point on the beam where the maximum BM will occur.

By looking at a typical BM diagram (Figure A1.9, Appendix 1), it will be evident that the bending moment increases initially and then peaks off at the maximum value, so its slope is 0 at that point. For calculating the BM at any point on the beam it is necessary to form an equation, which is then differentiated and equated to 0 as the slope of the line forming the bending moment diagram is 0 at the maximum value of the BM.

More details on the behaviour of beams can be found on the companion website.

Example 11.1

Find the position and magnitude of the maximum bending moment for the beam shown in Figure 11.1.

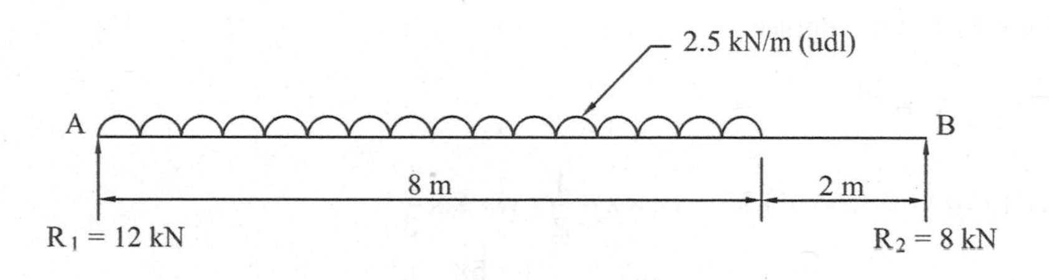

Figure 11.1

Solution:

Let us assume that the maximum BM occurs at point C, x metres from A, as shown in Figure 11.2. Bending moment at this point is given by:

$$M_x = R_1 \times x - 2.5 \times x \times \frac{x}{2}$$
$$= 12 \times x - 2.5 \times \frac{x^2}{2}$$

At maximum BM the slope of the curve is 0, therefore $\frac{dM_x}{dx} = 0$

$$\frac{dM_x}{dx} = 12 - 2.5 \times 2 \times \frac{x}{2} = 0$$

$$12 - 2.5\,x = 0 \quad \text{or} \quad x = \frac{12}{2.5} = \mathbf{4.8\ m}$$

$$M_x = 12 \times 4.8 - 2.5 \times \frac{(4.8)^2}{2} = \mathbf{28.8\ kNm}$$

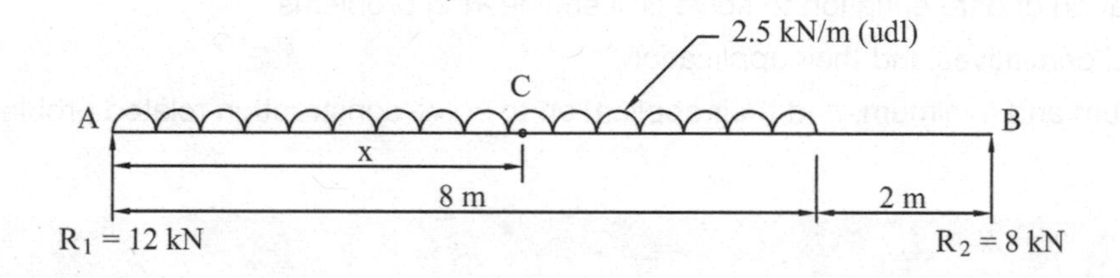

Figure 11.2

Example 11.2

A simply supported 8 m long beam carries a uniformly increasing load from 0 at 1 end to 5 kN/m at the other. Find the position and magnitude of the maximum bending moment on the beam.

Solution:

The beam and the load are shown in Figure 11.3.

$$\text{Total load on the beam} = \frac{1}{2} \times 8 \times 5 = 20\,\text{kN}$$

$$R_1 \times 8 = 20 \times \frac{8}{3} \quad \text{or} \quad R_1 = 6.67\ \text{kNm}$$

$$R_2 = 20\ \text{kN} - 6.67\ \text{kN} = 13.33\ \text{kN}$$

Triangles ADE and ACB are similar:

$$\frac{w}{x} = \frac{5}{8} \quad \text{or} \quad w = \frac{5x}{8}$$

$$\text{BM at point E (x metres from A)}, M_x = R_1 \times x - \frac{1}{2} \times w \times x \times \frac{x}{3}$$
$$= 6.67 \times x - \frac{1}{2} \times \frac{5x}{8} \times \frac{x^2}{3}$$

$$= 6.67 \times x - \frac{5x^3}{48}$$

At maximum BM the slope of the curve is zero, therefore $\frac{dM_x}{dx} = 0$

$$\frac{dM_x}{dx} = 6.67 - \frac{5 \times 3 \times x^2}{48} = 0$$

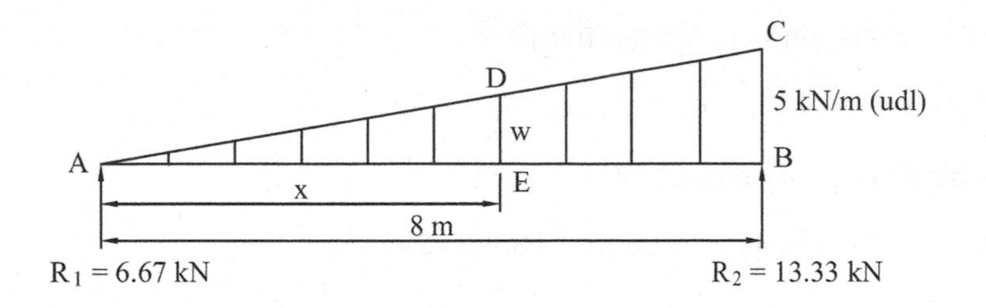

Figure 11.3

$$\frac{5 \times 3 \times x^2}{48} = 6.67;$$

$x^2 = 21.344$ or x = **4.62 m**

$$M_x = 6.67 \times 4.62 - \frac{5}{48} \times (4.62)^3 = \mathbf{20.54\ kNm}$$

11.2 Second derivatives

If $y = 2x^3$, then after differentiating:

$$\frac{dy}{dx} = 6x^{3-1} \text{ or } 6x^2,$$

$\frac{dy}{dx}$ is called the first derivative or the first differential coefficient.

On differentiating the above equation again, we have:

$\frac{d^2y}{dx^2} = 2 \times 6x^{2-1} = 12x$; $\frac{d^2y}{dx^2}$ is called the second derivative or the second differential coefficient of y with respect to x.

Example 11.3

Find $\frac{dy}{dx}$ and $\frac{d^2y}{dx^2}$, if:

i) $y = 2x^2 - 3x + 2$
ii) $y = 4 \sin 3x - \cos 4x$
iii) $y = e^{2x} - e^{-2x}$

Solution:

i) $y = 2x^2 - 3x + 2$

$$\frac{dy}{dx} = \mathbf{4x - 3}$$

$$\frac{d^2y}{dx^2} = \mathbf{4}$$

ii) $y = 4\sin 3x - \cos 4x$

$$\frac{dy}{dx} = 4 \times \cos 3x \times 3 - (-\sin 4x) \times 4 \text{(chain rule)}$$

$$= \mathbf{12\cos 3x + 4\sin 4x}$$

$$\frac{d^2y}{dx^2} = 12 \times (-\sin 3x) \times 3 + 4 \times \cos 4x \times 4$$

$$= -36\sin 3x + 16\cos 4x = \mathbf{16\cos 4x - 36\sin 3x}$$

iii) $y = e^{2x} - e^{-2x}$

$$\frac{dy}{dx} = 2e^{2x} - \left(-2e^{-2x}\right) = \mathbf{2e^{2x} + 2e^{-2x}}$$

$$\frac{d^2y}{dx^2} = 2 \times 2e^{2x} + 2 \times (-2)e^{-2x}$$

$$= \mathbf{4e^{2x} - 4e^{-2x}}$$

11.3 Velocity and acceleration

The rate of change of distance with respect to time is known as velocity. If the distance is denoted by s and time by t, then:

$$\text{Velocity} = \frac{ds}{dt}$$

Since acceleration is the rate of change of velocity with respect to time, therefore the second derivative, i.e. $\frac{d^2s}{dt^2}$, will give the acceleration.

Example 11.4

The distance travelled (s) by a vehicle in time t seconds is given by:

$s = t^3 - 6t^2 + 9t + 4$

Calculate:
a) the velocity of the vehicle at the end of 5 seconds.
b) the acceleration at the end of 5 seconds.
c) the time when the velocity of the vehicle is 0.

Solution:

a) $s = t^3 - 6t^2 + 9t + 4$

Differentiate with respect to s, $\frac{ds}{dt} = 3t^2 - 12t + 9$

The rate of change of distance (s) with respect to time (t) is known as velocity.

Therefore, velocity $= \frac{ds}{dt} = 3t^2 - 12t + 9$

When t = 5, $\frac{ds}{dt} = 3(5)^2 - 12(5) + 9$

$= 75 - 60 + 9 =$ **24 m/s**

b) Since acceleration is the rate of change of velocity with respect to time, therefore the second derivative, i.e. $\frac{d^2s}{dt^2}$, will give the acceleration.

From part (a), $\frac{ds}{dt} = 3t^2 - 12t + 9$

$\frac{d^2s}{dt^2} = 6t - 12$

Acceleration at the end of 5 seconds $= 6 \times 5 - 12 =$ **18 m/s²**.

c) $\frac{ds}{dt}$ is zero, when the velocity of the vehicle is zero.

Therefore, $\frac{ds}{dt} = 3t^2 - 12t + 9 = 0$

This is a quadratic equation, its solution will give the value of time t

$$t = \frac{-(-12) \pm \sqrt{(-12)^2 - 4 \times 3 \times 9}}{2 \times 3}$$

$$= \frac{12 \pm \sqrt{36}}{6} = \frac{12 \pm 6}{6}$$

$t = \frac{12+6}{6} =$ **3.0 seconds**

or $t = \frac{12-6}{6} =$ **1.0 second**

11.4 Maximum and minimum

There are many examples in construction and engineering where the maximum or minimum value of a quantity is required. These values are examples of turning points, as shown in Figure 11.4. The turning point at P is called the maximum turning point and the turning point at Q is called the minimum turning point; the turning points are not necessarily the maximum and the minimum values of y. The gradient of the tangents to the curve at the turning points is 0 as the tangents are horizontal lines. As the gradient of the curve is 0, $\frac{dy}{dx} = 0$

There are 2 methods of distinguishing between a maximum value and a minimum value. If $\frac{d^2y}{dx^2}$ is calculated, it is negative at maximum and positive at minimum. If the calculation of $\frac{d^2y}{dx^2}$ is difficult a second test may be used. At a maximum, $\frac{dy}{dx}$ changes from positive just before to negative just after, and at a minimum $\frac{dy}{dx}$ changes from negative just before, to positive just after.

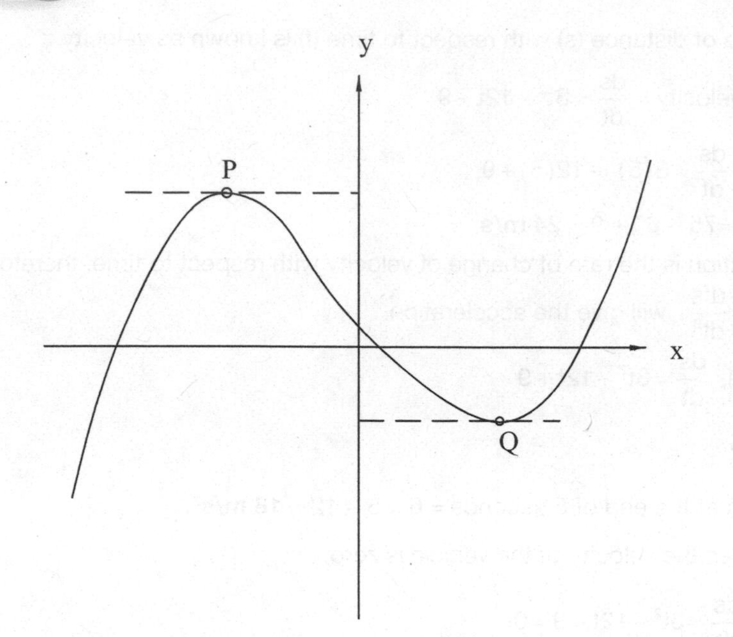

Figure 11.4

Example 11.5

Find the maximum and minimum values of y, if: $y = x^3 - 7x^2 - 8x + 15$.

Solution:

$$y = x^3 - 7x^2 - 8x + 15$$

Therefore $\frac{dy}{dx} = 3x^2 - 14x - 8$

$$\frac{d^2y}{dx^2} = 6x - 14$$

At a turning point $\frac{dy}{dx} = 0$

$$\frac{dy}{dx} = 3x^2 - 14x - 8 = 0$$

The above equation is a quadratic equation and may solved using the quadratic formula:

$$x = \frac{-(-14) \pm \sqrt{(-14)^2 - 4(3)(-8)}}{2 \times 3}$$

$$= \frac{14 \pm 17.09}{6}$$

Either $x = \frac{14 + 17.09}{6} = 5.18$, or $x = \frac{14 - 17.09}{6} = -0.52$

When $x = 5.18$, $\frac{d^2y}{dx^2} = 6x - 14 = 6 \times 5.18 - 14 = 17.08$

Since 17.08 is positive, the turning point at **$x = 5.18$ is a minimum.**

To find the minimum value of y, put x = 5.18 in the original equation.

y (minimum) = $(5.18)^3 - 7(5.18)^2 - 8 \times 5.18 + 15 =$ **−75.28**

When $x = -0.52$, $\frac{d^2y}{dx^2} = 6x - 14 = 6 \times (-0.52) - 14 = -17.12$

Since 17.12 is negative, the turning point at **$x = -0.52$ is a maximum.**

y (maximum) = $(-0.52)^3 - 7(-0.52)^2 - 8 \times -0.52 + 15 =$ **17.13**

Example 11.6

The cost (C) of producing x number of a building component is calculated by adding the fixed cost and the variable cost. If the fixed cost is £1000, and the variable cost is given by $0.05x^2 + 20x$, determine:

i) the output that will minimise the average cost per component.

ii) the average cost per component.

Solution:

Total cost of producing x number of building components = Fixed cost + Variable cost

$= 1000 + 0.05x^2 + 20x$

i) Average cost (C_{Av}) = Total cost ÷ x

$$= \frac{1000}{x} + \frac{0.05x^2}{x} + \frac{20x}{x}$$

$$= \frac{1000}{x} + 0.05x + 20 = 1000x^{-1} + 0.05x + 20$$

Differentiating the above equation, $\frac{dC_{Av}}{dx} = 1000(-1)x^{-2} + 0.05$

For maximum/minimum, $\frac{dC_{Av}}{dx}$ must be equal to 0

$1000(-1)x^{-2} + 0.05 = 0$

$0.05 = 1000x^{-2}$ or $0.05 = \frac{1000}{x^2}$

Therefore, $x^2 = \frac{1000}{0.05}$ or $x = \sqrt{20000} =$ **141 components**

$$\frac{d^2C_{Av}}{dx^2} = -1000(-2)x^{-3} = \frac{2000}{x^3}$$

This will be positive for all values of x, therefore, **x = 141 is a minimum.**

ii) Average cost / component $= \frac{1000}{x} + 0.05x + 20$

$$= \frac{1000}{141} + 0.05 \times 141 + 20$$

$= 7.09 + 7.05 + 20 =$ **£34.14**

Example 11.7

A 750 litres capacity rectangular tank, with a square base and open top, is to be made from a sheet of corrosion resistant metal. Find:

i) The dimensions of the tank so that the area of the metal sheet used is minimum.
ii) The area of the metal sheet.

Solution:

750 litres = 0.750 m^3

i) Let the length (and width) of the base be x metres, and the height be h metres

Volume of the tank = length × width × height = 0.750 m^3

$$x \times x \times h = 0.750, \quad \text{therefore, } h = \frac{0.750}{x^2} \quad \ldots\ldots\ldots\ldots\ldots (1)$$

Total area of the metal sheet, A = surface area of the tank

$$= 4 \times (x \times h) + x^2$$

$$= 4xh + x^2$$

$$\text{As } h = \frac{0.750}{x^2}, \text{Therefore } A = 4x \times \frac{0.750}{x^2} + x^2$$

$$\text{or } A = \frac{3.0}{x} + x^2 = 3.0x^{-1} + x^2$$

$$\text{Differentiating the above equation, } \frac{dA}{dx} = -1 \times 3.0x^{-2} + 2x = \frac{-3.0}{x^2} + 2x$$

$$\text{and } \frac{d^2A}{dx^2} = (-2)(-3.0x^{-3}) + 2 = \frac{6.0}{x^3} + 2$$

$$\text{For a turning point, } \frac{dA}{dx} = 0$$

$$\text{Therefore } \frac{-3.0}{x^2} + 2x = 0 \quad \text{or} \quad 2x = \frac{3.0}{x^2}$$

$$\text{After transposition, } 2x^3 = 3.0 \qquad \text{or } x = \sqrt[3]{\frac{3.0}{2}} = \mathbf{1.145\ m}$$

$$\text{Test for a minimum: } \frac{d^2A}{dx^2} = \frac{6.0}{x^3} + 2$$

$$= \frac{6.0}{(1.145)^3} + 2 = 6$$

As this value is positive, area of the metal sheet (A) is minimum.

ii) $$\text{From eq. 1, Height of the tank, } h = \frac{0.750}{x^2}$$

$$\text{As } x = 1.145\text{ m}, \quad h = \frac{0.750}{(1.145)^2} = \mathbf{0.572\ m}$$

Area of the metal sheet, A = 4xh + x^2

= 4 × 1.145 × 0.572 + $(1.145)^2$ = **3.931 m^2**

Exercise 11.1

1. Find $\frac{dy}{dx}$ and $\frac{d^2y}{dx^2}$, if:
   i) $y = 2x^3 - 3x^2 + 2x$
   ii) $y = 3 \sin 2x + \cos 5x$
   iii) $y = e^{2x} - \log_e 7x$
2. The distance travelled (s) by a vehicle in time t seconds is given by:
   $s = 0.5t^3 - 2t^2 + 2t + 5$
   Calculate: a) the velocity of the vehicle at the end of 4 seconds.
   b) the time t seconds, when the vehicle comes to rest.
   c) the time when the acceleration of the vehicle is 8 m/s^2.
3. Find the position and magnitude of the maximum bending moment on the beam shown in Figure 11.5.

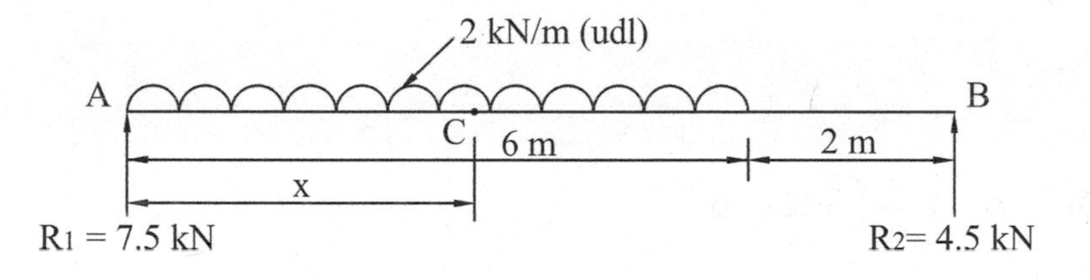

Figure 11.5

4. Find the maximum and minimum values of y, if: $y = x^3 - 6x^2 + 9x + 6$.
5. The profit (P) of a firm manufacturing hardwood doors is given by:
   $P = 20x - 0.04x^2$, where x is the number of doors produced (output). Calculate the best output for maximising the firm's profit.
6. The cost (C) of producing x number of a roof component is calculated by adding the fixed cost and the variable cost. If the fixed cost is £25000, and the variable cost is given by $0.05x^2 + 20x$, determine:
   i) the output that will minimise the average cost per component.
   ii) the average cost per component.
7. A steel plate, 700 mm wide, is used to manufacture a hollow beam having a width of b mm and depth of h mm in cross-section. If the strength of this beam is proportional to bh^3, find the dimensions of the strongest beam.
8. An open channel, made from 200 mm wide strip of steel, is rectangular in cross-section. Calculate the width and depth of the channel which provide the maximum area of cross-section.

9. An open channel, made from 400 mm wide strip of steel, is triangular in cross-section (see Figure 11.6). Calculate the depth of the channel and its width at the top so that the area of cross-section is maximum.

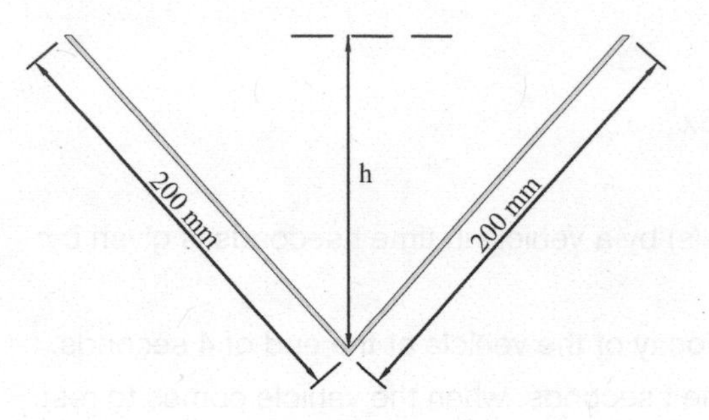

Figure 11.6 Open triangular channel

10. A pressurised hot water cylinder has flat top and base. Find the height and diameter of the cylinder if its capacity is 210 litres and has the least surface area.

Answers – Exercise 11.1

1. i) $\frac{dy}{dx} = 6x^2 - 6x + 2; \frac{d^2y}{dx^2} = 12x - 6$

   ii) $\frac{dy}{dx} = 6\cos 2x - 5\sin 5x; \frac{d^2y}{dx^2} = -12\sin 2x - 25\cos 5x$

   iii) $\frac{dy}{dx} = 2e^{2x} - x^{-1}; \frac{d^2y}{dx^2} = 4e^{2x} + x^{-2}$

2. a) 10 m/s b) 2 sec. or 0.67 sec. c) 4 sec.
3. Distance = 3.375 m from left support; Max. B.M. = 26.39 kNm
4. y (max) =10; y(min) = 6
5. 250
6. i) 707; ii) £90.71
7. b = 87.5 mm; h = 262.5 mm
8. Width = 100 mm; Depth = 50 mm
9. Width = 282.84 mm; Depth = 141.42 mm
10. Height = 644.22 mm; Diameter = 644.24 mm

CHAPTER 12

Integration

Topics covered in this chapter:

- Indefinite and definite integrals
- Integration by substitution
- Change of limits
- Integration by parts

12.1 Introduction

Integration is the reverse process of differentiation. If $y = x^2$, then after differentiation $\frac{dy}{dx} = 2x$ and after transposition $dy = 2x\,dx$. Here the power of x has been decreased by 1. As integration is the reverse process of differentiation, the power of x will be increased by 1. Integration is indicated by the sign ($\int$) shown in front of the differential; the integration of $dy = 2x\,dx$ is shown as:

$$\int dy = \int 2x\,dx$$

$$y = \frac{2x^{1+1}}{1+1} = \frac{2x^2}{2} = x^2 \text{ (increase the power by 1 and divide by the increased power)}$$

We can say this as the integral of 2x with respect to x is x^2.

So what is integration? we can understand this by considering Figure 12.1a that shows area ABCD. The area may be determined by first dividing the shape into 'n' number of strips of equal width δx. Let the area of one strip be δA:

$\delta A \approx y \times \delta x$ (δA is approximate as the strip is not a perfect rectangle)

Since δx is small, the shaded area at the top of each strip (Figure 12.1b) is small as compared to the area of the rectangle. The approximate area of figure ABCD is given by the sum of the areas of all the rectangles.

Area of shape ABCD = Area of all strips = $y_1\delta x + y_2\delta x + y_3\delta x + \ldots\ldots + y_n\delta x$

This can be written as: Area of ABCD $= \text{Area} = \sum_{x=a}^{x=b} y\,\delta x$ (Σ denotes the sum)

As δx approaches 0 (written as $\delta x \rightarrow 0$), the shaded area at the top of each strip becomes negligibly small and the above sum gives the exact value of the area. δx is replaced by dx and the Σ sign is replaced by the integral sign, i.e. $\int$. Thus the area is now written as: Area of ABCD $= \int_a^b y\,dx$, where a and b are the limits of integration.

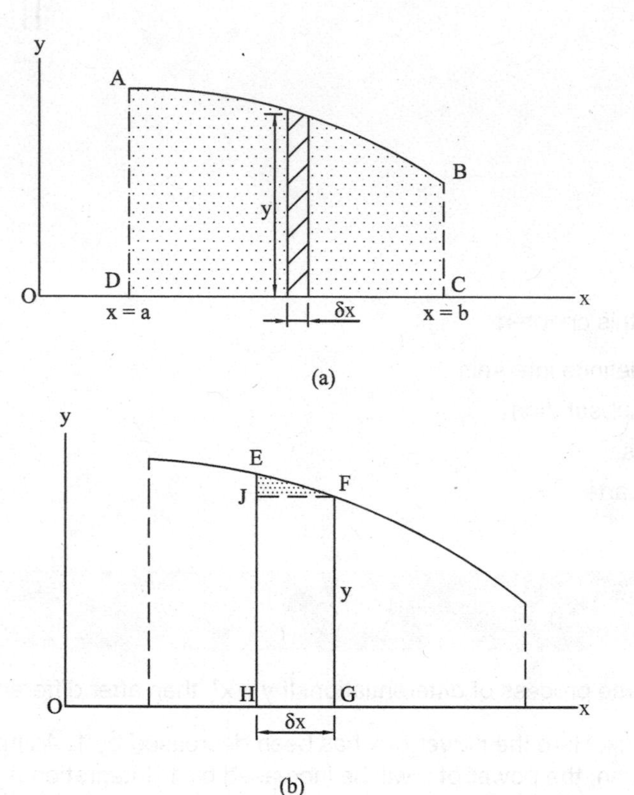

Figure 12.1

12.2 Indefinite integrals

If $y = x^2 + 3$, or $y = x^2 +$ any number, then:

$$\frac{dy}{dx} = 2x, \text{ or } dy = 2x\,dx$$

Integrating the above differential, we have: $\int dy = \int 2x\,dx$, or $y = x^2$

The answer must account for the number (constant) which was given in the original question. This is resolved by including a constant (c) in the answer.

Therefore, $\int 2x\,dx = x^2 + c$, where c is known as the arbitrary constant of integration and the result containing it is known as the indefinite integral. Table 12.1 shows a selection of the formulae used in integration:

Table 12.1

Formulae for integration:			
If $\frac{dy}{dx}=ax^n$, then $y=\int ax^n\,dx=\frac{ax^{n+1}}{n+1}$			
$\int \sin x=-\cos x$	$\int \cos x=\sin x$	$\int \sin ax=-\frac{1}{a}\cos ax$	
$\int \cos ax=\frac{1}{a}\sin ax$	$\int \sec^2 x=\tan x$	$\int \frac{1}{x}=\log_e x$	$\int e^{ax}=\frac{1}{a}e^{ax}$
$\int e^x=e^x$			

12.2.1 Integration of a sum

When an expression contains several terms, then the result will be the sum of the integrals of the separate terms. Only 1 arbitrary constant is used.

Example 12.1

Determine the following indefinite integrals:

i) $\int 5\,dx$

ii) $\int 6\,d\theta$

iii) $4x^3\,dx$

iv) $\int (2x^2-3x+2)\,dx$

v) $\int \left(\frac{1}{x}+e^{3x}\right)dx$

vi) $\int (\cos 2x+\sin 4x)\,dx$

Solution:

i) $\int 5\,dx=\mathbf{5x}$

ii) $\int 6\,d\theta=\mathbf{6\theta}$

(Note: Depending on the differential, a number after integration results in 5x, 6θ etc.)

iii) $\int 4x^3\,dx=\frac{4x^{3+1}}{3+1}+c$ (increase the power by one and divide by the increased power)

$$=\frac{4x^4}{4}+c=\mathbf{x^4+c}$$

iv) $\int (2x^2-3x+2)\,dx=\frac{2x^{2+1}}{2+1}-\frac{3x^{1+1}}{1+1}+2x+c$

$$=\frac{2x^3}{3}-\frac{3x^2}{2}+\mathbf{2x+c}$$

v) $\int \left(\frac{1}{x}+e^{3x}\right)dx=\mathbf{\log_e x}+\frac{e^{3x}}{3}+\mathbf{c}$

vi) $\int(\cos 2x+\sin 4x)dx=\frac{\sin 2x}{2}+\frac{-\cos 4x}{4}+c$

$=\frac{\sin 2x}{2}-\frac{\cos 4x}{4}+\mathbf{c}$

12.3 Definite integrals

An integral between 2 limits, the upper limit and the lower limit, is known as a definite integral. The indefinite integral is determined first and then the limits are used to find the definite value. The constant of integration disappears in this process. Let us assume that the upper and lower limits of an integral are b and a, then:

The value of the integral = the value of indefinite integral due to b – the value of indefinite integral due to a

Example 12.2

Evaluate $\int_1^3(x^2-1)dx$

Solution:

The first step is to go through the integration process and use the square brackets to show that this has been done. The second step involves the evaluation of the integral by putting x = 3 (upper limit) first and then putting x = 1 (lower limit), as shown:

$$\int_1^3(x^2-1)dx=\left[\frac{x^3}{3}-x\right]_1^3$$

$$=(\frac{3^3}{3}-3)-(\frac{1^3}{3}-1)$$

$$=(9-3)-(\frac{1}{3}-1)$$

$$=6-(-\frac{2}{3})=6+\frac{2}{3}$$

$$=6\frac{2}{3} \text{ or } \mathbf{6.67}$$

Example 12.3

Evaluate i) $\int_0^1 e^{3x}dx$

ii) $\int_0^1(\sin 2\theta-1)d\theta$

Solution:

$$\int_0^1 e^{3x}dx=\left[\frac{1}{3}e^{3x}\right]_0^1$$

$$=\frac{1}{3}e^{3\times 1}-\frac{1}{3}e^{3\times 0}$$

$$=\frac{1}{3}e^3-\frac{1}{3}e^0=\frac{1}{3}\times 20.086-\frac{1}{3}\times 1=\mathbf{6.36}$$

ii) $\int_0^1 (\sin 2\theta - 1)\,d\theta = \left[-\frac{1}{2}\cos 2\theta - \theta\right]_0^1$

$$= [-\frac{1}{2}\cos 2\times 1 - 1] - [-\frac{1}{2}\cos 2\times 0 - 0]$$

$$= -\frac{1}{2}\cos 2 - 1 + \frac{1}{2}\cos 0$$

(Your calculator must be set to give the values of angles in radians)

$$= -\frac{1}{2}\times(-0.416) - 1 + \frac{1}{2}\times 1$$

$$= 0.208 - 1 + 0.5 = -\mathbf{0.292}$$

12.4 Integration by substitution

The method of substitution may be used to simplify the integration of complex integrals. For example, the integration of $(2x - 3)^5$ may be done by expanding the expression first and then use the method described in section 12.1.1. However, this is time consuming process, and to solve this question quickly, integration by substitution may be used, as shown in example 12.4.

Example 12.4

Find $\int (2x - 3)^5\,dx$

Solution:

Let $u = 2x - 3$

Differentiating u, $\frac{du}{dx} = 2$, or $dx = \frac{du}{2}$

After substitution the question now becomes $\int (u)^5 \times \frac{du}{2}$

$$\int (u)^5 \times \frac{du}{2} = \frac{1}{2}\int u^5\,du$$

$$= \frac{1}{2}\times\frac{u^6}{6} + c = \frac{1}{12}\mathbf{(2x-3)^6 + c}$$

Example 12.5

Evaluate $\int_2^3 \frac{x\,dx}{(x^2 - 1)}$

Solution:

Let $u = x^2 - 1$

Differentiating u, $\frac{du}{dx} = 2x$ or $dx = \frac{du}{2x}$

After substitution, $\int_2^3 \frac{x\,dx}{(x^2-1)} = \int_2^3 \frac{x}{u}\times\frac{du}{2x}$

$$=\frac{1}{2}\int_2^3 \frac{1}{u}du=\frac{1}{2}\left[\log_e u\right]_2^3$$

$$=\frac{1}{2}\left[\log_e (x^2-1)\right]_2^3$$

$$=\frac{1}{2}\left[\log_e (3^2-1)-\log_e (2^2-1)\right]$$

$$=\frac{1}{2}\left[\log_e 8-\log_e 3\right]$$

$$=\frac{1}{2}\left[2.0794-1.0986\right]=\mathbf{0.49}$$

Example 12.6

Find $\int \frac{1}{2}\sin^2\theta\cos\theta d\theta$

Solution:

Let $u = \sin\theta$

After differentiation, $\frac{du}{d\theta}=\cos\theta$ or $d\theta=\frac{du}{\cos\theta}$

After substitution, $\int \frac{1}{2}\sin^2\theta\cos\theta d\theta=\int \frac{1}{2}u^2\cos\theta\frac{du}{\cos\theta}$

$$=\frac{1}{2}\int u^2 du=\frac{1}{2}\times\frac{u^3}{3}+c$$

$$=\frac{u^3}{6}+c=\frac{1}{6}\mathbf{sin^3\theta+c}$$

12.5 Change of limits

It is sometimes more convenient to change the limits while integrating definite integrals using the substitution method. The necessity of reintroducing the original variable is avoided by changing the limits.

Example 12.7

Evaluate $\int_1^2 \sqrt{(x^2+3)}\,3x\,dx$

Solution:

Let $u=x^2+3$; $\frac{du}{dx}=2x$ or $dx=\frac{du}{2x}$

Change limits: When $x = 2$, $u = 2^2 + 3 = 7$

When $x = 1$, $u = 1^2 + 3 = 4$

$$\int_1^2 \sqrt{(x^2+3)}\,3x\,dx = \int_4^7 (u)^{1/2}\,3x\frac{du}{2x} = \frac{3}{2}\int_4^7 (u)^{1/2}\,du$$

$$= \frac{3}{2}\left[\frac{u^{3/2}}{\frac{3}{2}}\right]_4^7 = \frac{3}{2}\times\frac{2}{3}\left[7^{3/2} - 4^{3/2}\right]$$

$$= 18.52 - 8.0 = \mathbf{10.52}$$

Example 12.8

Evaluate $\int_0^{\pi/3} \sin^2\theta\cos\theta\,d\theta$

Solution:

Let $u = \sin\theta$

After differentiation, $\frac{du}{d\theta} = \cos\theta$ or $d\theta = \frac{du}{\cos\theta}$

Change limits: When $\theta = \pi/3$, $u = \sin \pi/3 = 0.866$

When $\theta = 0$, $u = \sin 0 = 0$

$$\int_0^{\pi/3} \sin^2\theta\cos\theta\,d\theta = \int_0^{0.866} u^2\cos\theta\frac{du}{\cos\theta}$$

$$= \int_0^{0.866} u^2\,du = \left[\frac{u^3}{3}\right]_0^{0.866}$$

$$= \left[\frac{0.866^3}{3} - \frac{0^3}{3}\right] = \mathbf{0.216}$$

12.6 Integration by parts

Integration by parts is a special rule which can be used to integrate the product of 2 functions.

Let $y = u \times v$, where u and v are 2 functions

After differentiating, $\frac{dy}{dx} = \frac{d}{dx}(uv) = v\frac{du}{dx} + u\frac{dv}{dx}$

$$u\frac{dv}{dx} = \frac{d}{dx}(uv) - v\frac{du}{dx}$$

If we integrate both sides with respect to x, we have:

$$\int u\frac{dv}{dx}dx = \int\frac{d(uv)}{dx}dx - \int v\frac{du}{dx}dx$$

$$\int u\frac{dv}{dx}dx = uv - \int v\frac{du}{dx}dx$$

$\int u\,dv = uv - \int v\,du$ (integration by parts rule)

The formula replaces the integral on the left with another integral (on the right) that is simpler.

Example 12.9

Find $\int x \sin x \, dx$

Solution:

On comparing $\int x \sin x \, dx$ with the left side of the 'integration by parts rule'

$u = x$ and $dv = \sin x \, dx$

Differentiating $u = x$, we have $\frac{du}{dx} = 1$

Integrating $\int dv = \int \sin x \, dx$ we have $v = -\cos x$

Substituting the above in the 'integration by parts rule': $\int u \, dv = uv - \int v \, du$

$$\int x \sin x \, dx = x(-\cos x) - \int(-\cos x)(1) \, dx$$

$$= -x \cos x - (-\sin x)$$

$$\int x \sin x \, dx = \sin x - x \cos x + c$$

Example 12.10

Evaluate $\int_1^2 x e^x \, dx$

Solution:

On comparing $\int_1^2 x e^x \, dx$ with the left side of the ' integration by parts rule'

$u = x$ and $dv = e^x \, dx$

Differentiating $u = x$, we have $\frac{du}{dx} = 1$

Integrating $\int dv = \int e^x \, dx$, we have $v = e^x$

Substituting the above in the 'integration by parts rule': $\int u \, dv = uv - \int v \, du$

$$\int x e^x \, dx = x e^x - \int (e^x)(1) \, dx$$

$$= x e^x - e^x$$

$$\int_0^2 x e^x \, dx = \left[x e^x - e^x \right]_0^2$$

$$= [2 \times e^2 - e^2] - [0 \times e^0 - e^0]$$

$$= [14.778 - 7.389] - [0 - 1] = \mathbf{8.389}$$

Exercise 12.1

Integrate the following with respect to the variable:

1. $(2x + 4 - \frac{1}{x^2}) \, dx$
2. $(2x + 1)^2 \, dx$

3. $(\sin 3\theta - \cos 4\theta)\, d\theta$
4. $(e^x + \frac{1}{e^{3x}} - \frac{2}{x})\, dx$
5. $(\sqrt{x} + 3)^2 dx$

Evaluate the following (Q6 to Q11):

6. $\int_1^2 (2x^2 - 4x)\, dx$
7. $\int_1^2 (\sqrt{x} + \frac{1}{\sqrt{x}})\, dx$
8. $\int_0^{\pi/2} 2\sin 3\theta\, d\theta$
9. $\int_{0.5}^1 2\cos \frac{x}{2}\, dx$
10. $\int_{-1}^1 \frac{3}{2e^{3x}}\, dx$
11. $\int_2^3 (e^{x/3} + e^{-x/4})\, dx$

Use substitution for Q.12 to Q.15:

12. Find $\int 4x(\sqrt{(1-x^2)})dx$
13. Evaluate $\int_0^1 \frac{x\,dx}{\sqrt{x^2+1}}$
14. Evaluate $\int_0^1 \tan x \sec^2 x\, dx$
15. Evaluate $\int_1^2 \frac{2\log_e x}{x}\, dx$

Use substitution and change of limits for Q.16 to Q.18:

16. Evaluate $\int_1^2 2x e^{x^2-1} dx$
17. Evaluate $\int_0^{0.5} \sin(2x+1)dx$
18. Evaluate $\int_0^1 x(\sqrt{1+x^2})\, dx$

Use integration by parts to find/evaluate the following integrals:

19. $\int 2x \log_e x\, dx$
20. $\int x^3 \log_e x\, dx$
21. $\int x^2 \cos x\, dx$
22. $\int_1^2 \frac{\log_e x}{x^2} dx$
23. $\int_0^1 x(x+1)^2 dx$

Answers – Exercise 12.1

1. $x^2+4x+\frac{1}{x}+c$
2. $\frac{4x^3}{3}+2x^2+x+c$
3. $\frac{-\cos 3\theta}{3}-\frac{\sin 4\theta}{4}+c$
4. $e^x-\frac{1}{3e^{3x}}-2\log_e x+c$
5. $\frac{x^2}{2}+4x^{3/2}+9x+c$
6. – 1.33
7. 2.047
8. 0.67
9. 0.928
10. 10.018
11. 2.848
12. $-\frac{4}{3}(1-x^2)^{3/2}+c$
13. 0.414
14. 1.213
15. 0.48
16. 19.086
17. 0.478
18. 0.609
19. $x^2\log_e x-\frac{x^2}{2}+c$
20. $\frac{x^4}{4}\log_e x-\frac{x^4}{16}+c$
21. $(x^2-2)\sin x+2x\cos x+c$
22. 0.1534
23. 1.416

CHAPTER 13

Applications of integration

Topics covered in this chapter:

- Area under a curve
- Area enclosed between a curve and a straight line
- Volumes of revolution
- Civil engineering applications

13.1 Introduction

There are many applications of integration in civil engineering and building services engineering, but it is possible to consider only some of them here. Integration is used in several applications such as:

- Calculation of centroids and second moments of area of engineering components.
- Determination of areas under curves.
- Determination of volumes of revolution.
- Civil engineering processes in geotechnics, structural mechanics, fluid mechanics and other disciplines.

Some of these are dealt with in this chapter but centroids and second moment of areas are considered in another chapter.

13.2 Area under a curve

The area shown in Figure 13.1 may be determined by several methods depending on the degree of accuracy required. Integration may be used to give more accurate results as compared to methods such as the trapezoidal rule or the Simpson's rule. For determining the area of shape ABCD, shown in Figure 13.1, the shape is divided into 'n' number of strips of equal width δx. Let the area of 1 strip be δA, then:

$$\delta A \approx y \times \delta x$$

The approximate area of figure ABCD is given by the sum of the areas of all strips.

$$\text{Area of shape ABCD} = \text{Area of all strips} = \sum_{x=a}^{x=b} y\,\delta x$$

As δx approaches 0 the above sum gives the exact value of the area. δx is replaced by dx and the Σ sign is replaced by the integral sign. Thus area ABCD is given by: Area of ABCD $= \int_a^b y\,dx$ (a and b are the limits of integration)

For more details of the above process refer to section 12.1.

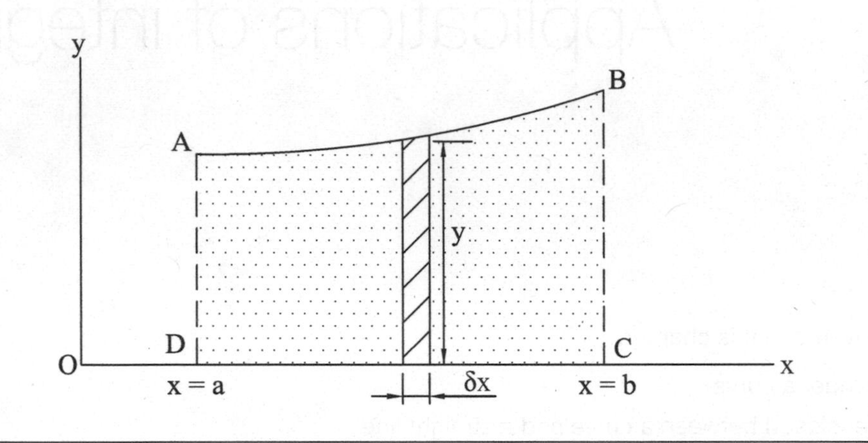

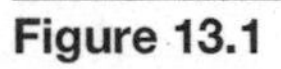
Figure 13.1

Example 13.1

Find the area (shaded) shown in Figure 13.2.

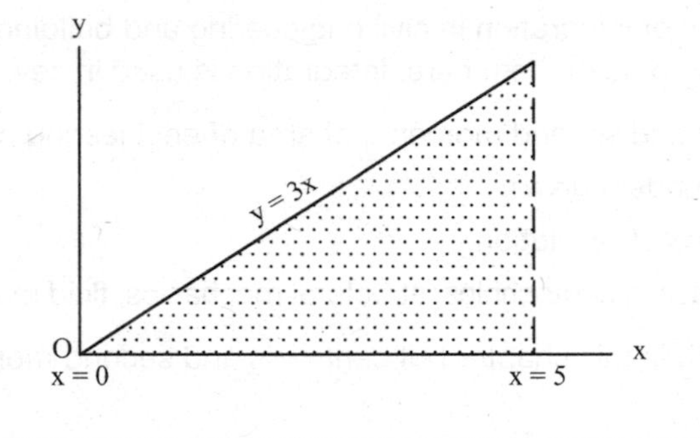

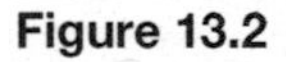
Figure 13.2

Solution:

$$\text{Area, } A = \int_0^5 y\,dx$$

$$= \int_0^5 3x\,dx \qquad (y = 3x\text{, is the equation of the line})$$

$$= \left[\frac{3x^2}{2}\right]_0^5 = \frac{3}{2}\left[x^2\right]_0^5$$

Therefore, Area $= \frac{3}{2}\left[5^2 - 0^2\right] =$ **37.5 square units**

Example 13.2

Find the area enclosed by the curve $y = x^3$, the x-axis and ordinates $x = 1$ m and $x = 3$ m.

Solution:

$$\text{Area, } A = \int_1^3 y\,dx$$

$$= \int_1^3 x^3\,dx \qquad \left(y = x^3, \text{ is the equation of the curve}\right)$$

$$= \left[\frac{x^4}{4}\right]_1^3 = \frac{1}{4}\left[x^4\right]_1^3$$

Therefore, Area $= \frac{1}{4}\left[3^4 - 1^4\right] =$ **20 m²**

13.3 Area enclosed between a curve and a straight line

Figure 13.3 shows a straight line intersecting the curve at 2 points and hence encloses the shaded area. The shaded area is the difference between the area under the curve and the area under the straight line.

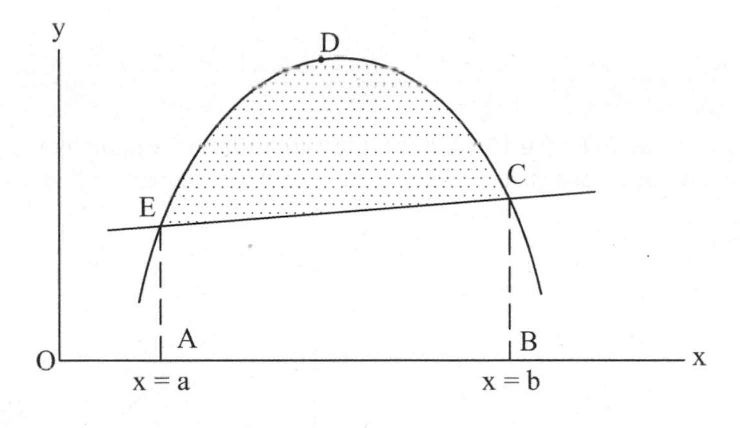

Figure 13.3

Example 13.3

Find the area between the curve $y = 5x - x^2$ and the line $y = x + 3$

Solution:

Refer to Figure 13.3.

The points of intersection of the 2 graphs are found by solving the equation:

$5x - x^2 = x + 3$ or $x^2 - 4x + 3 = 0$

$(x - 1)(x - 3) = 0$ Therefore $x = 1$ or $x = 3$

The above values of x are the limits of integration.

$$\text{Area AEDCBA} = \int_1^3 y\,dx = \int_1^3 (5x - x^2)\,dx$$

$$= \left[\frac{5x^2}{2} - \frac{x^3}{3}\right]_1^3$$

$$= \left[\frac{5\times 3^2}{2} - \frac{3^3}{3}\right] - \left[\frac{5\times 1^2}{2} - \frac{1^3}{3}\right]$$

$$= [22.5 - 9] - [2.5 - 0.33] = 11.33$$

$$\text{Area AECBA} = \int_1^3 y\,dx = \int_1^3 (x + 3)\,dx$$

$$= \left[\frac{x^2}{2} + 3x\right]_1^3$$

$$= \left[\frac{3^2}{2} + 3\times 3\right] - \left[\frac{1^2}{2} + 3\times 1\right]$$

$$= [4.5 + 9] - [0.5 + 3] = 10.0$$

Enclosed area (shaded) = 11.33 – 10.0 = **1.33 square units** (2 d.p.)

13.4 Volumes of revolution

Consider the triangle shown in Figure 13.4a. If this triangle is rotated 1 complete revolution about the x-axis, then it will trace out a cone which is known as the solid of revolution (Figure 13.4b).

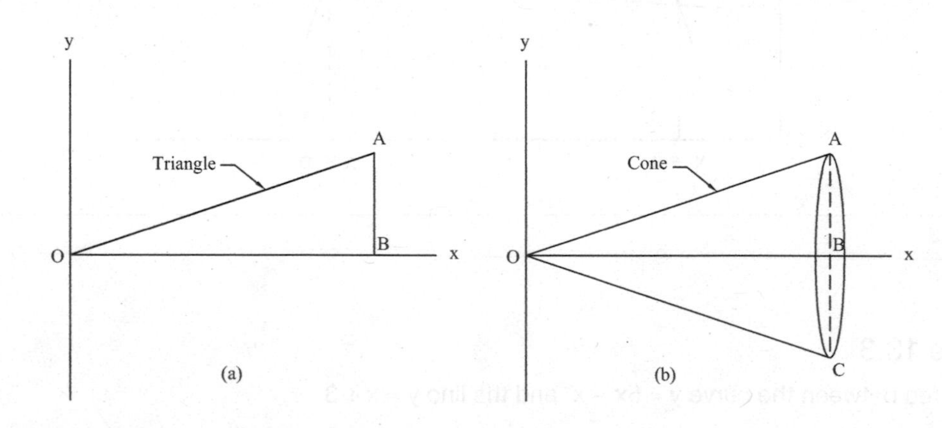

Figure 13.4

Figure 13.5 shows an area under the curve y = f(x) and a thin strip of thickness δx. If this area is rotated about the x-axis, the strip will produce a disc of nearly uniform area of cross-section. Thus the approximate volume of the disc is $\pi y^2 \delta x$. The volume of revolution is made up of several such elementary discs, hence the complete volume of revolution is the sum of all the elementary discs between the values of x = a, and x = b.

$$\text{Volume of revolution} = \sum_{x=a}^{x=b} \pi y^2\, \delta x \,(\text{approximately})$$

As the thickness of the discs approaches zero ($\delta x \to 0$) they can be considered as discs with uniform area of cross-section.

$$\text{Volume of disc} = \pi y^2\, dx$$

The accurate volume can be calculated by finding the sum of infinite number of discs. This can be achieved by integrating the volume of discs between the limits x = a and x = b

$$\text{Volume of revolution} = \int_a^b \pi y^2\, dx$$

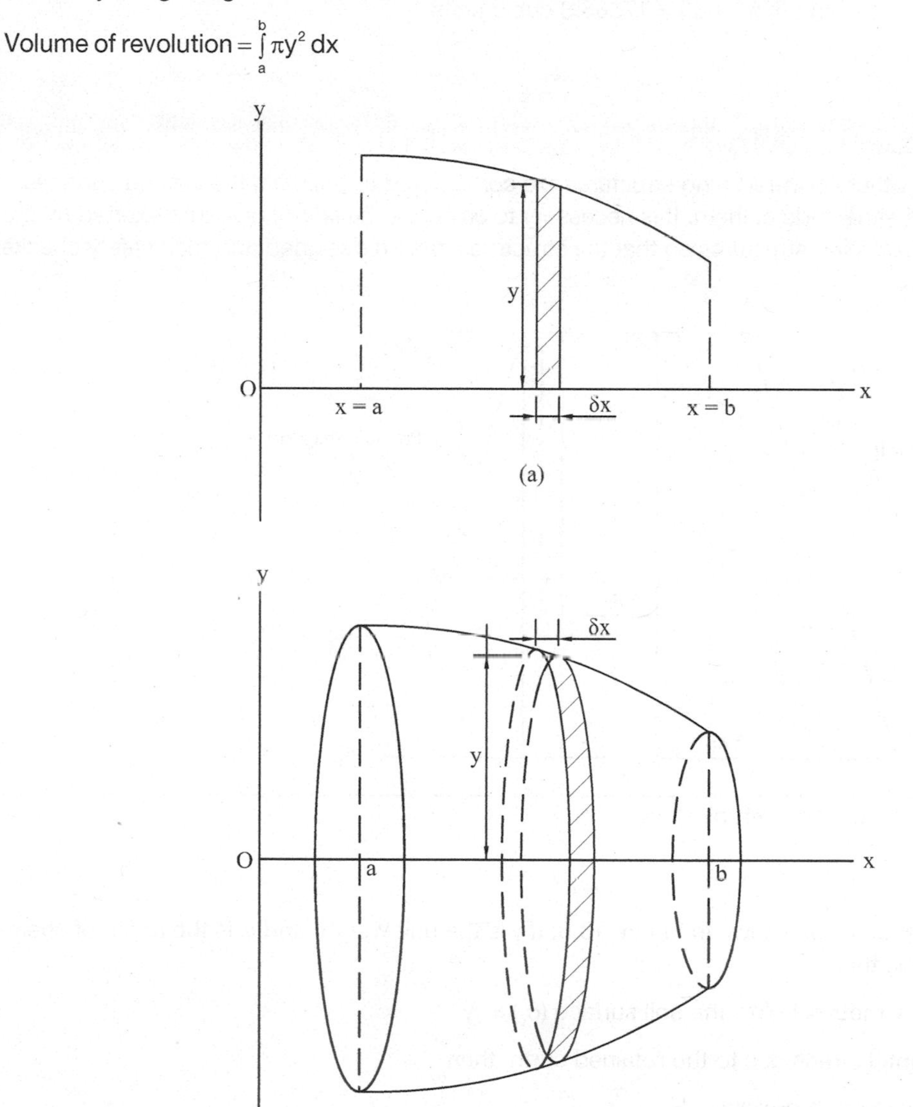

Figure 13.5

Example 13.4

Find the volume of revolution between the limits x = 1 and x = 5, when the curve $y = 3x^2$ is rotated about the x-axis.

Solution:

$$\text{Volume of revolution} = \int_1^5 \pi y^2\,dx = \int_1^5 \pi(3x^2)^2\,dx$$

$$= \int_1^5 \pi(3x^2)^2\,dx = \int_1^5 9\pi\,x^4\,dx$$

$$= 9\pi\left[\frac{x^5}{5}\right]_1^5 = 9\pi\frac{5^5}{5} - 9\pi\frac{1^5}{5}$$

$$= 9\pi\,(625 - 0.2) = 17665.8 \text{ cubic units}$$

13.5 Earth pressure on retaining walls

Retaining walls and other earth-retaining structures are constructed to maintain the ground surfaces at different levels on either side of them. It is necessary to calculate the lateral pressure exerted by the retained soil on the retaining structures so that the structures can be designed and their safety checked.

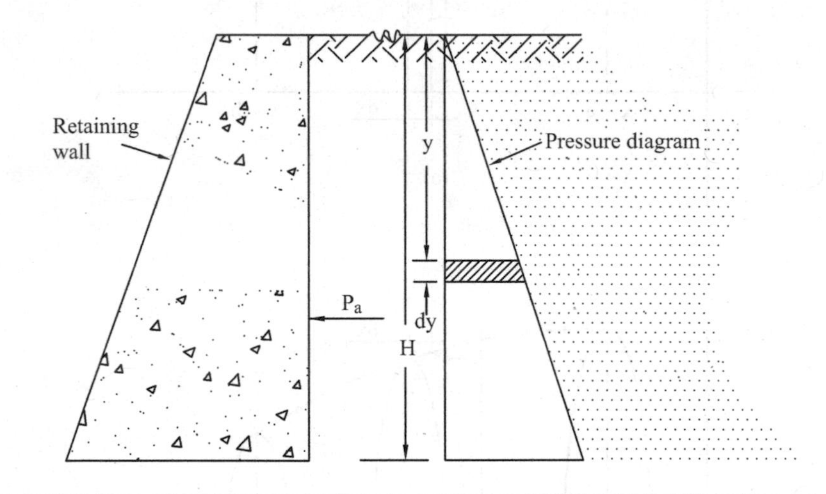

Figure 13.6 Earth pressure on a retaining wall

Consider the retaining wall shown in Figure 13.6. If γ is the unit weight and φ is the angle of shearing resistance of the soil, then

Vertical stress at y metres below the soil surface $(\sigma_v) = \gamma y$ (1)

If σ_h is the horizontal stress due to the retained earth, then

Coefficient of active earth pressure, $K_a = \dfrac{\sigma_h}{\sigma_v}$ (2)

From equations 1 and 2, $\sigma_h = K_a \gamma y$

Consider a strip of unit length and very small height dy

Area of the strip = 1 × dy = dy

Therefore force on the strip = $\sigma_h \times dy = K_a\gamma y\, dy$

Total thrust (force) on the wall, P_a, can be calculated by integrating $K_a\gamma y\, dy$

within the limits 0 and H (H = height of the wall).

$$P_a = \int_0^H K_a \gamma y \, dy$$

$$= \left[K_a \gamma \frac{y^2}{2} \right]_0^H$$

$$P_a = \left[K_a \gamma \frac{H^2}{2} - K_a \gamma \frac{0^2}{2} \right] = \frac{1}{2} K_a \gamma H^2$$

The value of K_a depends on the value of the angle of shearing resistance (φ) and is given by $K_a = \frac{1-\sin\varphi}{1+\sin\varphi}$

$K_a = \frac{1}{3}$, when $\varphi = 30°$

Example 13.5

A retaining wall is to be designed to retain a 5 m deep soil deposit. Calculate the total force on the wall due to the soil deposit if the unit weight of soil is 18.0 kN/m³ and the coefficient of active earth pressure is 0.33.

Solution:

$$\text{Total force on the wall, } P_a = \int_0^H K_a \gamma y \, dy = \int_0^5 0.33 \times 18.0 \, y \, dy$$

$$= 5.94 \int_0^5 y \, dy = 5.94 \left[\frac{y^2}{2} \right]_0^5$$

$$= 5.94 \left[\frac{5^2}{2} - \frac{0^2}{2} \right] = 74.25 \, \text{kN}$$

13.6 Permeability of soils

Darcy (circa 1856) proposed that the flow of water through a soil could be expressed as

$$v = ki$$

where v is the velocity of water flow

k is the coefficient of permeability of the soil

i is the hydraulic gradient $\left(i = \frac{\text{loss of head}}{\text{length of soil sample}} = \frac{\Delta h}{L} \right)$

This equation is considered to be one of the most important equations in soil mechanics. In order to determine the coefficient of permeability in a fine grained soil, a falling head permeameter (Figure 13.7) is used. A standpipe of cross-sectional area 'a' mm² is fitted over the mould that contains the soil sample. Water is allowed to flow through the permeameter, and the head of water (h) at any time (t) is the difference in the water level in the standpipe and the bottom tank.

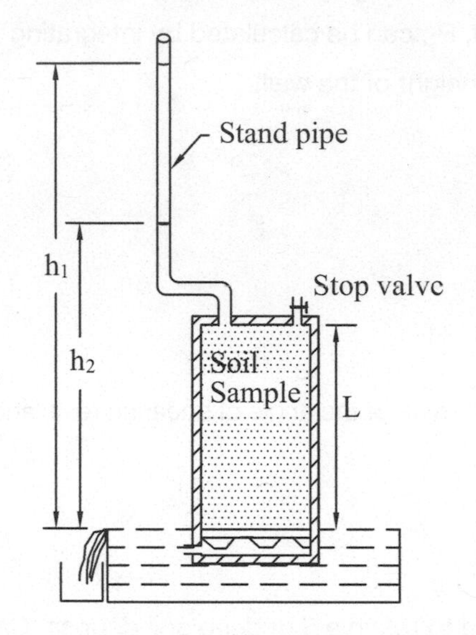

Figure 13.7 Falling head permeameter

Let h_1 and h_2 be the heads at time intervals t_1 and t_2. Let h be the head at time interval t and –dh be the change in the head in a small time interval dt.

Velocity of flow, $v = -\dfrac{dh}{dt}$

(minus sign shows that the head decreases with time)

Flow into the soil sample, $q_{in} = -a\dfrac{dh}{dt}$

Flow through and out of the sample, $q_{out} = Av = Aki$

(A is the cross-sectional area of the sample)

As $q_{out} = q_{in}$, Therefore $Aki = -a\dfrac{dh}{dt}$

$$Ak\frac{h}{L} = -a\frac{dh}{dt}$$

After transposition, $\dfrac{Ak}{aL}dt = -\dfrac{dh}{h}$

Integrating between the limits t_1 and t_2, and h_1 and h_2

$$\int_{t_1}^{t_2}\frac{Ak}{aL}dt = \int_{h_1}^{h_2}-\frac{dh}{h}$$

$$\frac{Ak}{aL}\int_{t_1}^{t_2}dt = \int_{h_2}^{h_1}\frac{dh}{h}$$

Example 13.6

The results of falling head permeability test on a fine grained soil are given below:

Initial head of water in standpipe = 1500 mm

Final head of water in standpipe = 700 mm

Sample length = 150 mm

Sample diameter = 100 mm

Standpipe diameter = 5 mm

Duration of test = 300 seconds

Calculate the permeability of the soil.

Solution:

Cross-sectional area of the sample, $A = \frac{\pi}{4} \times 100^2 = 7853.982 \text{ mm}^2$

Cross-sectional area of the standpipe, $a = \frac{\pi}{4} \times 5^2 = 19.635 \text{ mm}^2$

$$\frac{7853.982 \times k}{19.635 \times 150} \int_0^{300} dt = \int_{700}^{1500} \frac{dh}{h}$$

$$2.667 \times k \left[t\right]_0^{300} = \left[\log_e h\right]_{700}^{1500}$$

$2.667 \times k \times [300 - 0] = \log_e 1500 - \log_e 700$

$k \times 800.1 = 0.762$

k (coefficient of permeability) = 9.53×10^{-4} mm/s

13.7 Bending moment and shear force in beams

When structural engineers prepare the structural design of beams, they have to consider the forces and their effects on the beams. Shear forces and bending moments are caused by the forces and they affect the beams in several ways. Shear forces, as the name says, tend to cause shearing of the fibres of the material from which the beam has been manufactured. The forces also try to cause deflection in a beam which is due to their turning or rotational effect. This turning effect is known as the moment of a force or bending moment. For more information refer to Appendix 1.

The magnitude of bending moments for simple cases may be determined from shear force diagrams by integration. The area of shear force diagram gives the magnitude of bending moment, as shown in the following example.

Example 13.7

The bending moment in a beam may be calculated from the area enclosed by the shear force diagram. Calculate area of the shear force diagram and hence the bending moment at 2.5 m from the beam end. Use 0 m and 2.5 m as the limits of integration.

Solution:

The shear force diagram (SFD) and the bending moment diagram (BMD) are shown in Figure 13.8.

Slope of the line forming the SFD (line CD), $m = 11 \div 4 = -2.75$ (negative slope)

Intercept $c = 11$ kN, therefore equation of line CD is: $y = -2.75x + 11$

(the standard equation of a straight line is: $y = mx + c$)

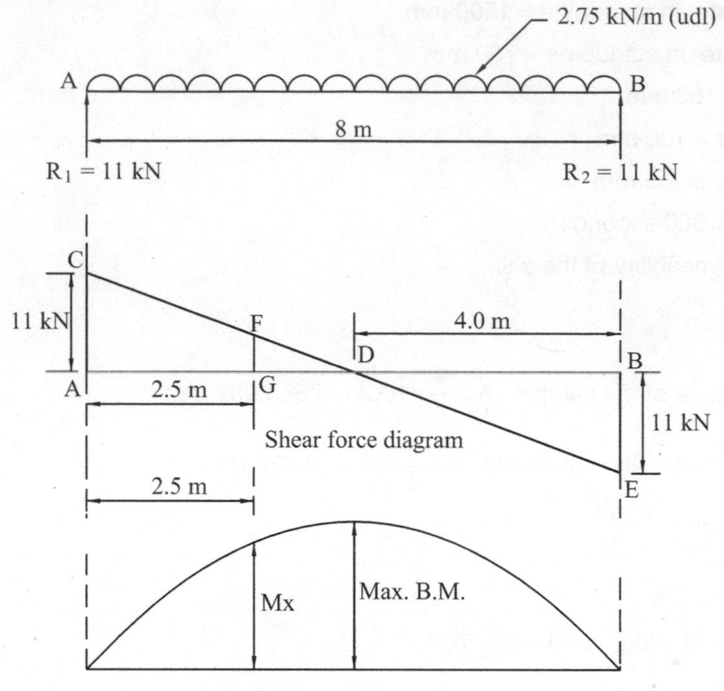

Figure 13.8

As the bending moment is equal to the area of the SFD, therefore:

Let M_x be the bending moment at 2.5 m from A

$$M_x = \text{Area ACFG of the SFD} = \int_0^{2.5} (-\ 2.75\,x\ +\ 11)\,dx$$

$$= \left[\frac{-2.75x^2}{2} + 11x\right]_0^{2.5}$$

$$= \left[\frac{-2.75 \times 2.5^2}{2} + 11 \times 2.5\right] - \left[\frac{0}{2} + 0\right]$$

$$= -\ 8.59 + 27.5 = \mathbf{18.91\ kNm}$$

Exercise 13.1

1. Determine the area enclosed by the line $y = 3x + 1$, the x-axis and the ordinates $x = 1$ and $x = 3$.
2. Calculate the area enclosed by the curve $y = 2x^2 + 5$, the x-axis and the ordinates $x = 1$ and $x = 4$.
3. Find the area under the curve $y = \sin\theta$, between $\theta = 0$ and $\theta = \frac{\pi}{2}$.
4. Find the area between the curve $y = 7x - x^2$ and the line $y = x + 5$.

For questions 5 to 7, find the volume of revolution between the given limits when the following curves are rotated about the x-axis:

5. $y = 3$; limits: $x = 0$ and $x = 4$.
6. $y = 0.6x$; limits: $x = 0$ and $x = 10$.
7. $y = 2x^2 + 3x$; limits: $x = 2$ and $x = 5$.
8. Find the volume of a cone by integration if its base diameter and the height are 20 cm and 30 cm, respectively.
9. The dimensions of the slump cone which is used for testing fresh concrete are shown in Figure 13.9. Use the method of integration to find the volume of the slump cone in mm³ and litres.

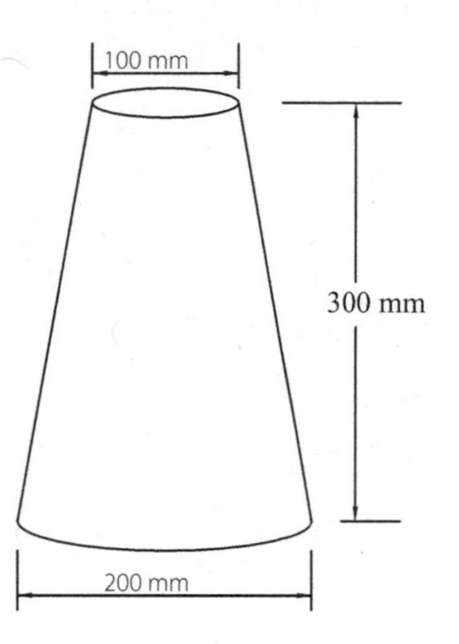

Figure 13.9 The slump cone

10. A retaining wall is to be designed to retain a 6 m deep soil deposit. If the unit weight of the retained soil is 18.0 kN/m³ and the coefficient of active earth pressure is 0.33, calculate the force on the wall due to the soil deposit at a depth of: a) 4 m b) 6 m.
11. The bending moment in a beam may be calculated from the area enclosed by the shear force diagram. If the equation of the straight line forming the shear force diagram is:

    $y = -2.5x + 10$ (y represents the shear force and x is the distance along the beam)

    use integration to calculate the bending moment at 2.2 m from the beam end. Use 0 m and 2.2 m as the limits of integration.

Answers – Exercise 13.1

1. 14.0 square units
2. 57.0 square units
3. 1.0 square unit
4. 10.66 square units
5. 113.1 cubic units
6. 376.99 cubic units

7. 14615.95 cubic units
8. 3141.59 cm^3
9. 5497840.55 mm^3 or 5.498 litres
10. a) 47.52 kN b) 106.92 kN
11. 15.95 kNm

CHAPTER 14

Properties of sections

Topics covered in this chapter:

- Centroids of simple shapes
- Centroids of simple/complex shapes by integration
- Second moment of area and radius of gyration
- Parallel axis theorem

14.1 Centroids of simple shapes

If an object is suspended by a string, it will only balance when its centre of gravity is vertically in line with the string. The entire mass of the object may be considered to act at its centre of gravity. When an area is considered, rather than the mass, the term centre of area or centroid is used for the point where the centre of gravity of the area would be situated. The area may be represented by a lamina i.e. a thin flat sheet of negligible thickness and mass.

The positions of the centre of gravity of common shapes are well known, for example, the centre of gravity of a thin flat circular disc is at the centre of the disc, therefore we may say that the centroid of a circular area is at the centre of the area. Similarly, the centroid of a rectangular area is at the intersection of the centre lines, as shown in Figure 14.1.

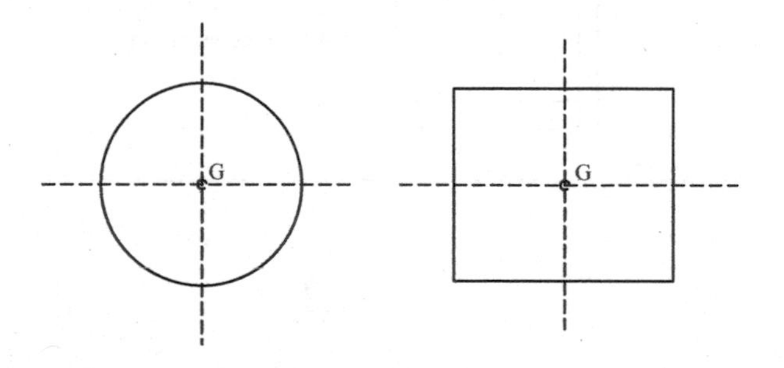

Figure 14.1 Position of the centroid (G)

The first moment of area, defined as the product of the area and the perpendicular distance of its centroid from a given axis in the plane of the area, is used to determine the position of the centroid (Figure 14.2a). Let $\bar{x}$ and $\bar{y}$ be the co-ordinates of the centroid, as shown in Figure 14.2b:

$$\bar{x} = \frac{\Sigma Ax}{\Sigma A} \quad \text{and} \quad \bar{y} = \frac{\Sigma Ay}{\Sigma A} \tag{14.1}$$

where ΣAx and ΣAy are the sum of the first moments of the component areas about the axes Oy and Ox respectively.

ΣA is the sum of the component areas

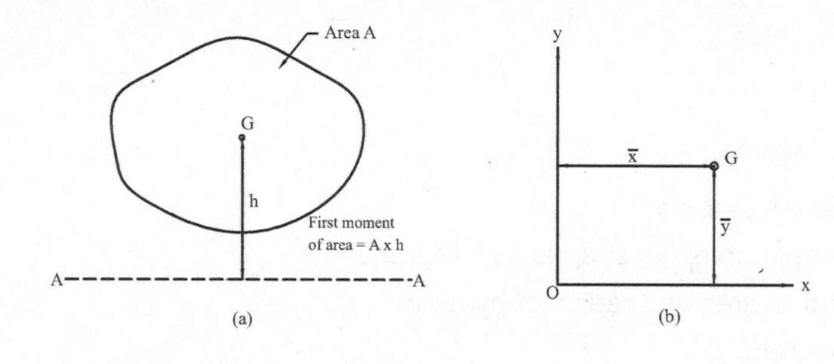

Figure 14.2

Example 14.1

Find the position of the centroid of the shape shown in Figure 14.3.

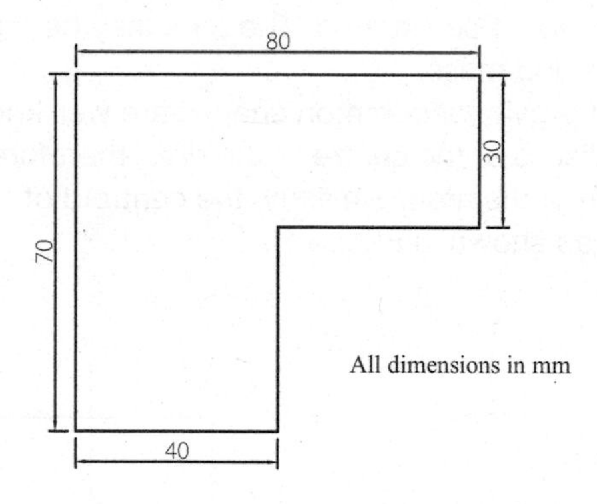

Figure 14.3

Solution:

Let Ox and Oy be the reference axes.

The shape has been divided into two parts, as shown in Figure 14.4. The centroids of both parts (G1 and G2) are also shown in the figure. Let G be the centroid of the whole section; situated at $\bar{x}$ from axis Oy and $\bar{y}$ from axis Ox.

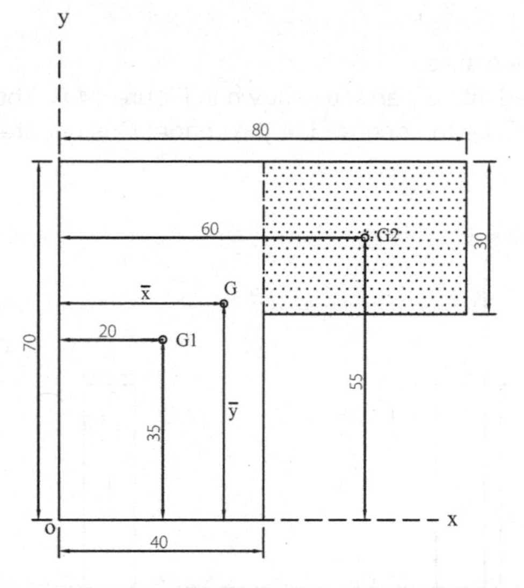

Figure 14.4

	Area	Distance to centroid		Moment of area	
		From y-axis	From x-axis	About Oy	About Ox
	A (mm²)	x (mm)	y (mm)	Ax	Ay
1	70 × 40 = 2800	20	35	2800 × 20 = 56 000	2800 × 35 = 98 000
2	40 × 30 – 1200	60	55	72 000	66 000
	ΣA = 4000			ΣAx – 128 000	ΣAy = 164 000

$$\bar{x} = \frac{\Sigma Ax}{\Sigma A} = \frac{128000}{4000} = \mathbf{32.0\ mm}$$

$$\bar{y} = \frac{\Sigma Ay}{\Sigma A} = \frac{164000}{4000} = \mathbf{41.0\ mm}$$

Example 14.2

Find the position of the centroid of the area shown in Figure 14.5.

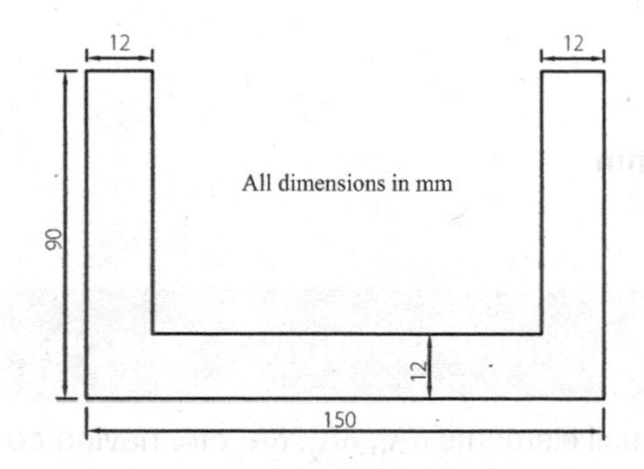

Figure 14.5

Solution

Let Ox and Oy be the reference axes.

The shape has been divided into 3 parts, as shown in Figure 14.6. The centroids of all 3 parts are also shown in the figure. Let G be the centroid of the shape; G is situated at $\bar{x}$ from Oy and at $\bar{y}$ from Ox.

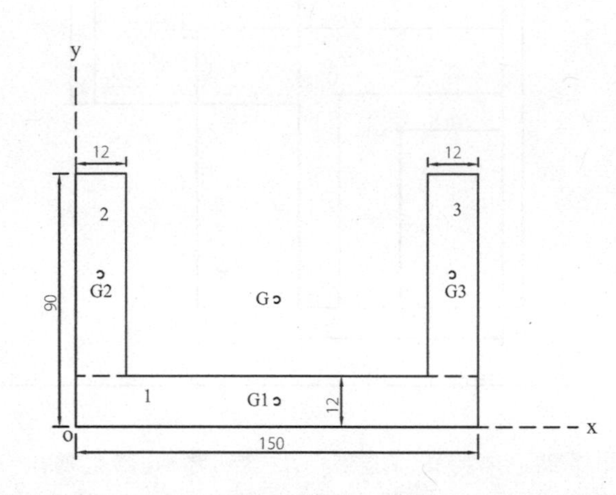

Figure 14.6

	Area	Distance to centroid		Moment of area	
		From y-axis	From x-axis	About Oy	About Ox
	A (mm²)	x (mm)	Y (mm)	Ax	Ay
1	150 × 12 = 1800	75	6	1800 × 75 = 135 000	1800 × 6 = 10 800
2	12 × 78 = 936	6	51	936 × 6 = 5616	936 × 51 = 47 736
3	12 × 78 = 936	144	51	936 × 144 = 134 784	936 × 51 = 47 736
	ΣA = 3672			ΣAx = 275 400	ΣAy = 106 272

$$\bar{x} = \frac{\Sigma Ax}{\Sigma A} = \frac{275400}{3672} = \mathbf{75.0\ mm}$$

$$\bar{y} = \frac{\Sigma Ay}{\Sigma A} = \frac{106272}{3672} = \mathbf{28.94\ mm}$$

14.2 Centroids of simple/complex shapes by integration

The shapes are split up into small elements δA_1, δA_2, δA_3 etc. having co-ordinates (x_1, y_1), (x_2, y_2), (x_3, y_3) etc. from the reference axes.

From equations 14.1, we have

$$\bar{x} = \frac{\Sigma \delta A\, x}{\Sigma \delta A}, \qquad \text{and} \qquad \bar{y} = \frac{\Sigma \delta A\, y}{\Sigma \delta A}$$

As $\delta A \to 0$, the summations tend to integrals, (also see section 12.1)

$$\bar{x} = \frac{\int x\, dA}{A}, \qquad \text{and} \qquad \bar{y} = \frac{\int y\, dA}{A} \tag{14.2}$$

where A is the total area of the shape.

Example 14.3

Find the position of the centroid of a rectangle, by integration.

Solution:

Figure 14.7 shows rectangle ABCD drawn on x-axis and y-axis. An elementary strip of length d and thickness δx is situated at a distance of x from Oy. Let the co-ordinates of the centroid be $\bar{x}, \bar{y}$.

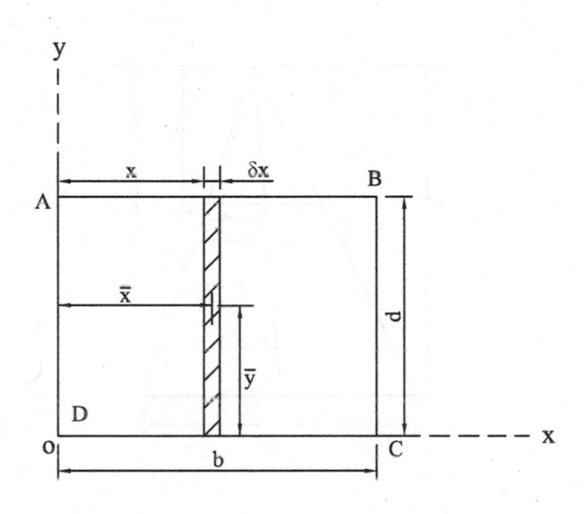

Figure 14.7

$$\text{By integration,}\ \bar{x} = \frac{\int_0^b x(d\, dx)}{\int_0^b d\, dx}$$

$$= \frac{\left[d\frac{x^2}{2}\right]_0^b}{\left[dx\right]_0^b} = \frac{\left[d\frac{b^2}{2}\right]}{\left[db\right]}$$

$$= \frac{b}{2}$$

$$\bar{y} = \frac{\int_0^b \frac{d}{2}(d\,dx)}{\int_0^b d\,dx} \qquad \text{(dx after integration becomes x)}$$

$$= \frac{\left[d^2 \frac{x}{2}\right]_0^b}{\left[dx\right]_0^b} = \frac{\left[d^2 \frac{b}{2}\right]}{\left[db\right]}$$

$$= \frac{d}{2}$$

The centroid lies at $(\frac{b}{2}, \frac{d}{2})$

Example 14.4

Find, by integration, the position of the centroid of the triangle, shown in Figure 14.8.

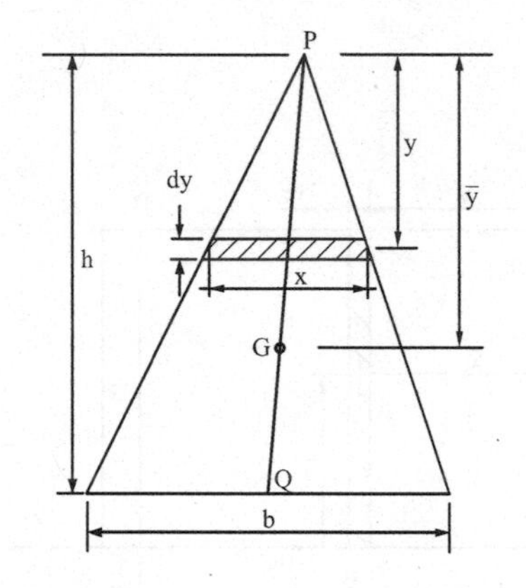

Figure 14.8

Solution

Consider an elementary strip of thickness dy at a distance y from the apex. Let the length of the strip be x.

Area of the strip (dA) = x dy (14.3)

From similar triangles, $\frac{x}{y} = \frac{b}{h}$

or $x = \frac{b}{h}y$ (14.4)

From (14.3) and (14.4), $dA = \frac{b}{h}y\,dy$

From equations (14.2), $\bar{y} = \frac{\int y\,dA}{A}$

$$= \frac{\int_0^h y \frac{b}{h} y\,dy}{\frac{b \times h}{2}}$$

$$= \frac{\frac{2b}{h}\int_0^h y^2\,dy}{b \times h}$$

$$= \frac{2}{h^2}\left[\frac{y^3}{3}\right]_0^h = \frac{2}{h^2}\left[\frac{h^3}{3}\right]$$

$$\bar{y} = \frac{2h}{3}$$

The centroid of the triangle (G) lies on the median (PQ), $\frac{2h}{3}$ from the apex or $\frac{1}{3}h$ from the base.

Example 14.5

Find, by integration, the position of the centroid of the sector of a circle of radius r and an angle of α radians subtended at the centre.

Solution:

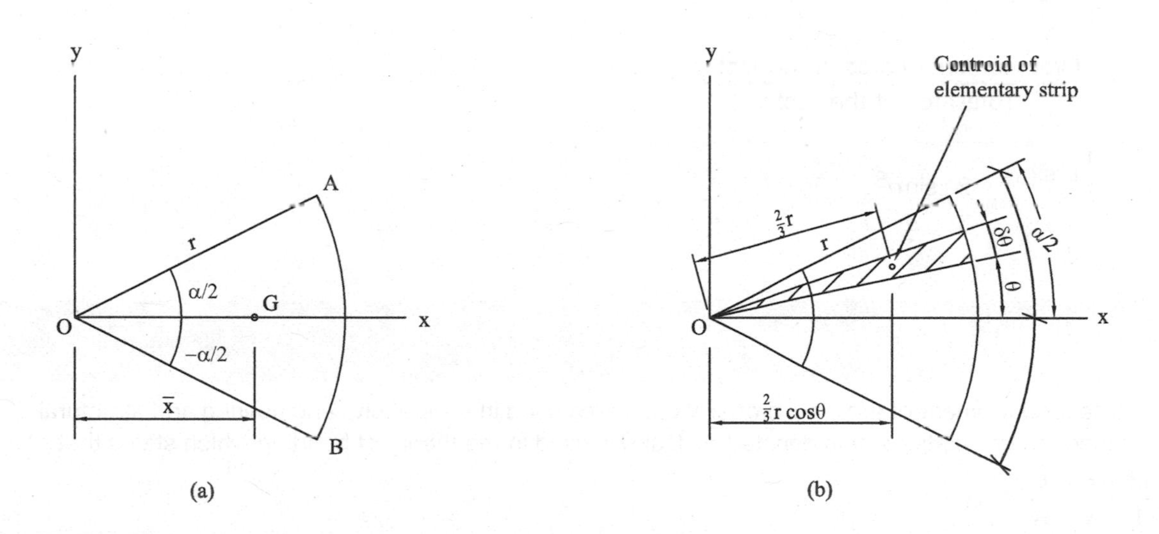

Figure 14.9

The sector is shown in Figure 14.9a. By symmetry, the centroid (G) of the sector lies on the x-axis. The sector can be divided into elementary strips, 1 of which is shown in Figure 14.9b. The angle of the strip is δθ, and by considering the strip to be a triangle, its area is given by:

Area of the element $= \frac{1}{2} \times \text{base} \times \text{height}$

$$= \frac{1}{2} \times r\delta\theta \times r = \frac{1}{2} \times r^2\delta\theta$$

The centroid of the strip is at a distance of $\frac{2}{3}r$ from O. The perpendicular distance from the centroid to Oy is $\frac{2}{3}r\cos\theta$.

Area of sector $= \pi r^2 \times \frac{\alpha}{2\pi} = \frac{1}{2}r^2\alpha$

First moment of area of the element $= \frac{1}{2} \times r^2\,\delta\theta \times \frac{2}{3}r\cos\theta$

First moment of area of the sector = Sum of the first moment of area of all elementary strips

$$= \sum_{\theta=-\alpha/2}^{\theta=\alpha/2} \frac{1}{2}r^2\,\delta\theta \times \frac{2}{3}r\cos\theta$$

$$= \int_{-\alpha/2}^{\alpha/2} \frac{1}{3}r^3\cos\theta\,d\theta$$

$$= \frac{1}{3}r^3\left[\sin\theta\right]_{-\alpha/2}^{\alpha/2}$$

$$= \frac{1}{3}r^3\left[\sin\alpha/2 - \sin(-\alpha/2)\right]$$

$$= \frac{1}{3}r^3\sin\alpha$$

$$\bar{x} = \frac{\text{First moment of area of the sector}}{\text{Total area of the sector}}$$

$$= \frac{\frac{1}{3}r^3\sin\alpha}{\frac{1}{2}r^2\alpha} = \frac{2}{3}r\frac{\sin\alpha}{\alpha}$$

14.3 Second moment of area

The second moment of area is a property of an area used in several civil engineering and structural engineering calculations. It is denoted by I, and is used in the theory of bending which states that $\frac{M}{I} = \frac{\sigma}{y} = \frac{E}{R}$

Figure 14.10 shows a strip of very small width at distance x_1 from axis yy. Let the area of the strip be δA_1. Its first moment of area about axis yy is $\delta A_1\,x_1$. Its second moment of area about yy is given by $\delta A_1\,x_1^2$. The second moment of area I_y of the whole section about yy is the sum of all the second moments of the elementary areas, i.e.

$$I_{yy} = \delta A_1\,x_1^2 + \delta A_2\,x_2^2 + \ldots\ldots\ldots\ldots\ldots\ldots\ldots$$

$$= \Sigma\,\delta A\,x^2 \qquad (14.5)$$

As $\delta A \to 0$, the summation becomes an integral, $I_{yy} = \int x^2 \, dA$

The above formula may be applied for determining the second moment of area of a rectangle, as shown in example 14.6.

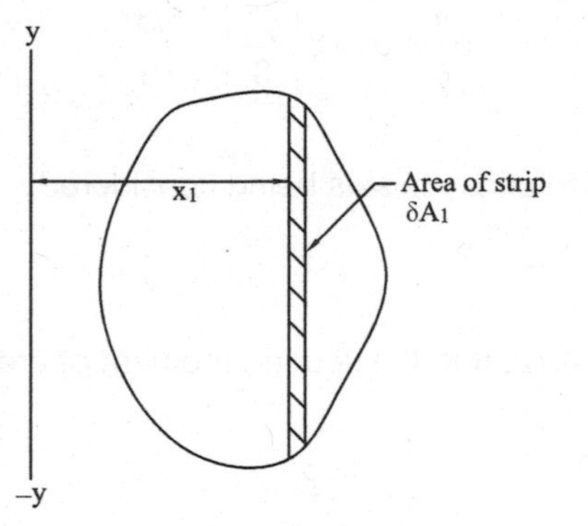

Figure 14.10

14.4 Radius of gyration

Several elemental areas δA_1, δA_2 etc. at distances x_1, x_2 etc. from a fixed axis may be replaced by a single area A ($A = \delta A_1 + \delta A_2 + ...$) at distance k from the axis, so that $Ak^2 = \Sigma\, \delta A\, x^2$, where k is known as the radius of gyration of area A about the given axis.

From 14.5, $\Sigma\, \delta A\, x^2 = I$, therefore $Ak^2 = I$

Radius of gyration, $k = \sqrt{\dfrac{I}{A}}$

Example 14.6

Find the second moment of area of the rectangle shown in Figure 14.11.

Solution:

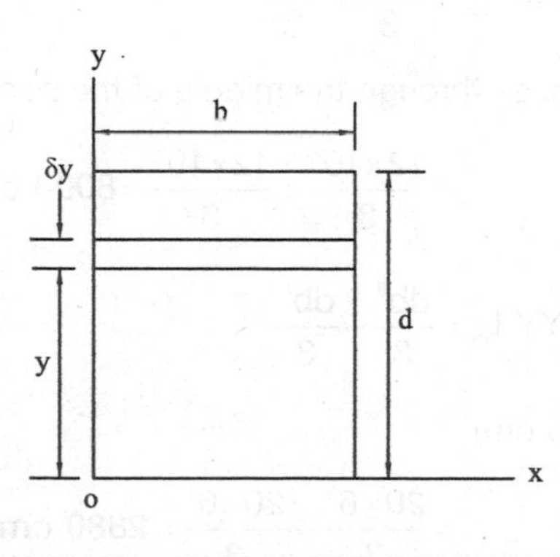

Figure 14.11

Consider a small strip of width δy, at a distance y from Ox

Area of the strip, $\delta A = b \times \delta y$

Second moment of area about $Ox = \sum_{x=0}^{x=d} (b \times \delta y)\, y^2$

$$= b\int_0^d y^2\, dy = b\left[\frac{y^3}{3}\right]_0^d = \frac{bd^3}{3}$$

(Note: depth is always perpendicular to the axis being considered)

Example 14.7

For the beam shown in Figure 14.12, find the second moment of area about the axes AA, XX and YY.

Solution:

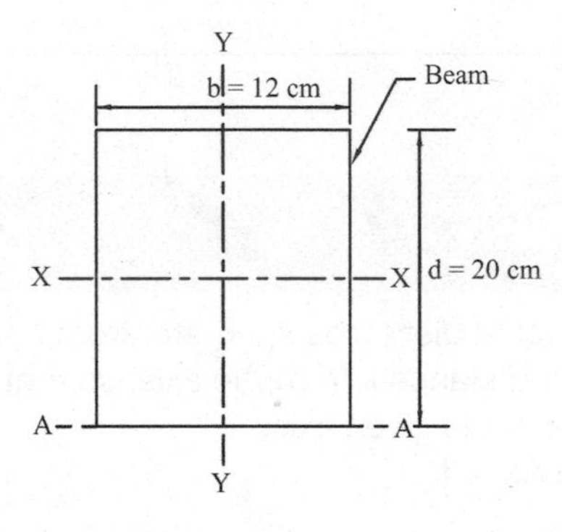

Figure 14.12

Second moment of area about AA, $I_{AA} = \frac{bd^3}{3}$

$$= \frac{12 \times 20^3}{3} = \mathbf{32000\ cm^4}$$

Second moment of area about XX, $I_{XX} = \frac{bd^3}{3} + \frac{bd^3}{3}$

(d here is 20 ÷ 2, as Axis XX passes through the middle of the beam)

$$= \frac{12 \times 10^3}{3} + \frac{12 \times 10^3}{3} = \mathbf{8000\ cm^4}$$

Second moment of area about YY, $I_{YY} = \frac{db^3}{3} + \frac{db^3}{3}$

(As before b here is 12 ÷ 2, i.e. 6 cm)

$$= \frac{20 \times 6^3}{3} + \frac{20 \times 6^3}{3} = \mathbf{2880\ cm^4}$$

14.5 The parallel axis theorem

Figure 14.13 shows a plane area whose centroid is G. There are 2 parallel axes, 1 passes through the centroid and the other is at a distance h from the first. According to the parallel axis theorem:

$I_H = I_G + Ah^2$ where I_H is the second moment of area about the parallel axis

I_G is the second moment of area about the axis passing through the centroid

A is the cross-sectional area of the section

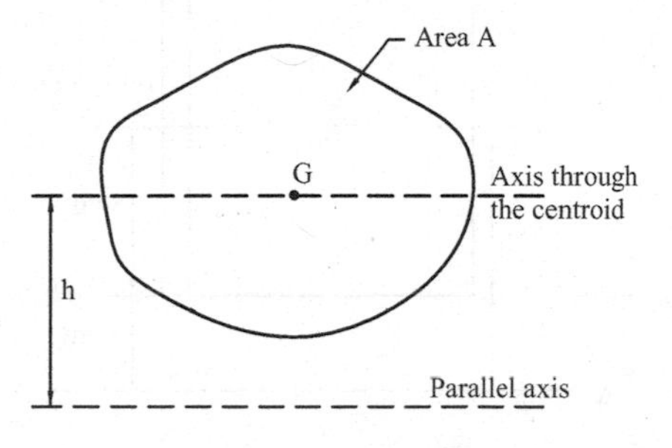

Figure 14.13

The parallel axis theorem can be used to determine the second moment of area of a rectangle about the axis through its centroid. Figure 14.14 shows a rectangle of width b and depth d. If I is the second moment of area about HH, then:

$I_H = I_G + Ah^2$ (A = area = bd)

$$\frac{bd^3}{3} = I_G + bd \times h^2 \quad (h = \frac{d}{2})$$

$$= I_G + bd \times \frac{d^2}{4} = I_G + \frac{bd^3}{4}$$

$$I_G = \frac{bd^3}{3} - \frac{bd^3}{4} = \frac{bd^3}{12} \qquad (14.6)$$

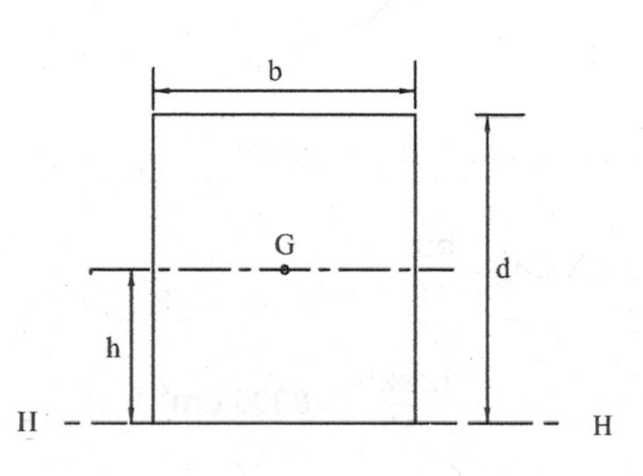

Figure 14.14

Example 14.8

Figure 14.15 shows the cross-section of a timber beam. Find its second moment of area about axis AA.

Solution:

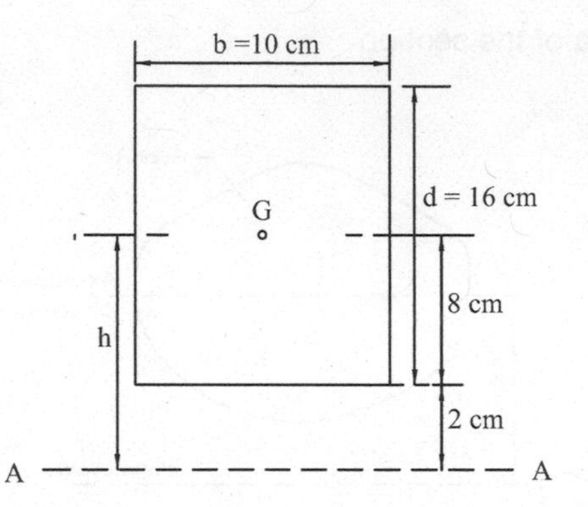

Figure 14.15

$I_{AA} = I_G + Ah^2$

$$= \frac{bd^3}{12} + bd \times h^2$$

$$= \frac{10 \times 16^3}{12} + 10 \times 16 \times 10^2$$

$= 3413.33 + 16\,000 =$ **19 413.33 cm**^{4}

Example 14.9

Use equation 14.6 to determine I_{XX} and I_{YY} of the beam shown in Figure 14.12.

Solution:

Axes XX and YY pass through the centroid of the beam.

Second moment of area about XX, $I_{XX} = \frac{bd^3}{12}$

$$= \frac{12 \times 20^3}{12} = \mathbf{8000\ cm^4}$$

$$\text{Second moment of area about YY}, I_{YY} = \frac{db^3}{12}$$

$$= \frac{20 \times 12^3}{12} = \mathbf{2880\ cm^4}$$

Example 14.10

Determine the second moment of area of the section shown in Figure 14.16 about axes AA, XX and YY.

Solution:

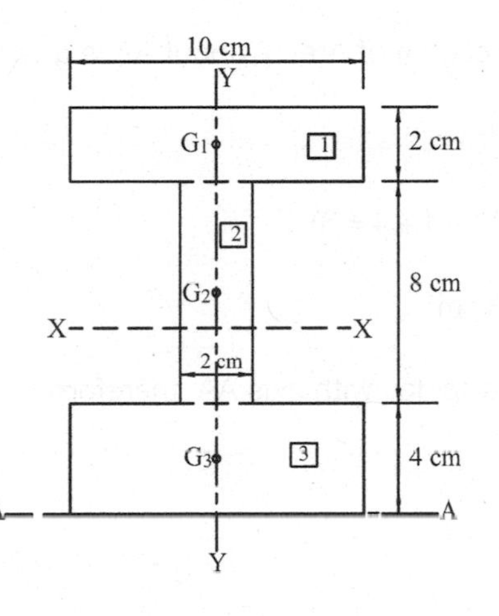

Figure 14.16

The section is divided into 3 rectangles, 1, 2 and 3, and their centroids denoted by G1, G2 and G3. The centroid of the whole section will lie on the axis of symmetry YY but its exact position needs to be determined as the section is not perfectly symmetrical.

	Area	Distance of centroid From axis AA	Moment of area About AA
	A (cm²)	y (cm)	Ay
1	20	13	20 × 13 = 260
2	16	8	16 × 8 = 128
3	40	2	40 × 2 = 80
	ΣA = 76		ΣAy = 468

$$\bar{y} = \frac{\Sigma Ay}{\Sigma A} = \frac{468}{76} = \mathbf{6.16\ cm}$$

The centroid is positioned on axis YY at a distance of 6.16 cm from axis AA.

Second moment of area about axis AA

Area 1: The second moment of area of Area 1 about AA is given by:

$$I_{AA} = I_{G1} + A_1 h_1^2$$

$$= \frac{10 \times 2^3}{12} + 20 \times 13^2 \ (A_1 = 20;\ h_1 = 4 + 8 + 1 = 13)$$

$$= 6.67 + 3380 = 3386.67\ cm^4$$

Area 2: The second moment of area of Area 2 about AA is given by:

$$I_{AA} = I_{G2} + A_2 h_2^2$$

$$= \frac{2 \times 8^3}{12} + 16 \times 8^2 \ (A_2 = 16;\ h_2 = 4 + 4 = 8)$$

$$= 85.33 + 1024 = 1109.33\ cm^4$$

Area 3: The base of Area 3 coincides with axis AA, therefore:

$$I_{AA} = \frac{bd^3}{3}$$

$$= \frac{10 \times 4^3}{3} = 213.33\ cm^4$$

Therefore, $I_{AA} = 3386.67 + 1109.33 + 213.33 = \mathbf{4709.3\ cm^4}$

Second moment of area about axis XX

$$I_{AA} = I_{XX} + Ah^2$$

$$I_{XX} = I_{AA} - A \times h^2$$

$$= 4709.33 - 76 \times 6.16^2 = 4709.33 - 2883.87 = \mathbf{1825.5\ cm^4}$$

Second moment of area about axis YY

Axis YY passes through the centroids of all three parts

$$I_{YY} = \frac{b_1 d_1^3}{12} + \frac{b_2 d_2^3}{12} + \frac{b_3 d_3^3}{12}$$

$$= \frac{2 \times 10^3}{12} + \frac{8 \times 2^3}{12} + \frac{4 \times 10^3}{12} = 166.67 + 5.33 + 333.33 = \mathbf{505.3\ cm^4}$$

Exercise 14.1

1. Find the position of the centroid of the areas (a to d) shown in Figure 14.17.

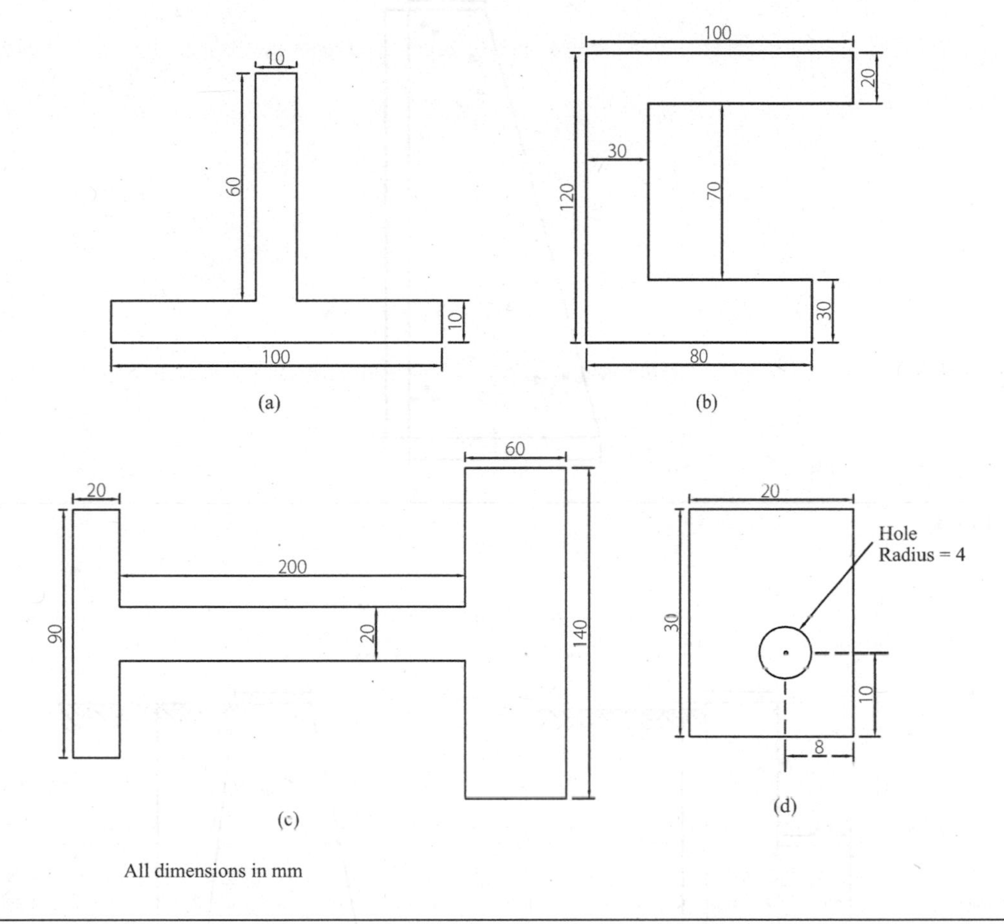

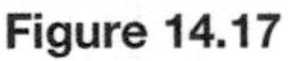
Figure 14.17

2. Figure 14.18 shows the cross-section of a concrete dam. Find the position of the centroid of the cross-section.
3. Figure 14.19 shows 2 retaining walls, a and b, constructed of stone and concrete respectively. Find the position of their centroids.
4. A timber beam is 8 cm wide and 12 cm deep. Calculate the second moment of area of the beam about the axes AA, XX and YY. Axis AA passes through the base and axes XX and YY pass through the centroid of the beam.
5. Figure 14.20 shows the cross-section of a timber beam. Find its second moment of area about axis AA.

6. Find I_{XX} and I_{YY} of the angle section shown in Figure 14.21.
7. Determine I_{XX} and I_{YY} of the channel section shown in Figure 14.22.

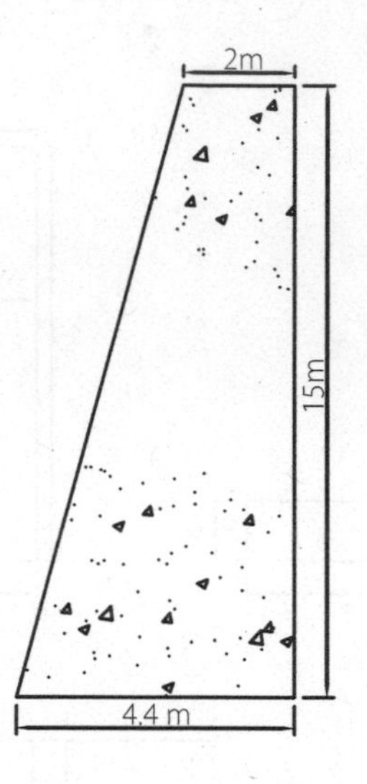

Figure 14.18

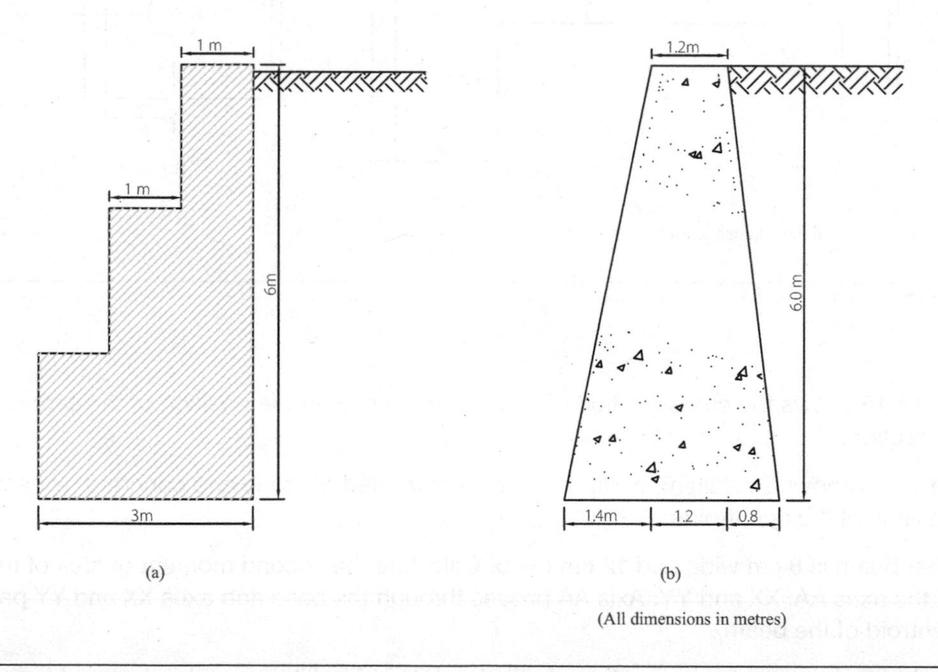

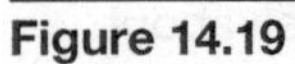

Figure 14.19

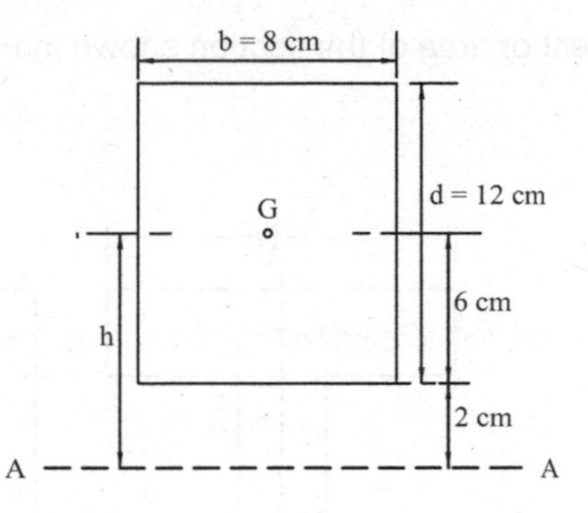

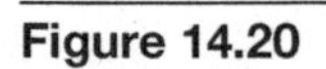
Figure 14.20

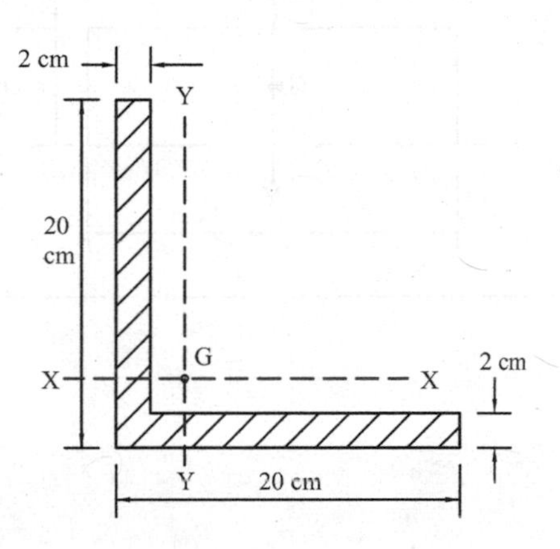

Figure 14.21

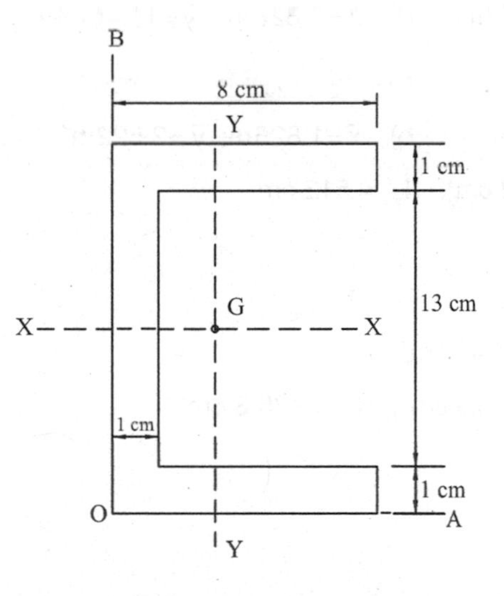

Figure 14.22

8. Determine the second moment of area of the section shown in Figure 14.23 about axes AA, XX and YY.

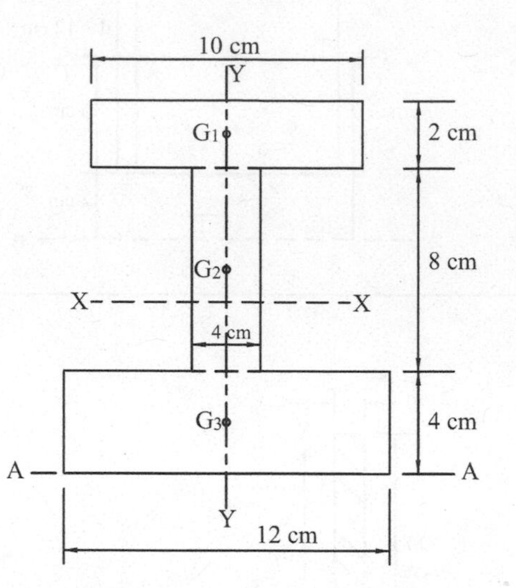

Figure 14.23

Answers – Exercise 14.1

1. a) $\bar{x}=50.0\text{mm}$; $\bar{y}=18.13\text{mm}$ b) $\bar{x}=35.0\text{mm}$; $\bar{y}=60.38\text{mm}$

   c) $\bar{x}=182.96\text{mm}$; $\bar{y}=70.0\text{mm}$ d) $\bar{x}=9.82\text{mm}$; $\bar{y}=15.46\text{mm}$
2. $\bar{x}=2.725\text{m}$; $\bar{y}=6.563\text{m}$
3. a) $\bar{x}=1.833\text{m}$; $\bar{y}=2.333\text{m}$ b) $\bar{x}=1.826\text{m}$; $\bar{y}=2.522\text{m}$
4. $I_{AA} = 4608\ \text{cm}^4$, $I_{XX} = 1152\ \text{cm}^4$, $I_{YY} = 512\ \text{cm}^4$
5. $I_{AA} = 7296\ \text{cm}^4$
6. $I_{XX} = I_{YY} = 2877.3\ \text{cm}^4$
7. $I_{XX} = 968.42\ \text{cm}^4$, $I_{YY} = 174.43\ \text{cm}^4$
8. $I_{AA} = 5861.3\ \text{cm}^4$, $I_{XX} = 2115.9\ \text{cm}^4$, $I_{YY} = 785.3\ \text{cm}^4$

CHAPTER 15

Matrices and determinants

Topics covered in this chapter:

- Addition and subtraction of matrices
- Determinants and their properties
- Multiplication of matrices
- Application of matrices in solving simultaneous equations

15.1 Introduction

A matrix is an array of numbers enclosed in brackets. If there are a large number of equations to be solved then matrices (plural of matrix) make their solution much easier as compared to the method of substitution or elimination.

Each number or symbol in a matrix is known as the element of the matrix. The number of rows and columns in a matrix determines its order, for example, $\begin{pmatrix} 10 & 4 \\ 9 & 7 \end{pmatrix}$ is a 2×2 matrix; the first value is used for the number of rows and the second value used for the number of columns.

$\begin{pmatrix} 3 & 2 \\ 6 & 7 \\ 8 & 7 \\ 5 & 4 \end{pmatrix}$ This is a 4×2 matrix, i.e. 4 rows and 2 columns.

15.2 Square matrix

In a square matrix the number of rows is equal to the number of columns. For example,

$\begin{pmatrix} 1 & 6 & -5 \\ -2 & 1 & 0 \\ 5 & 3 & 3 \end{pmatrix}$ is a square matrix.

15.3 Diagonal matrix

A square matrix in which all elements are 0, except along the diagonal, is known as a diagonal matrix. For example:

$$\begin{pmatrix} 3 & 0 & 0 \\ 0 & 6 & 0 \\ 0 & 0 & 7 \end{pmatrix}$$

15.4 Unit matrix

In a unit matrix all main diagonals are 1, and all the other elements are 0. A matrix remains unchanged when it is multiplied by a unit matrix, which means that the unit matrix behaves like number 1.

15.5 Addition and subtraction

Matrices can be added/subtracted if they are conformable, i.e. are of the same size.

Example 15.1

a) Add $\begin{pmatrix} 2 & 5 \\ 4 & 1 \end{pmatrix}$ and $\begin{pmatrix} 3 & 9 \\ 5 & 7 \end{pmatrix}$

b) Subtract $\begin{pmatrix} 3 & 9 \\ 5 & 7 \end{pmatrix}$ from $\begin{pmatrix} 2 & 5 \\ 4 & 1 \end{pmatrix}$

Solution:

a) $$\begin{pmatrix} 2 & 5 \\ 4 & 1 \end{pmatrix} + \begin{pmatrix} 3 & 9 \\ 5 & 7 \end{pmatrix} = \begin{pmatrix} 2+3 & 5+9 \\ 4+5 & 1+7 \end{pmatrix} = \begin{pmatrix} 5 & 14 \\ 9 & 8 \end{pmatrix}$$

b) $$\begin{pmatrix} 2 & 5 \\ 4 & 1 \end{pmatrix} - \begin{pmatrix} 3 & 9 \\ 5 & 7 \end{pmatrix} = \begin{pmatrix} 2-3 & 5-9 \\ 4-5 & 1-7 \end{pmatrix} = \begin{pmatrix} -1 & -4 \\ -1 & -6 \end{pmatrix}$$

15.6 Multiplication/division by a scalar

If a matrix is to be multiplied/divided by a scalar (a number), then each element is multiplied/divided by that number, as shown in example 15.2.

Example 15.2

If $\mathbf{A} = \begin{pmatrix} 3 & 5 \\ 7 & 1 \end{pmatrix}$ Find 10**A**

Solution:

$$10\mathbf{A} = 10 \times \begin{pmatrix} 3 & 5 \\ 7 & 1 \end{pmatrix} = \begin{pmatrix} 3\times 10 & 5\times 10 \\ 7\times 10 & 1\times 10 \end{pmatrix} = \begin{pmatrix} 30 & 50 \\ 70 & 10 \end{pmatrix}$$

Multiplication of matrices is more complex, as shown in example 15.3. 2 matrices can only be multiplied if they are conformable, i.e. the number of columns in the first matrix is equal to the number of rows in the second matrix. The following process is used:

$$\begin{pmatrix} a & b \\ c & d \end{pmatrix} \times \begin{pmatrix} e & f \\ g & h \end{pmatrix} = \begin{pmatrix} ae+bg & af+bh \\ ce+dg & cf+dh \end{pmatrix}$$

Example 15.3

Evaluate $\begin{pmatrix} 2 & 4 \\ 3 & 5 \end{pmatrix} \times \begin{pmatrix} 6 & 2 \\ 0 & 5 \end{pmatrix}$

Solution:

$$\begin{pmatrix} 2 & 4 \\ 3 & 5 \end{pmatrix} \times \begin{pmatrix} 6 & 2 \\ 0 & 5 \end{pmatrix} = \begin{pmatrix} (2\times 6)+(4\times 0) & (2\times 2)+(4\times 5) \\ (3\times 6)+(5\times 0) & (3\times 2)+(5\times 5) \end{pmatrix}$$

$$= \begin{pmatrix} (12+0) & (4+20) \\ (18+0) & (6+25) \end{pmatrix} = \begin{pmatrix} 12 & 24 \\ 18 & 31 \end{pmatrix}$$

Example 15.4

Evaluate $\begin{pmatrix} 2 & 5 \\ 3 & 6 \end{pmatrix} \times \begin{pmatrix} 4 \\ 6 \end{pmatrix}$

Solution:

$$\begin{pmatrix} a & b \\ c & d \end{pmatrix} \times \begin{pmatrix} e \\ g \end{pmatrix} = \begin{pmatrix} ae+bg \\ ce+dg \end{pmatrix}$$

$$\begin{pmatrix} 2 & 5 \\ 3 & 6 \end{pmatrix} \times \begin{pmatrix} 4 \\ 6 \end{pmatrix} = \begin{pmatrix} (2\times 4)+(5\times 6) \\ (3\times 4)+(6\times 6) \end{pmatrix}$$

$$= \begin{pmatrix} 8+30 \\ 12+36 \end{pmatrix} = \begin{pmatrix} 38 \\ 48 \end{pmatrix}$$

Example 15.5

If $\mathbf{A} = \begin{pmatrix} -2 & 0 \\ 6 & -3 \end{pmatrix}$ and $\mathbf{B} = \begin{pmatrix} 2 & -2 \\ -6 & 3 \end{pmatrix}$, find $2\mathbf{A} + \mathbf{B}$

Solution:

$$2\mathbf{A} = 2\begin{pmatrix} -2 & 0 \\ 6 & -3 \end{pmatrix} = \begin{pmatrix} -4 & 0 \\ 12 & -6 \end{pmatrix}$$

$$2\mathbf{A} + \mathbf{B} = \begin{pmatrix} -4 & 0 \\ 12 & -6 \end{pmatrix} + \begin{pmatrix} 2 & -2 \\ -6 & 3 \end{pmatrix} = \begin{pmatrix} -4+2 & 0+(-2) \\ 12+(-6) & -6+3 \end{pmatrix} = \begin{pmatrix} -2 & -2 \\ 6 & -3 \end{pmatrix}$$

15.7 Transpose

The transpose of a matrix **A** (denoted as **A′** or $\mathbf{A}^T$) is obtained by writing its rows as columns in order, i.e. row 1 becomes column 1. For example:

$$\text{If } \mathbf{A} = \begin{pmatrix} 3 & 2 & 4 \\ 3 & 4 & 1 \\ 2 & 3 & 3 \end{pmatrix}, \text{ then } \mathbf{A}^T = \begin{pmatrix} 3 & 3 & 2 \\ 2 & 4 & 3 \\ 4 & 1 & 3 \end{pmatrix}$$

15.8 Determinants

If we have a 2 × 2 matrix, $\mathbf{A} = \begin{pmatrix} a & b \\ c & d \end{pmatrix}$, then its determinant (denoted by det **A** or $|A|$) is given by:

$$\det \mathbf{A} \text{ or } |A| = \begin{vmatrix} a & b \\ c & d \end{vmatrix} = ad - cb$$

For a 3 × 3 determinant

$$\det \mathbf{A} \text{ or } |A| = \begin{vmatrix} a & b & c \\ d & e & f \\ g & h & i \end{vmatrix} = a\begin{vmatrix} e & f \\ h & i \end{vmatrix} - b\begin{vmatrix} d & f \\ g & i \end{vmatrix} + c\begin{vmatrix} d & e \\ g & h \end{vmatrix}$$

Each element of the top row is multiplied by the 2 × 2 determinant from the remaining rows and columns not containing the element. Also, the sign of the middle element of the top row is changed. These 2 × 2 determinants are known as minors of the corresponding element, for example the minor of element **a** is $\begin{vmatrix} e & f \\ h & i \end{vmatrix}$

15.9 Cofactors

The cofactor of an element is the minor determinant formed from the rows and columns not containing that element with due regard for the appropriate sign, as shown below:

For 2 × 2 determinant 3 × 3 determinant 4 × 4 determinant

$$\begin{vmatrix} + & - \\ - & + \end{vmatrix} \qquad \begin{vmatrix} + & - & + \\ - & + & - \\ + & - & + \end{vmatrix} \qquad \begin{vmatrix} + & - & + & - \\ - & + & - & + \\ + & - & + & - \\ - & + & - & + \end{vmatrix}$$

Example 15.6

Evaluate the following determinants:

a) $\begin{vmatrix} 3 & 4 \\ 1 & 2 \end{vmatrix}$ b) $\begin{vmatrix} 4 & 1 & 5 \\ 2 & -1 & 2 \\ -1 & 2 & 0 \end{vmatrix}$

Solution:

a) Refer to section 15.8, $\begin{vmatrix} 3 & 4 \\ 1 & 2 \end{vmatrix} = 3 \times 2 - 1 \times 4 = \mathbf{2}$

b) $\begin{vmatrix} 4 & 1 & 5 \\ 2 & -1 & 2 \\ -1 & 2 & 0 \end{vmatrix} = 4\begin{vmatrix} -1 & 2 \\ 2 & 0 \end{vmatrix} - 1\begin{vmatrix} 2 & 2 \\ -1 & 0 \end{vmatrix} + 5\begin{vmatrix} 2 & -1 \\ -1 & 2 \end{vmatrix}$

$= 4(0 - 4) - 1(0 + 2) + 5(4 - 1)$

$= -16 - 2 + 15 = \mathbf{-3}$

15.10 Inverse of a square matrix

The inverse of a square matrix **A** is denoted as $\mathbf{A}^{-1}$, so that $\mathbf{A} \times \mathbf{A}^{-1} = 1$

Consider a 2 × 2 matrix, $\mathbf{A} = \begin{pmatrix} a & b \\ c & d \end{pmatrix}$, the inverse matrix $\mathbf{A}^{-1}$ is given by:

$$\mathbf{A}^{-1} = \frac{1}{|A|}\begin{pmatrix} d & -b \\ -c & a \end{pmatrix} = \mathbf{I}$$

A matrix that does not have an inverse is called a singular matrix. In this case the determinant of the matrix is 0.

For example, the determinant of $\begin{pmatrix} 2 & 8 \\ 1 & 4 \end{pmatrix} = \begin{vmatrix} 2 & 8 \\ 1 & 4 \end{vmatrix} = 2 \times 4 - 1 \times 8 = 0$

Therefore $\begin{pmatrix} 2 & 8 \\ 1 & 4 \end{pmatrix}$ is a singular matrix.

Example 15.7

If $\mathbf{A} = \begin{pmatrix} 4 & 2 \\ 6 & 4 \end{pmatrix}$, find $\mathbf{A}^{-1}$, and verify the result.

Solution:

$|A| = 4 \times 4 - 6 \times 2 = 4$

$$\mathbf{A}^{-1} = \frac{1}{4}\begin{pmatrix} 4 & -2 \\ -6 & 4 \end{pmatrix} = \begin{pmatrix} \frac{4}{4} & \frac{-2}{4} \\ \frac{-6}{4} & \frac{4}{4} \end{pmatrix} = \begin{pmatrix} 1 & -0.5 \\ -1.5 & 1 \end{pmatrix}$$

$$\text{Verification: } \mathbf{AA}^{-1} = \begin{pmatrix} 4 & 2 \\ 6 & 4 \end{pmatrix}\begin{pmatrix} 1 & -0.5 \\ -1.5 & 1 \end{pmatrix}$$

$$= \begin{pmatrix} 4\times1+2\times(-1.5) & 4\times(-0.5)+2\times1 \\ 6\times1+4\times(-1.5) & 6\times(-0.5)+4\times1 \end{pmatrix}$$

$$= \begin{pmatrix} 4-3 & -2+2 \\ 6-6 & -3+4 \end{pmatrix} = \begin{pmatrix} 1 & 0 \\ 0 & 1 \end{pmatrix} = \mathbf{I}$$

15.11 Properties of determinants

a) The rows and columns in determinants may be changed either to alter the value of the determinant or leave it unchanged:

   i) The value of a determinant does not change if rows become columns and columns become rows.

   ii) Changing a row with another row or a column with another column causes the determinant to be multiplied by −1. For example:

$$\begin{vmatrix} 2 & 5 & 7 \\ 3 & 2 & 6 \\ 1 & 8 & 4 \end{vmatrix} = 44 \text{ and } \begin{vmatrix} 3 & 2 & 6 \\ 2 & 5 & 7 \\ 1 & 8 & 4 \end{vmatrix} = -44$$

   iii) If 2 rows and 2 columns are identical, the value of the determinant is 0.

   iv) If a determinant is multiplied by a number, then each element of 1 row or 1 column only is multiplied. For example:

$$\det A = \begin{vmatrix} 2 & 5 & 7 \\ 3 & 2 & 6 \\ 1 & 8 & 4 \end{vmatrix}, \text{ then } 2\times\det A = \begin{vmatrix} 4 & 10 & 14 \\ 3 & 2 & 6 \\ 1 & 8 & 4 \end{vmatrix} = \begin{vmatrix} 4 & 5 & 7 \\ 6 & 2 & 6 \\ 2 & 8 & 4 \end{vmatrix},$$

Each element of the first row, or each element of the first column is multiplied by 2.

A common factor in any row or column can be taken outside the determinant by following the reverse process. In the following example, the second row has a common factor of 3; the determinant factorises to:

$$\begin{vmatrix} 1 & 3 & 8 \\ 6 & 9 & 15 \\ 3 & 7 & 4 \end{vmatrix} = 3\begin{vmatrix} 1 & 3 & 8 \\ 2 & 3 & 5 \\ 3 & 7 & 4 \end{vmatrix}$$

Example 15.8

Factorise $\begin{vmatrix} 2 & 6 & 8 \\ 3 & -7 & 8 \\ 1 & 5 & -12 \end{vmatrix}$

Solution:

The first row has a common factor of 2 and the third column a common factor of 4.

$$\begin{vmatrix} 2 & 6 & 8 \\ 3 & -7 & 8 \\ 1 & 5 & -12 \end{vmatrix} = 2\begin{vmatrix} 1 & 3 & 4 \\ 3 & -7 & 8 \\ 1 & 5 & -12 \end{vmatrix} = 2\times 4\begin{vmatrix} 1 & 3 & 1 \\ 3 & -7 & 2 \\ 1 & 5 & -3 \end{vmatrix}$$

15.12 Application of matrices

Determinants may be used to solve simultaneous equations involving 2 unknowns; the equations are written as:

$$a_1x + b_1y = c_1$$

$$a_2x + b_2y = c_2$$

Let **D** be the determinant comprising the coefficients of x and y, and $\mathbf{D_x}$ and $\mathbf{D_y}$ respectively be the determinants obtained from **D** by replacing its first and second columns by the column (c_1, c_2) as shown below:

$$\det \mathbf{D} = \begin{vmatrix} a_1 & b_1 \\ a_2 & b_2 \end{vmatrix}; \mathbf{D}_x = \begin{vmatrix} c_1 & b_1 \\ c_2 & b_2 \end{vmatrix}; \mathbf{D}_y = \begin{vmatrix} a_1 & c_1 \\ a_2 & c_2 \end{vmatrix}$$

$$x = \frac{D_x}{D} \quad y = \frac{D_y}{D}$$

For 3 unknowns, 3 equations are required which may be written as:

$$a_1x + b_1y + c_1z = d_1$$

$$a_2x + b_2y + c_2z = d_2$$

$$a_3x + b_3y + c_3z = d_3$$

$$\det \mathbf{D} = \begin{vmatrix} a_1 & b_1 & c_1 \\ a_2 & b_2 & c_2 \\ a_3 & b_3 & c_3 \end{vmatrix}; \mathbf{D}_x = \begin{vmatrix} d_1 & b_1 & c_1 \\ d_2 & b_2 & c_2 \\ d_3 & b_3 & c_3 \end{vmatrix}; \mathbf{D}_y = \begin{vmatrix} a_1 & d_1 & c_1 \\ a_2 & d_2 & c_2 \\ a_3 & d_3 & c_3 \end{vmatrix}$$

$$\mathbf{D}_z = \begin{vmatrix} a_1 & b_1 & d_1 \\ a_2 & b_2 & d_2 \\ a_3 & b_3 & d_3 \end{vmatrix}$$

$$x = \frac{D_x}{D}; y = \frac{D_y}{D}; z = \frac{D_z}{D}$$

Example 15.9

Solve the simultaneous equations: $x + 2y = -3$ and $2x - 2y = 6$

Solution:

$a_1 = 1 \quad b_1 = 2 \quad a_2 = 2 \quad b_2 = -2 \; c_1 = -3 \; c_2 = 6$

$$\det \mathbf{D} = \begin{vmatrix} 1 & 2 \\ 2 & -2 \end{vmatrix} = -2 - 4 = -6$$

$$\det \mathbf{D}_x = \begin{vmatrix} -3 & 2 \\ 6 & -2 \end{vmatrix} = 6 - 12 = -6$$

$$\det \mathbf{D}_y = \begin{vmatrix} 1 & -3 \\ 2 & 6 \end{vmatrix} = 6 - (-6) = 12$$

$$\mathbf{x} = \frac{D_x}{D} = \frac{-6}{-6} = \mathbf{1} \qquad \mathbf{y} = \frac{D_y}{D} = \frac{12}{-6} = \mathbf{-2}$$

Example 15.10

Solve the simultaneous equations:

$x + 2y + z = 2$

$2x + 2y - z = 8$

$3x - y + 2z = 1$

Solution:

$$\det \mathbf{D} = \begin{vmatrix} a_1 & b_1 & c_1 \\ a_2 & b_2 & c_2 \\ a_3 & b_3 & c_3 \end{vmatrix} = a_1 \begin{vmatrix} b_2 & c_2 \\ b_3 & c_3 \end{vmatrix} - b_1 \begin{vmatrix} a_2 & c_2 \\ a_3 & c_3 \end{vmatrix} + c_1 \begin{vmatrix} a_2 & b_2 \\ a_3 & b_3 \end{vmatrix}$$

$$= 1 \begin{vmatrix} 2 & -1 \\ -1 & 2 \end{vmatrix} - 2 \begin{vmatrix} 2 & -1 \\ 3 & 2 \end{vmatrix} + 1 \begin{vmatrix} 2 & 2 \\ 3 & -1 \end{vmatrix}$$

$$= 1(4 - 1) - 2(4 + 3) + 1(-2 - 6) = -19$$

$$\mathbf{D}_x = \begin{vmatrix} d_1 & b_1 & c_1 \\ d_2 & b_2 & c_2 \\ d_3 & b_3 & c_3 \end{vmatrix} = d_1 \begin{vmatrix} b_2 & c_2 \\ b_3 & c_3 \end{vmatrix} - b_1 \begin{vmatrix} d_2 & c_2 \\ d_3 & c_3 \end{vmatrix} + c_1 \begin{vmatrix} d_2 & b_2 \\ d_3 & b_3 \end{vmatrix}$$

$$= 2 \begin{vmatrix} 2 & -1 \\ -1 & 2 \end{vmatrix} - 2 \begin{vmatrix} 8 & -1 \\ 1 & 2 \end{vmatrix} + 1 \begin{vmatrix} 8 & 2 \\ 1 & -1 \end{vmatrix}$$

$$= 2(4 - 1) - 2(16 + 1) + 1(-8 - 2) = -38$$

$$\mathbf{D}_y = \begin{vmatrix} a_1 & d_1 & c_1 \\ a_2 & d_2 & c_2 \\ a_3 & d_3 & c_3 \end{vmatrix} = a_1\begin{vmatrix} d_2 & c_2 \\ d_3 & c_3 \end{vmatrix} - d_1\begin{vmatrix} a_2 & c_2 \\ a_3 & c_3 \end{vmatrix} + c_1\begin{vmatrix} a_2 & d_2 \\ a_3 & d_3 \end{vmatrix}$$

$$= 1\begin{vmatrix} 8 & -1 \\ 1 & 2 \end{vmatrix} - 2\begin{vmatrix} 2 & -1 \\ 3 & 2 \end{vmatrix} + 1\begin{vmatrix} 2 & 8 \\ 3 & 1 \end{vmatrix}$$

$$= 1(16 + 1) - 2(4 + 3) + 1(2 - 24) = -19$$

$$\mathbf{D}_z = \begin{vmatrix} a_1 & b_1 & d_1 \\ a_2 & b_2 & d_2 \\ a_3 & b_3 & d_3 \end{vmatrix} = a_1\begin{vmatrix} b_2 & d_2 \\ b_3 & d_3 \end{vmatrix} - b_1\begin{vmatrix} a_2 & d_2 \\ a_3 & d_3 \end{vmatrix} + d_1\begin{vmatrix} a_2 & b_2 \\ a_3 & b_3 \end{vmatrix}$$

$$= 1\begin{vmatrix} 2 & 8 \\ -1 & 1 \end{vmatrix} - 2\begin{vmatrix} 2 & 8 \\ 3 & 1 \end{vmatrix} + 2\begin{vmatrix} 2 & 2 \\ 3 & -1 \end{vmatrix}$$

$$= 1(2 + 8) - 2(2 - 24) + 2(-2 - 6) = 38$$

$$x = \frac{D_x}{D} = \frac{-38}{-19} = 2;\ y = \frac{-19}{-19} = 1;\ z = \frac{38}{-19} = -2$$

Exercise 15.1

1. Add a) $\begin{pmatrix} 2 & -3 \\ 6 & 4 \end{pmatrix}$ and $\begin{pmatrix} 2 & 0 \\ -3 & 5 \end{pmatrix}$

   b) $\begin{pmatrix} 3 & 2 & -4 \\ 3 & 4 & 1 \\ 2 & 3 & -3 \end{pmatrix}$ and $\begin{pmatrix} 1 & 0 & -5 \\ -2 & 1 & 0 \\ 5 & 3 & 3 \end{pmatrix}$

2. Subtract a) $\begin{pmatrix} -2 & 0 \\ -3 & 5 \end{pmatrix}$ from $\begin{pmatrix} 2 & -3 \\ 6 & 4 \end{pmatrix}$

   b) $\begin{pmatrix} 1 & 6 & -5 \\ -2 & 1 & 0 \\ 5 & 3 & 3 \end{pmatrix}$ from $\begin{pmatrix} 3 & 2 & -4 \\ 3 & 4 & 1 \\ 2 & 3 & -3 \end{pmatrix}$

3. If $A = \begin{pmatrix} 2 & -2 \\ -6 & 3 \end{pmatrix}$ and $B = \begin{pmatrix} -2 & 0 \\ 6 & -3 \end{pmatrix}$, find $A - 2B$

4. If $A = \begin{pmatrix} 2 & 5 \\ 4 & 1 \end{pmatrix}$ and $B = \begin{pmatrix} 3 & 1 \\ 2 & 0 \end{pmatrix}$, find $3A + B^2$

5. Solve the equation $\mathbf{X} - \begin{pmatrix} 5 & 3 \\ 7 & 0 \end{pmatrix} = \begin{pmatrix} 2 & 4 \\ 1 & -8 \end{pmatrix}$ where $\mathbf{X}$ is a 2×2 matrix

6. Evaluate $\begin{pmatrix} 3 & 2 \\ 6 & 1 \end{pmatrix} \times \begin{pmatrix} 5 & 3 \\ 6 & 0 \end{pmatrix}$

7. Evaluate $\begin{pmatrix} 3 & 4 \\ 7 & 2 \end{pmatrix} \times \begin{pmatrix} 5 \\ 6 \end{pmatrix}$

8. Evaluate the following determinants:

   a) $\begin{vmatrix} -5 & -6 \\ 4 & 3 \end{vmatrix}$ b) $\begin{vmatrix} -2 & 3 & 0 \\ -3 & 1 & 1 \\ -2 & 3 & 4 \end{vmatrix}$

9. If $\mathbf{A} = \begin{pmatrix} 3 & 2 \\ 1 & 2 \end{pmatrix}$, find $\mathbf{A}^{-1}$, and verify the result

10. Solve the simultaneous equations: $2x + 3y = 4$ and $x - y = -3$

11. Solve the simultaneous equations: $2x + 2y = -2$ and $3x - y = 9$

12. Solve the simultaneous equations:

    $x + 2y + z = 6$

    $2x + 2y - z = 9$

    $3x - y + 2z = -2$

Answers – Exercise 15.1

1. a) $\begin{pmatrix} 0 & -3 \\ 3 & 9 \end{pmatrix}$

   b) $\begin{pmatrix} 4 & 8 & -9 \\ 1 & 5 & 1 \\ 7 & 6 & 0 \end{pmatrix}$

2. a) $\begin{pmatrix} 4 & -3 \\ 9 & -1 \end{pmatrix}$

   b) $\begin{pmatrix} 2 & -4 & 1 \\ 5 & 3 & 1 \\ -3 & 0 & -6 \end{pmatrix}$

3. $\begin{pmatrix} 6 & -2 \\ -18 & 9 \end{pmatrix}$

4. $\begin{pmatrix} 17 & 18 \\ 18 & 5 \end{pmatrix}$

5. $\begin{pmatrix} 7 & 7 \\ 8 & -8 \end{pmatrix}$

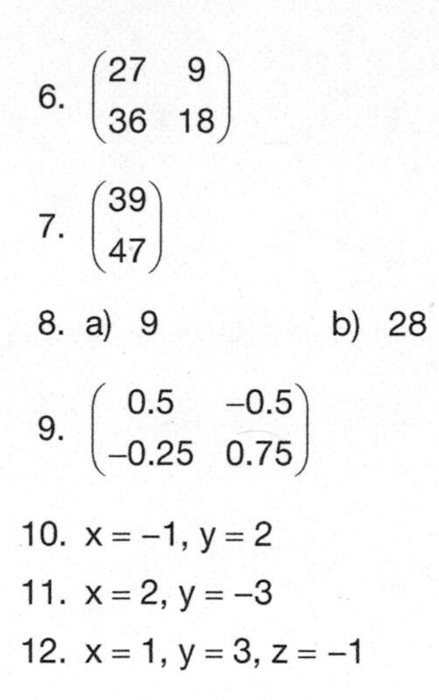

6. $\begin{pmatrix} 27 & 9 \\ 36 & 18 \end{pmatrix}$

7. $\begin{pmatrix} 39 \\ 47 \end{pmatrix}$

8. a) 9 b) 28

9. $\begin{pmatrix} 0.5 & -0.5 \\ -0.25 & 0.75 \end{pmatrix}$

10. $x = -1$, $y = 2$

11. $x = 2$, $y = -3$

12. $x = 1$, $y = 3$, $z = -1$

CHAPTER 16

Vectors

Topics covered in this chapter:

- Vectors and their addition/subtraction
- Determination of the resultant of two or more vectors
- Resolution of vectors

16.1 Introduction

Physical quantities which have magnitude but no direction are called scalar quantities or scalars. For example mass, temperature and energy are scalar quantities. Physical quantities which are described by both magnitude and direction are called vector quantities or vectors, typical examples being force, velocity, acceleration and displacement. A vector may be represented by a line whose direction is parallel to the actual vector and the length of the line is proportional to the magnitude of the vector. An arrow, shown at the end of the vector, is used to show the direction of the vector (Figure 16.1a). A vector is written in bold print such as **v** or as OA, where OA is the line representing the vector and the arrow above letters OA means that the direction of the vector is from O to A. The magnitude of the vector **v** is written as $|\mathbf{v}|$.

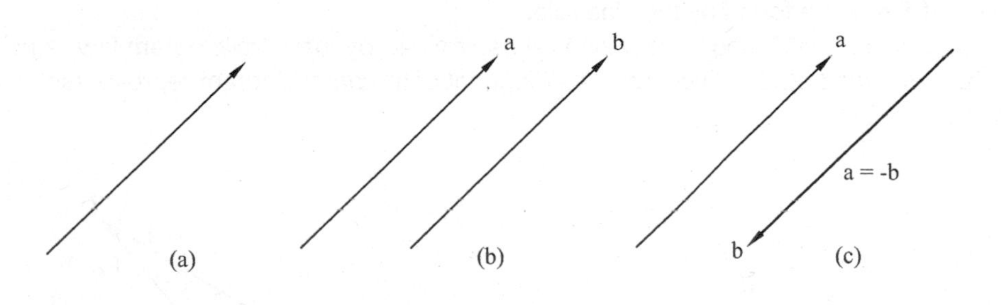

Figure 16.1

2 vectors **a** and **b** are equal if they have the same magnitude and the same direction, as shown in Figure 16.1b. Vectors **a** and **b** are equal and opposite (Figure 16.1c) if they have the same magnitude but act in opposite direction.

Example 16.1

A 50 N force acts upwards and is inclined at an angle of 30° to the horizontal. Draw the vector.

Solution:

The first step is to choose a suitable scale, say 1 mm = 1 Newton. Using this scale, 50 N = 50 mm. Next, draw a horizontal line as the reference line. Draw line AB = 50 mm, inclined at 30° to the reference line, as shown in Figure 16.2. Draw an arrow at point B to show the direction of the force. AB is the required vector.

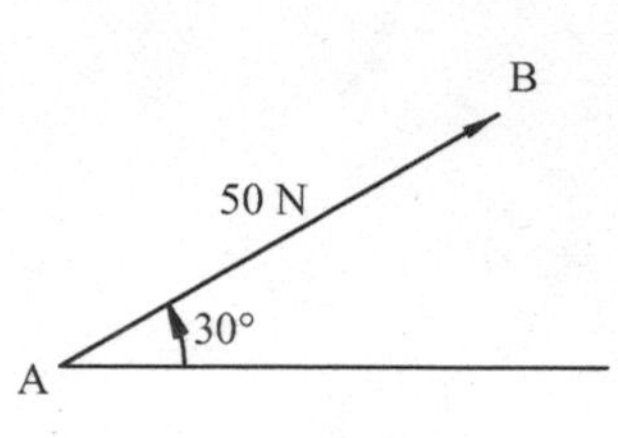

Figure 16.2

16.2 Addition of vectors

Figure 16.3a shows 2 vectors $\mathbf{F_1}$ and $\mathbf{F_2}$ acting at point O. The resultant of adding vectors $\mathbf{F_1}$ and $\mathbf{F_2}$ can be obtained by drawing OP to represent F_1, and PQ to represent F_2 as shown in Figure 16.3b. OP$\|$ F_1, and PQ$\|$ F_2. Line OQ, which represents the vector **F**, is drawn to complete the triangle QOP.

$$\mathbf{F} = \mathbf{F_1} + \mathbf{F_2}$$

F is the resultant vector that replaces $\mathbf{F_1}$ and $\mathbf{F_2}$. The magnitude of F (written as $|\mathbf{F}|$) and its direction are obtained from triangle QOP, either by drawing the triangle to scale or by calculations using the sine and cosine rules. The magnitude of **F** in Figure 16.3b is given by:

$$|\mathbf{F}|^2 = |\mathbf{F_1}|^2 + |\mathbf{F_2}|^2 - 2\,|\mathbf{F_1}|\,|\mathbf{F_2}|\cos\theta$$

The direction of **F** may be found by the sine rule.

The resultant of vectors $\mathbf{F_1}$ and $\mathbf{F_2}$ may also be determined by the parallelogram law. A parallelogram is drawn which is based on the 2 vectors. The diagonal of the parallelogram represents the resultant, as illustrated in example 16.2.

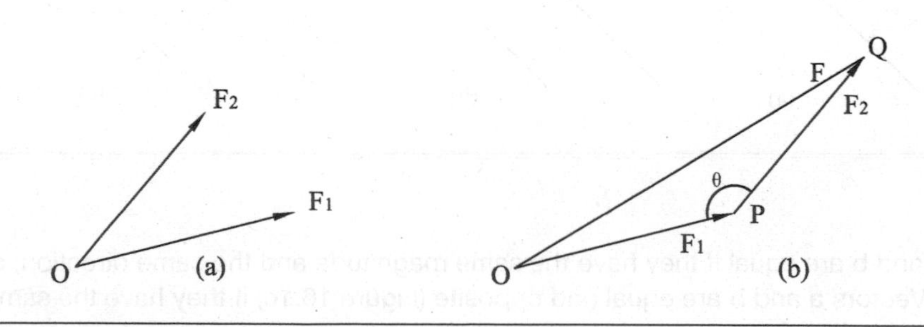

Figure 16.3

If there are more than 2 vectors the resultant of the vectors may be determined graphically. Consider 4 vectors **a**, **b**, **c** and **d** as shown in Figure 16.4a. Select a suitable scale and draw line OM to scale and parallel to vector **a,** as shown in Figure 16.4b. At the end of vector **a**, i.e. at point M, line MN is drawn to scale and parallel to vector **b** of Figure 16.4a. Similarly the remaining 2 vectors are drawn and finally from point Q line QO is drawn to represent the resultant of the 4 vectors. Figure 16.4b is called the polygon of vectors.

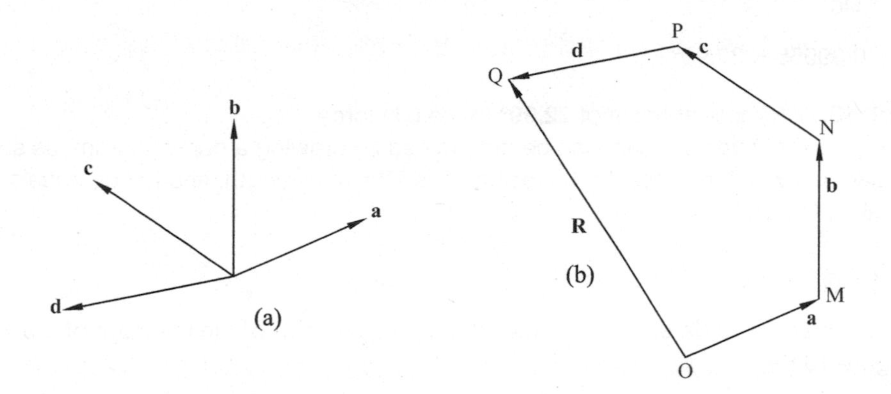

Figure 16.4 Polygon of vectors

Example 16.2

2 forces act at a point as shown in Figure 16.5a. Find the magnitude and the direction of the resultant of the 2 forces.

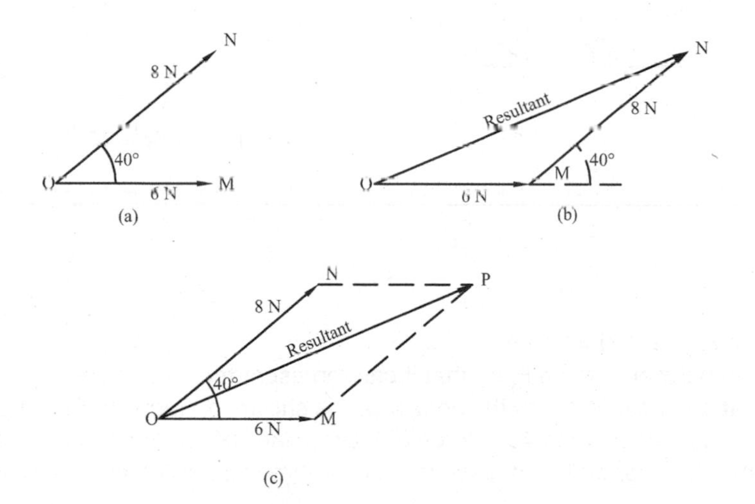

Figure 16.5

Solution:

Select a suitable scale, i.e. 1 N = 6 mm (1 mm = 1 ÷ 6 = 0.1666 N).

Draw, line OM = 36 mm, and parallel to the 6 N force, as shown in Figure 16.5b. At point M draw line MN = 48 mm and parallel to the 8 N force. Join ON to represent the resultant. Measure the length of line ON:

ON = 79 mm. 79 mm converted into Newtons equal **13.17 N.**

Use the sine rule to determine ∠O: $\frac{79}{\sin 140^\circ} = \frac{48}{\sin O} = \frac{36}{\sin N}$ (∠M =140°)

$\frac{79}{\sin 140^\circ} = \frac{48}{\sin O}$ which gives, $\sin O = \frac{48 \times \sin 140^\circ}{79} = 0.39055$

∠O = $\sin^{-1} 0.39055$ = **22.99°**

The resultant (**13.17 N**) acts at angle of **22.99°** to the 6 N force.

The resultant of the 2 vectors can also be determined by drawing a parallelogram, as shown in Figure 16.5c. The diagonal which represents the resultant, is 79 mm in length, and hence equals 13.17 N after using the scaling factor.

Example 16.3

Use a graphical method to determine the magnitude and direction of the resultant of the 3 forces shown in Figure 16.6a.

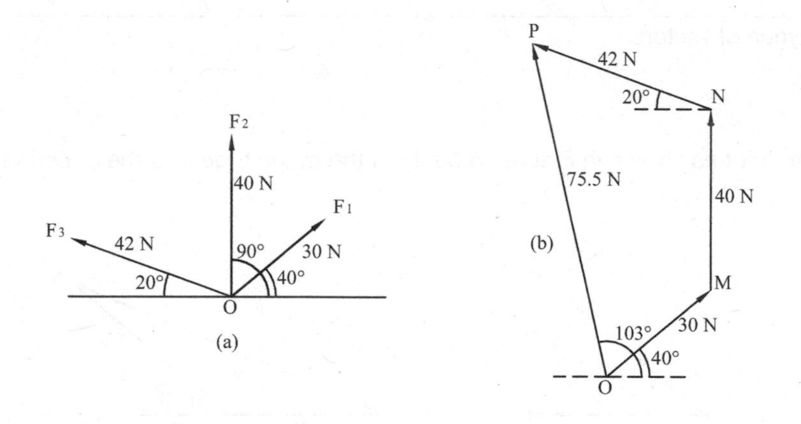

Figure 16.6

Solution:

Select a suitable scale, i.e. 1 N = 1 mm.

Line OM is drawn parallel to force F_1 so that it can represent the force. Therefore, draw line OM, 30 mm long, at an angle of 40° to the horizontal, as shown in Figure 16.6b. At point M draw, line MN = 40 mm and parallel to the 40 N force (F_2). Draw line NP, 42 mm long and parallel to the 42 N force (F_3). Draw line PO to represent the resultant of the three forces. Measure the length of line PO:

Resultant PO = 75.5 mm. 75.5 mm converted into Newtons equal 75.5 N.

The resultant is inclined at 103° to the horizontal or 63° to the 30N force.

16.3 Subtraction of vectors

Consider 2 vectors **a** and **b** shown in Figure 16.7a. To find the difference between **b** and **a** (i.e. find **b** – **a**) vector **b** is drawn from point O, as shown in Figure 16.7b. Then at the end of vector **b**, vector **a** is drawn, its direction being opposite to the direction shown in Figure 16.7a.

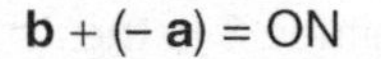

$\mathbf{b} + (-\mathbf{a}) = \text{ON}$

ON is equal to vector PQ, drawn across the ends of the 2 vectors, as illustrated in Figure 16.7c. Therefore, subtraction of 2 vectors may also be achieved by drawing a vector across their ends. The direction of the resultant vector is away from the subtracted vector.

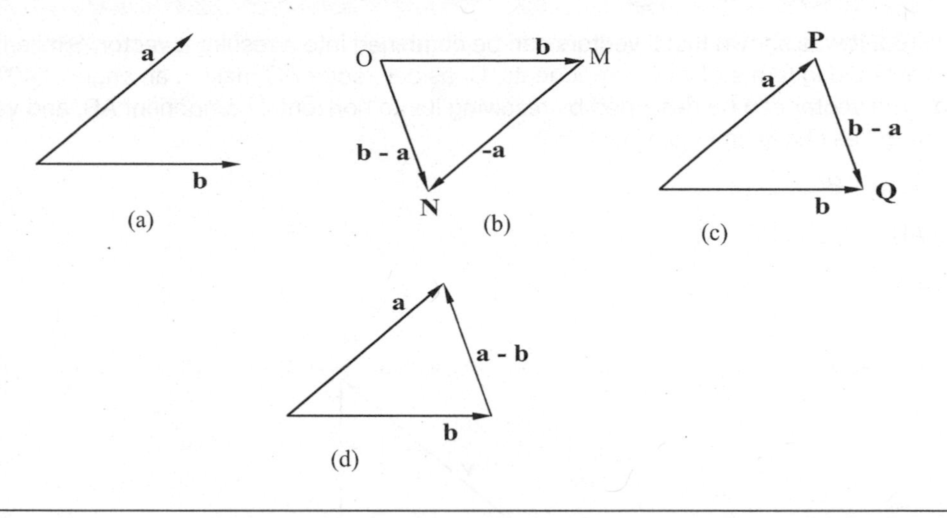

Figure 16.7

To find **a** – **b**, the process is similar to the determination of **b** – **a**, the only difference is that the direction of vector **a** – **b** is opposite to the direction of vector **b** – **a** (Figure 16.7d).

16.4 Unit vectors

A vector whose magnitude is 1 is known as a unit vector. In Cartesian axis, as shown in Figure 16.8, the unit vector along x-direction is called i. Similarly, the unit vectors along y-direction and z-direction are called j and k respectively.

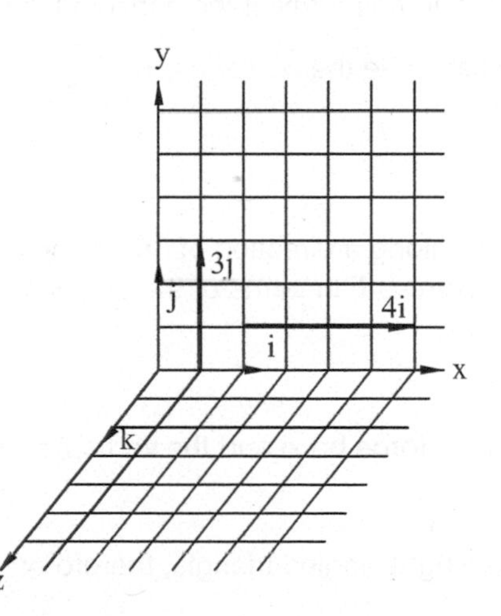

Figure 16.8

A vector of magnitude 4 along the x-direction will be 4i, similarly, a vector of magnitude 3 in the y-direction is 3j.

16.5 Resolution of vectors

In section 16.2 it was shown that 2 vectors can be combined into a resultant vector. Similarly a vector may be expressed in terms of its 2 components. Consider vector AC making an angle of 40° to the horizontal. The vector can be described by resolving it into horizontal component AB, and vertical component BC as shown in Figure 16.9.

$AC = \mathbf{v} = AB + BC$

Writing $AB = xi$, and $BC = yj$:

$\mathbf{v} = AC = xi + yj$

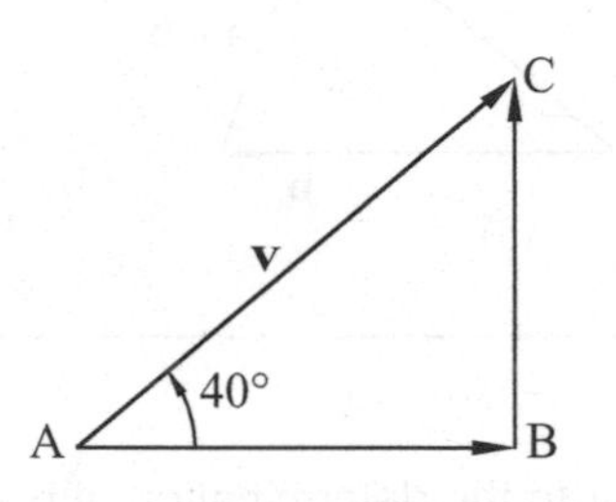

Figure 16.9

We can obtain the magnitude of '**v**' in terms of x and y by Pythagorus' theorem:

$|\mathbf{v}|^2 = x^2 + y^2$

$|v| = \sqrt{x^2 + y^2}$

A vector can also be written as a column matrix. If the horizontal and the vertical components of a vector (**u**) are 4 and 3 respectively, we can write the vector as: $\mathbf{u} = \begin{pmatrix} 4 \\ 3 \end{pmatrix}$

Example 16.4

A force (**F**) of magnitude 50 N acts along a direction of 38° to the horizontal. Calculate its horizontal and vertical components and hence express **F** in terms of the unit vectors **i** and **j**.

Solution:

Let the horizontal component of the force be xi and the vertical component be yj.

$\mathbf{F} = x\mathbf{i} + y\mathbf{j}$

Triangle AOB (Figure 16.10) is a right-angled triangle, therefore:

$|OB| = x = |F| \cos 38°$

$= 50 \times \cos 38° = 39.4$ N

$|BA| = y = |F| \sin 38°$

$= 50 \times \sin 38° = 30.78$ N

Hence, **F** = 39.4**i** + 30.78**j**

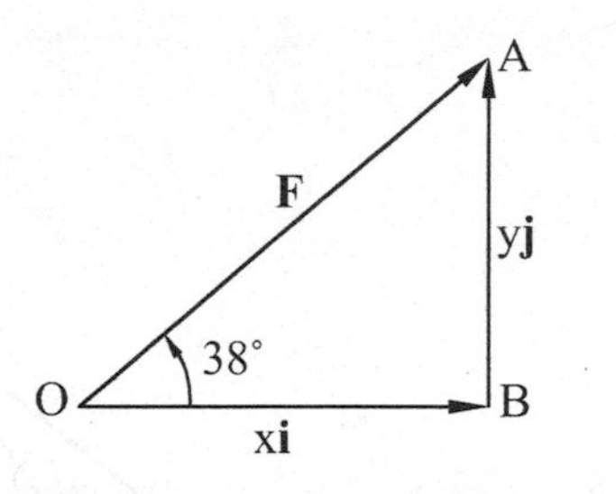

Figure 16.10

Example 16.5

With reference to the x – axis, find the magnitude and direction of **a** = 3**i** + 4**j**.

Solution:

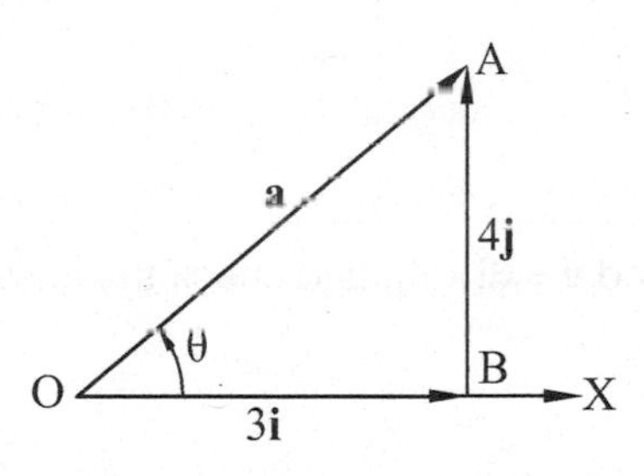

Figure 16.11

$x = 3; y = 4; |a| = \sqrt{x^2 + y^2}$

Refer to Figure 16.11, $|a| = \sqrt{3^2 + 4^2} = \sqrt{25} = \mathbf{5N}$

In ΔAOB, $\tan\theta = \frac{4}{3}$, therefore $\theta = \tan^{-1}\left(\frac{4}{3}\right) = \mathbf{53.13^{o}}$

The resultant vector **a** has a magnitude of **5.0 N**, and acts at an angle of **53.13°** to the horizontal.

16.6 Addition and subtraction of two vectors in Cartesian form

Consider 2 vectors u and v; their resultant (w) is: **w = u + v** (Figure 16.12). In terms of Cartesian components, this becomes:

$\mathbf{u} = x_1\mathbf{i} + y_1\mathbf{j}$, and $\mathbf{v} = x_2\mathbf{i} + y_2\mathbf{j}$

$\mathbf{w} = \mathbf{u} + \mathbf{v} = (x_1 + x_2)\mathbf{i} + (y_1 + y_2)\mathbf{j}$

Similarly, $\mathbf{u} - \mathbf{v} = (x_1 - x_2)\mathbf{i} + (y_1 - y_2)\mathbf{j}$

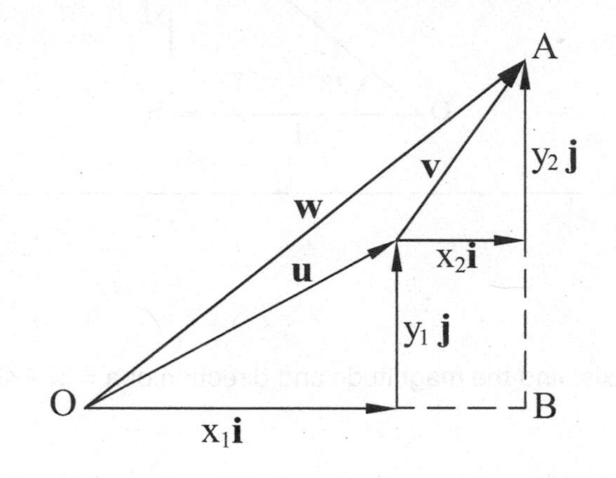

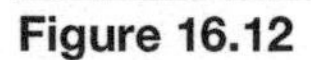
Figure 16.12

Example 16.6

Find the resultant of: **u** = 5**i** + 3**j**, and **v** = 3**i** + 4**j**, and check the answer from graphical solution.

Solution:

Resultant $\mathbf{w} = \mathbf{u} + \mathbf{v} = (x_1 + x_2)\mathbf{i} + (y_1 + y_2)\mathbf{j}$

$= (5 + 3)\mathbf{i} + (3 + 4)\mathbf{j}$

$= 8\mathbf{i} + 7\mathbf{j}$

$$|w| = \sqrt{8^2 + 7^2} = \sqrt{113} = \mathbf{10.63}$$

$$\tan\theta = \frac{7}{8}, \text{ therefore } \theta = \tan^{-1}\left(\frac{7}{8}\right) = \mathbf{41.19^\circ}$$

The magnitude of the resultant is **10.63** units, and it acts at **41.19°** to the horizontal
Figure 16.13 illustrates the problem. From this figure:

Magnitude of the resultant, $|\mathbf{w}|$ = 10.6 units; θ = **41°**

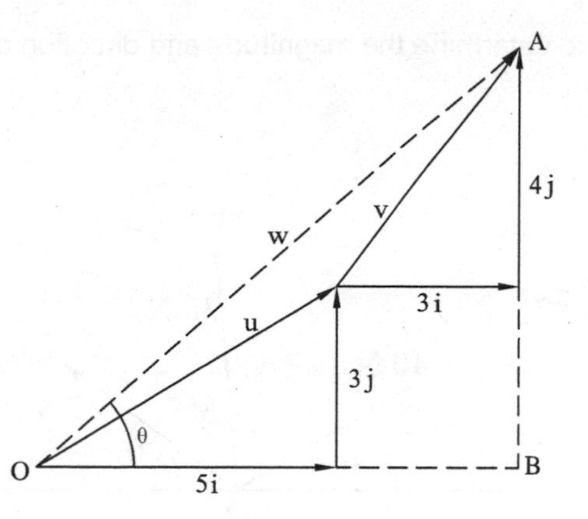

Figure 16.13

Exercise 16.1

1. 2 forces act at a point as shown in Figure 16.14. Find the magnitude and the direction of the resultant of the 2 forces.

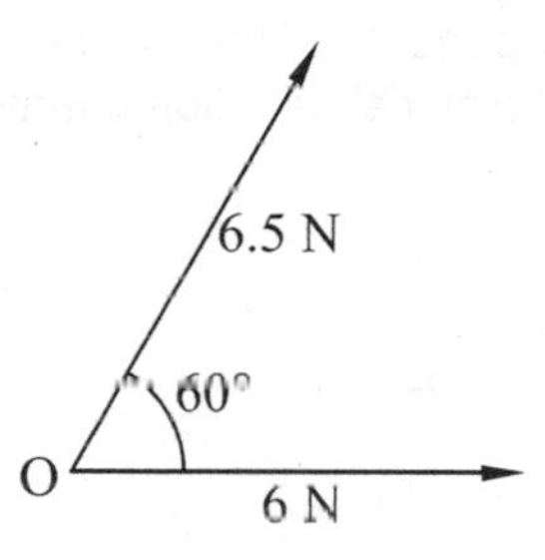

Figure 16.14

2. Use the parallelogram law to determine the magnitude and direction of the resultant of the forces shown in Figure 16.15.

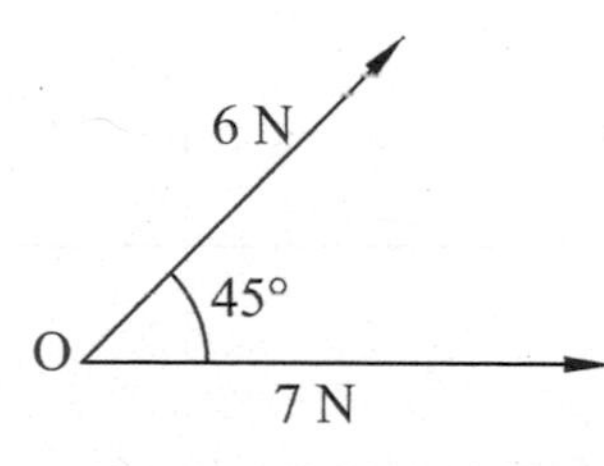

Figure 16.15

3. Use a graphical method to determine the magnitude and direction of the resultant of the forces shown in Figure 16.16.

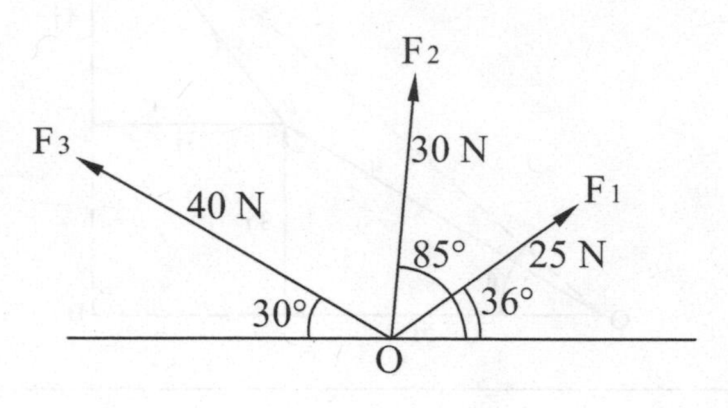

Figure 16.16

4. Find the magnitude and the direction of the vector, $\mathbf{v} = \begin{pmatrix} 6 \\ 4 \end{pmatrix}$.
5. A force (**F**) of magnitude 55 N acts along a direction of 46° to the horizontal. Calculate its horizontal and vertical components and hence express **F** in terms of the unit vectors **i** and **j**.
6. Find the magnitude and direction of the vector: $\mathbf{a} = 5\mathbf{i} + 6\mathbf{j}$, with reference to the x-axis.
7. Find the horizontal and vertical components of the vectors shown in Figure 16.17, and write each of them in the equivalent component form, i.e. as column matrices.

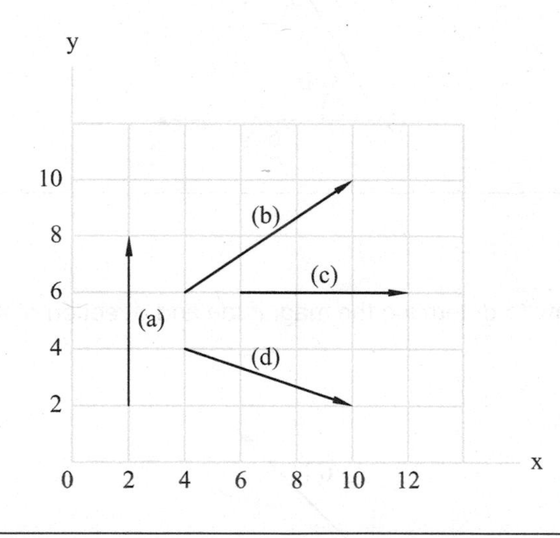

Figure 16.17

8. Find the magnitude and direction of the vector: $\mathbf{v} = 5\mathbf{i} - 3\mathbf{j}$.

Answers – Exercise 16.1

1. 10.83 N acts at 31.31° to 6N force
2. 12 N acts at 21° to the horizontal
3. 65.5 N acts at 65° to 25N force
4. 7.21 units acts at 33.69° to the horizontal
5. $\mathbf{F} = 38.21\mathbf{i} + 39.56\mathbf{j}$
6. 7.81 units acts at 50.19° to the horizontal
7. a) $\begin{pmatrix} 0 \\ 6 \end{pmatrix}$ b) $\begin{pmatrix} 6 \\ 4 \end{pmatrix}$ c) $\begin{pmatrix} 6 \\ 0 \end{pmatrix}$ d) $\begin{pmatrix} 6 \\ -2 \end{pmatrix}$
8. 5.83 units acts at –30.96° to the horizontal

Statistics

Topics covered in this chapter:

- Averages: mean, mode and median of the given data
- Statistical diagrams
- Variance and standard deviation
- Binomial, Poisson and normal distributions

17.1 Introduction

Statistics is a branch of mathematics that involves the collection, preparation, analysis, presentation and interpretation of data. Whenever we want to carry out an investigation we can do that by acquiring primary or secondary data. Primary data is obtained from people by carrying out surveys via questionnaires and interviews. Primary data is published in many newspapers, journals, magazines and websites; the Office for National Statistics (ONS) publishes data on construction related activities. If primary data is extracted from websites, journals and other sources then it is known as secondary data.

17.2 Types of data

Depending on the subject of the investigation, the data collected may be either discrete data or continuous data.

17.2.1 Discrete data

Data collected as integers (whole numbers) is known as discrete data.

For example, number of employees in a company, number of bedrooms in buildings, number of children per family, number of cars per family etc.

17.2.2 Continuous data

Unlike discrete data, continuous data does not increase in jumps, but can have any value between the given limits. For example, cost of building materials, labour costs, daily temperatures etc.

17.2.3 Raw data

Data shown in the manner in which it was collected is known as raw data. For example, the following data shows the crushing strength of 50 concrete cubes and is shown as it was obtained; it has not been re-arranged into any order:

46	40	37	40	35	40	42	43	40	45	39	38
46	45	44	50	45	35	39	38	35	48	37	42
50	39	46	41	44	41	51	42	47	49	36	47
53	48	49	50	37	44	44	43	51	54	53	42
38	41 N/mm^2										

17.2.4 Grouped data

Sometimes we have to deal with a very large set of data, therefore, the data may be arranged in groups or classes so that the analysis and then the presentation of results become convenient. The data given in section 17.2.3 show that the crushing strengths of concrete cubes vary between 35 N/mm^2 and 54 N/mm^2. In order to present and understand the information easily, the data may be arranged in groups. The number of groups will depend on the amount of the data; for a small set of data, i.e. 25 to 50 items, the number of groups may be 5 to 10. For larger amounts of data the upper limit should be 20. The following groups may be used for the data in section 17.2.3:

35–37, 38–40, 41–43, 44–46, 47–49, 50–52, 53–55

A group is called a class and each class is specified by 2 limits, the lower class limit and the upper class limit. For the first class, 35 is the lower and 37 the upper class limit.

Example 17.1

The crushing strengths of 50 concrete cubes are given in section 17.2.3. Group the data into 7 classes and find the frequency of each class.

Solution:

The minimum and the maximum strengths are 35 N/mm^2 and 54 N/mm^2 respectively. The 7 classes and their frequencies are shown in Table 17.1. A tally is prepared using a vertical bar for each value and for the fifth value a crossing diagonal/horizontal bar is used.

Table 17.1

Class interval	Tally	Frequency
35–37	𝍸 II	7
38–40	𝍸 𝍸	10
41–43	𝍸 III	8
44–46	𝍸 𝍸	10
47–49	𝍸 I	6
50–52	𝍸 I	6
53–55	III	3
		Total = 50

17.3 Averages

Having collected the data, it is often necessary to calculate the average result. In statistics there are 3 types of average, i.e. mean, mode and median.

The mean, also called the arithmetic mean, is calculated by dividing the sum of all values of the data by the number of values:

$$\text{Mean} = \frac{\text{Sum of all values of data}}{\text{Number of values in the data}}$$

This can also be written as: $\bar{x} = \frac{\Sigma x}{n}$

Where $\bar{x}$ (read as x bar) is the mean value

Σx means the sum of all values of the data.

n denotes the number of values in the data set.

In the case of grouped data it is necessary to determine the class mid-point for each class. The class mid-point is the average of the lower class boundary and the upper class boundary. For grouped data:

$$\text{Mean, } \bar{x} = \frac{\Sigma fx}{\Sigma f}$$

Where, x = class mid-point

f = frequency of each class.

The number that occurs most often in a set of data is called its mode. For grouped data, the class having the highest frequency is called the modal class. In some cases the data may not have a clear mode or there may be more than one mode.

If the data is arranged in ascending or descending order, the middle number is called the median.

17.3.1 Comparison of mean, mode and median

The mean is the most commonly used average and is the only average that involves all values in the data set, but its value is easily affected if some of the values are extremely high or low as compared to the rest of the data. It can also give an answer that may be impossible, e.g. 2.4 children per family.

The median is not affected by the extreme values as its calculation does not involve all data values. However, a large data set is required for the median value to be reliable.

The mode is easy to determine and is useful where the mean gives meaningless results. Some data sets may show no mode or more than 1 mode, and like the median, the mode does not involve all values of the data.

Example 17.2

The compressive strengths of 13 bricks are given below. Calculate the mean, median and mode strengths.

33, 31, 29, 30, 40, 36, 39, 32, 41, 37, 35, 36 and 38 N/mm^2

Calculate the mean, mode and median strengths.

Solution:

a) Mean, $\bar{x} = \frac{\Sigma x}{n}$

$$= \frac{33+31+29+30+40+36+39+32+41+37+35+36+38}{13}$$

$$= \frac{457}{13} = \mathbf{35.15\ N/mm^2}$$

The mean can be calculated using the calculator functions. The procedure given here is based on Casio fx-83GT PLUS, for other models the user manual should be referred to.

Press Mode and select 2 for statistics. Select 1 for '1-VAR' and then start inputting the numbers (data). Remember to press '=' after each number.

Now press AC to clear the table. Press shift and 1. From the list of options, press 4 for VAR and 3 for $\bar{x}$. $\bar{x}$ is the mean.

Answer from calculator operation = **35.15 N/mm^2**

b) The data has been arranged in an ascending order to find the median (middle number):

29, 30, 31, 32, 33, 35, 36, 36, 37, 38, 39, 40, 41

The middle number is 36, therefore the median = **36 N/mm^2**

c) Number 36 occurs twice, all other numbers occur once. Therefore, the mode = **36 N/mm^2**

Example 17.3

Consider the grouped data of example 17.1 and calculate the mean compressive strength.

Solution:

In the case of grouped data the actual values are not known, therefore, it is necessary to determine the class mid-point for each class, as shown in Table 17.2. The class mid-point gives the average value of the data within a class.

For grouped data: Mean, $\bar{x} = \frac{\Sigma fx}{\Sigma f}$, where x = class mid-point

f = frequency of each class.

Table 17.2

Class interval	Class mid-point (x)	Frequency (f)	F × x
35–37	36	7	36 × 7 = 252
38–40	39	10	39 × 10 = 390
41–43	42	8	42 × 8 = 336
44–46	45	10	45 × 10 = 450
47–49	48	6	48 × 6 = 288
50–52	51	6	51 × 6 = 306
53–55	54	3	54 × 3 = 162
		$\Sigma f = 50$	$\Sigma fx = 2184$

$$\text{Mean, } \bar{x} = \frac{\Sigma fx}{\Sigma f} = \frac{2184}{50} = \mathbf{43.68\ N/mm^2}$$

17.4 Statistical diagrams

After the collection and analysis of data it becomes necessary to present the results. Different types of statistical diagrams may be used to present the results so that the technical as well as non-technical readers can understand the information and the obvious trends. Some commonly used diagrams are: pictograms, bar charts, pie charts and line graphs.

Pictograms use different types of pictures to show the information. The pictures should be able to represent all features of the data.

Bar charts consist of data represented in the form of vertical or horizontal bars of equal width with some space in-between 2 bars. Their heights or lengths depend on the quantity they represent. A bar chart may consist of single bars, multiple bars, component bars, Gantt-charts and back-to-back bars.

A pie chart is a circle of convenient diameter which is divided into a number of sectors, each sector representing 1 portion of the data. The angle of each sector is calculated in accordance with the quantity they represent; the total angle of all portions of the data should be 360°. Several types of pie charts are used, e.g. simple pie, exploded pie, 3-dimensional pie etc.

Line graphs are invariably used to present the data from experimental work, and have already been dealt with in chapter 5.

Example 17.4

A civil engineering company employs 4 directors, 10 civil engineers, 5 structural engineers, 10 technician engineers, 6 technicians and 5 administrative staff. Represent the data as:

a) a vertical bar chart, b) a horizontal bar chart c) a pie chart

Solution:

a) In Figure 17.1a the x-axis represent the staff and the y-axis represent their numbers or frequency. The width of a bar is not important, but to avoid confusion all bars in a chart should be of the same width.

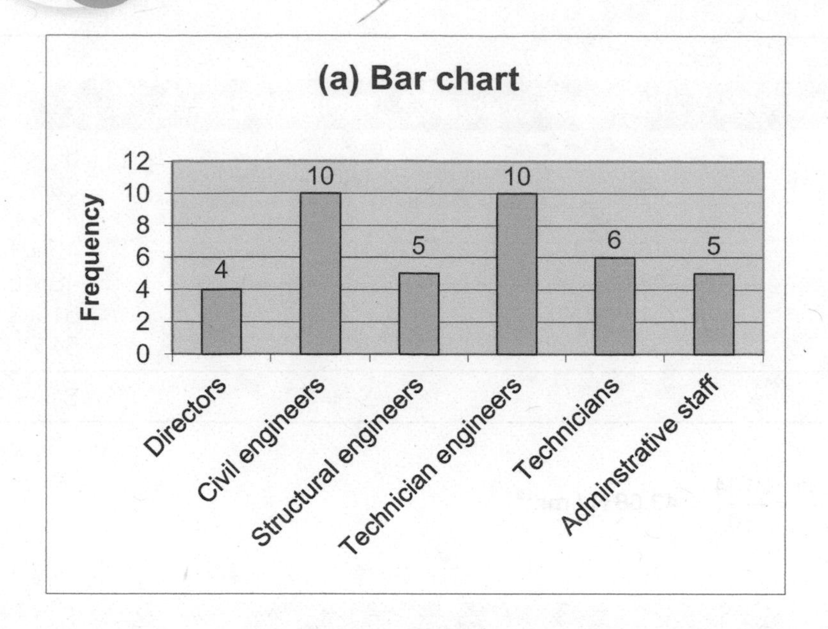

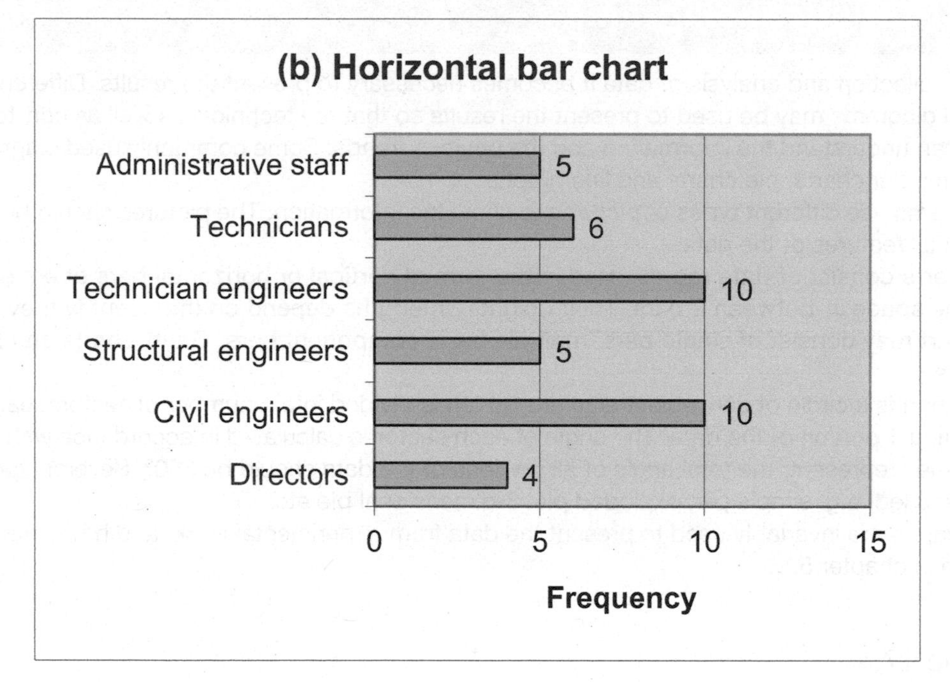

Figure 17.1a and 17.1b

b) In a horizontal bar chart (Figure 17.1b) the length of a bar is proportional to the number or frequency it represents.

c) The total number of employees is 40, which the pie (circle) will represent. The total angle at the centre of a circle is 360°, therefore 9° will represent one employee $\left(\frac{360}{40} = 9°\right)$. Table 17.3 shows the angles representing the different categories of employees.

Table 17.3

Category of employee	Number of employees	Angle representing each category
Directors	4	$= 9 \times 4 = 36°$
Civil engineer	10	$= 9 \times 10 = 90°$
Structural engineers	5	$= 9 \times 5 = 45°$
Technician engineers	10	$= 9 \times 10 = 90°$
Technicians	6	$= 9 \times 6 = 54°$
Administrative staff	5	$= 9 \times 5 = 45°$

The pie chart can be drawn manually using the angles for the relevant sector, however the chart shown in Figure 17.1c has been produced with the Microsoft Excel program.

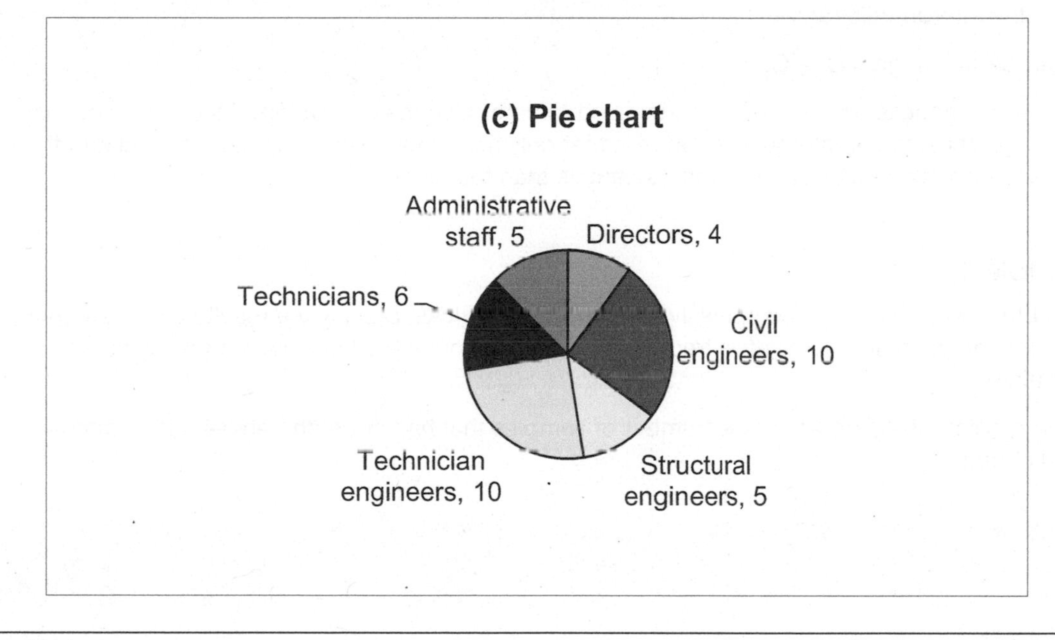

Figure 17.1c

17.5 Frequency distributions

The frequency of a particular observation is the number of times it occurs in the survey. The frequency distribution of the data may be represented by frequency tables, histograms, frequency polygons and cumulative frequency polygons. The frequency tables are the same as tally charts, as shown in Table 17.1. A histogram is similar to a vertical bar chart except that the bars are without any gaps. In a histogram (Figure 17.2) scales are used for both axes, which means that it can be used to display continuous data. The area of a bar is equal to the frequency multiplied by the class interval, hence the bars of a histogram can be of different widths. The mode of grouped data can be determined from a histogram.

A frequency polygon can also be used to show a frequency distribution. It is drawn by plotting the frequencies at the mid-points of the class intervals, and joining the points by straight lines (Figure 17.3). As a polygon should be closed, extra class intervals are added and the straight lines extended to meet the x-axis.

A cumulative frequency curve, also known as an ogive, may also be used to represent a frequency distribution. The cumulative frequency for each class is obtained by adding all frequencies up to that class and plotted against corresponding upper class boundaries. The points are joined by a smooth curve to get the cumulative frequency curve or ogive, as shown in example 17.5. Cumulative frequency curves may be used to determine the median and the dispersion of the data. The values, which divide the data into 4 equal parts, are called quartiles and denoted by Q_1, Q_2 and Q_3.

Q_1 = the lower quartile or first quartile

Q_2 = the middle quartile or second quartile

Q_3 = the upper quartile or third quartile. The difference between the upper and lower quartiles is called the interquartile range:

Interquartile range = $Q_3 - Q_1$

Interquartile range is one of the measures used to determine the dispersion of the data. The main advantage of using the interquartile range is that only the middle half of the data is used and thus this measure is considered to be more representative than the range.

Example 17.5

The table below shows the compressive strength of 52 bricks. Display the results as a histogram, frequency polygon and cumulative frequency polygon. From the cumulative frequency curve, determine:

a) the median of the data b) the number of samples that had strength between 41 N/mm^2 and 50 N/mm^2.

Compressive strength (N/mm^2)	35–37	38–40	41–43	44–46	47–49	50–52	53–55
Frequency	5	10	13	10	6	5	3

Solution:

Figure 17.2 shows the histogram in which the compressive strength is taken on the x-axis and the frequency on the y-axis. As the class intervals are the same, bars of equal width have been used. The first bar is produced by drawing vertical lines at the lower class boundary and the upper class boundary of the 35–37 class and drawing horizontal line at frequency equal to 5. Other bars are produced by repetition of this process.

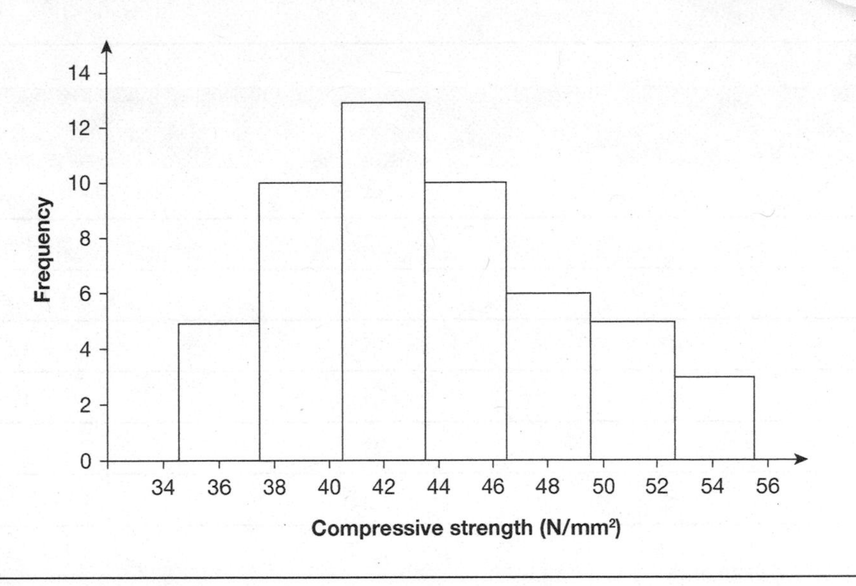

Figure 17.2 Histogram

For the frequency polygon the points are plotted between class mid-point and the frequency for each class. The points are joined by straight lines. Extra class intervals are added so that a closed polygon is produced by drawing 2 lines, as shown by dashed lines in Figure 17.3.

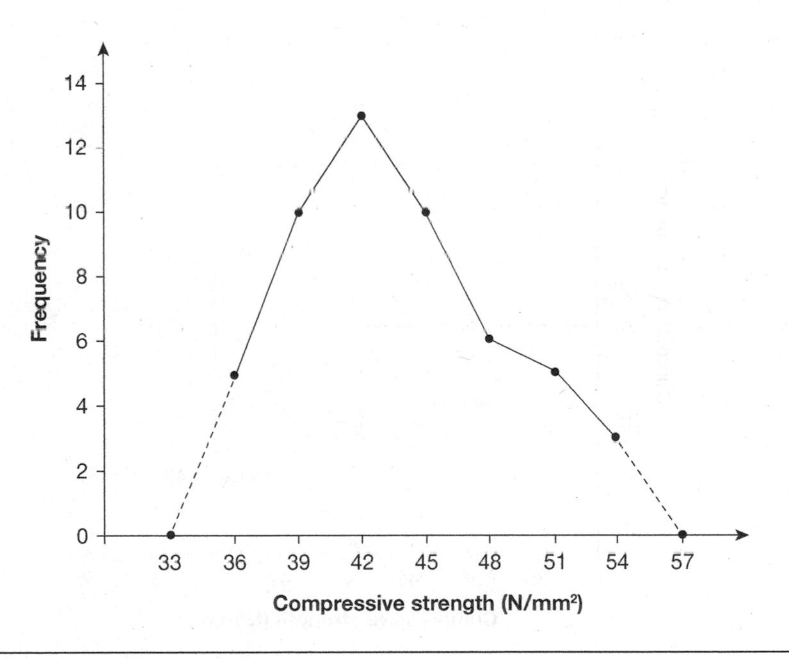

Figure 17.3 Frequency polygon

For producing the cumulative frequency curve we need to find the cumulative frequency for each class, as shown in Table 17.4.

Table 17.4

Class interval (Moisture content)	Frequency	Moisture content – less than	Cumulative Frequency
		34.5	0
35–37	5	37.5	5
38–40	10	40.5	5 + 10 = 15
41–43	13	43.5	15 + 13 = 28
44–46	10	46.5	28 + 10 = 38
47–49	6	49.5	38 + 6 = 44
50–52	5	52.5	44 + 5 = 49
53–55	3	55.5	49 + 3 = 52

The cumulative frequencies are plotted against the upper class boundaries and a smooth curve is drawn through all points (Figure 17.4). The cumulative frequency is divided into 4 parts by Q_1, Q_2 and Q_3.

a) Q_2 = Median = **43.1 N/mm²**

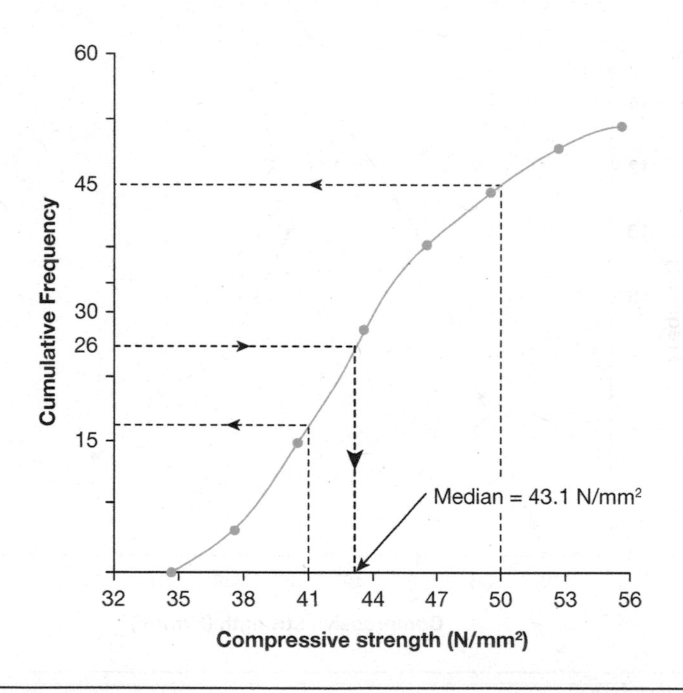

Figure 17.4 Cumulative frequency curve

b) The cumulative frequencies corresponding to 41 N/mm² and 50 N/mm² are 16.9 and 45 respectively. Therefore, the number of samples having strength between 41 and 50 N/mm² are:

45 – 16.9 = 28.1, say **28.**

17.6 Measures of dispersion

The range is the simplest method to find the spread or dispersion of data and is defined as the difference between the highest and the lowest values.

Range = maximum value – minimum value

The interquartile range contains the middle 50% of the data, therefore the minimum and maximum values which are not representative, are ignored. For grouped data the interquartile range can be determined from the value of Q_3 and Q_1 that can be obtained from the cumulative frequency curve.

Interquartile range = $Q_3 - Q_1$

The mean deviation uses all values of the data and is determined by finding the mean of the deviations from the mean, ignoring the negative signs:

$$\text{Mean deviation} = \frac{\Sigma(x - \bar{x})}{\Sigma f}$$

The standard deviation overcomes the problem of signs encountered in the calculation of mean deviation. The values of $(x - \bar{x})$ are squared to get rid of the negative signs of some of the values. Finally the square root is determined to cancel out the squaring done earlier.

$$\text{Variance} = \text{mean of the squared deviations} = \frac{\Sigma(x - \bar{x})^2}{n}$$

$$\text{Standard deviation} = \sqrt{(\text{variance})} = \sqrt{\frac{\Sigma(x - \bar{x})^2}{n}}$$

$$\text{For grouped data, standard deviation} - \sqrt{\frac{\Sigma f(x - \bar{x})^2}{\Sigma f}}$$

For grouped data, the mid-class value is taken as representative of each group, as done in example 17.3. A low value of standard deviation means that the data points are very close to the mean whereas high standard deviation indicates that the data points are dispersed over a large range of values.

Example 17.6

Consider the grouped data of example 17.5, which has 35 N/mm² and 54 N/mm² as the minimum and maximum compressive strengths. Calculate the range and the interquartile range.

Solution:

Range = maximum value – minimum value

= 54 – 35 = **19 N/mm²**

For the interquartile range the data from Table 17.4 been used to produce Figure 17.5. From Figure 17.5:

Interquartile range = $Q_3 - Q_1$

$Q_1 = 40$ and $Q_3 = 46.8$

Hence interquartile range = 46.8 – 40 = **6.8 N/mm²**

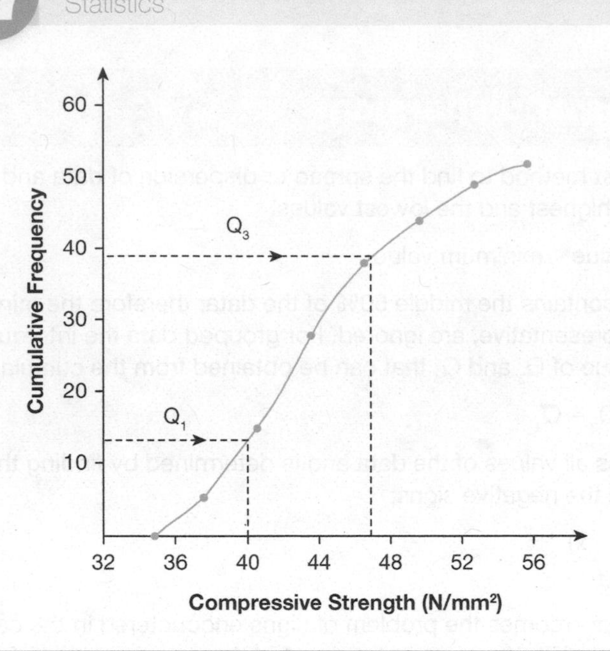

Figure 17.5

17.6.1 Standard deviation

Standard deviation is one of the methods to determine the dispersion of data, however, it measures the spread of the data about the mean value.

Consider 6 bricks having strengths of 40, 44, 45, 47, 50 and 50 N/mm². The mean strength of these bricks is 46 mm². The deviation of each value from the mean is shown in Figure 17.6.

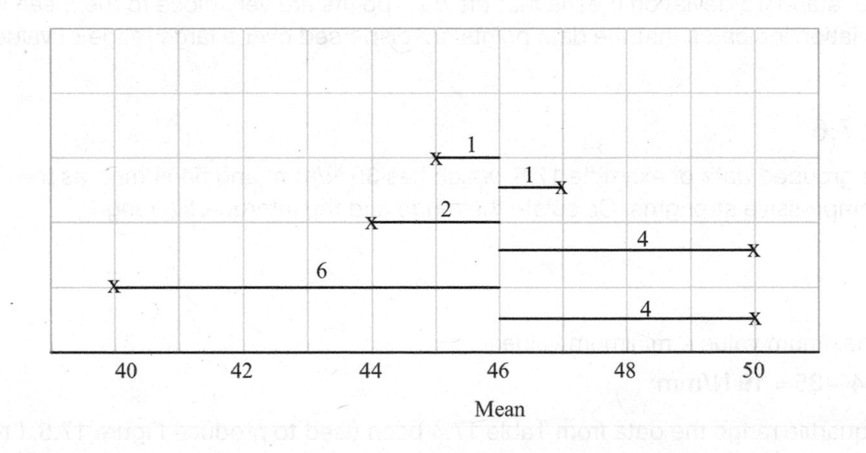

Figure 17.6

The data and their deviation from the mean are also shown in Table 17.5.

Table 17.5

Strength (mm²)	Mean ($\bar{x}$)	Deviation ($x - \bar{x}$)
40	46	40 – 46 = – 6
44	46	44 – 46 = – 2
45	46	45 – 46 = – 1
47	46	47 – 46 = 1
50	46	50 – 46 = 4
50	46	50 – 46 = 4

The average of the deviations from the mean, known as **mean deviation**, is calculated by ignoring their negative signs:

$$\text{Mean deviation} = \frac{6+2+1+1+4+4}{6} = 3$$

Another method to get positive values of the deviations is to square these values and find their mean. The mean of the squares of the deviations is called the **variance**.

$$\text{Variance} = \frac{\Sigma(x-\bar{x})^2}{n}$$

A column is added to Table 17.5 to show the squares of the deviations in Table 17.6:

Table 17.6

Strength (N/mm²)	Mean ($\bar{x}$)	Deviation ($x - \bar{x}$)	$(x - \bar{x})^2$
40	46	40 – 46 = – 6	36
44	46	44 – 46 = – 2	4
45	46	45 – 46 = – 1	1
47	46	47 – 46 = 1	1
50	46	50 – 46 = 4	16
50	46	50 – 46 = 4	16
			$\Sigma = 74$

$$\text{Variance} = \frac{\Sigma(x-\bar{x})^2}{n} = \frac{74}{6} = 12.33$$

This result is totally different from the mean deviation. However if we take square root of 12.33, the answer is 3.5 which is close to the mean deviation. The square root of variance is called the **standard deviation** (σ).

$$\text{Standard deviation, } \sigma = \sqrt{12.33} = \mathbf{3.51\ N/mm^2}$$

Example 17.7

10 bricks were tested to determine their crushing strength. The results are:

40, 35, 31, 32, 36, 38, 38, 40, 34, 36 N/mm²

a) Calculate the standard deviation of the results.

b) Use the functions of a scientific calculator to verify the answer.

Solution:

a) Mean strength = 36 N/mm²

The deviations from the mean and their squares are shown in Table 17.7.

Table 17.7

Sample No.	Strength (x)	Mean ($\bar{x}$)	$(x-\bar{x})$	$(x-\bar{x})^2$
1	40	36	4	16
2	35	36	−1	1
3	31	36	−5	25
4	32	36	−4	16
5	36	36	0	0
6	38	36	2	4
7	38	36	2	4
8	40	36	4	16
9	34	36	−2	4
10	36	36	0	0
				$\Sigma = 86$

Standard deviation, $\sigma = \sqrt{\frac{\Sigma(x-\bar{x})^2}{n}} = \sqrt{\frac{86}{10}} = \mathbf{2.93 N/mm^2}$

b) The procedure given here is based on Casio fx-83GT PLUS.

Press Mode and select 2 for statistics. Select 1 for '1-VAR' and then start inputting the numbers (data). Remember to press '=' after each number.

Now press AC to clear the table. Press shift and 1. From the list of options, press 4 for VAR and 3 for σx. σx is the standard deviation.

Answer from calculator operation = **2.93 N/mm²**

Example 17.8

Find the standard deviation of the data given in example 17.5, and compare the answer to the interquartile range and range.

Solution:

Table 17.8 shows the different values that are needed for calculating the standard deviation.

Table 17.8

Moisture content (class interval	Frequency (f)	Class mid-point (x)	$f \times x$	Mean ($\bar{x}$)	$(x - \bar{x})$	$(x - \bar{x})^2$	$f \times (x - \bar{x})^2$
35–37	5	36	180	43.7	– 7.7	59.29	296.45
38–40	10	39	390	43.7	– 4.7	22.09	220.9
41–43	13	42	546	43.7	– 1.7	2.89	37.57
44–46	10	45	450	43.7	1.3	1.69	16.9
47–49	6	48	288	43.7	4.3	18.49	110.94
50–52	5	51	255	43.7	7.3	53.29	266.45
53–55	3	54	162	43.7	10.3	106.09	318.27
	$\sum f = 52$		$\sum fx = 2271$				$\Sigma f \times (x - \bar{x})^2 = 1267.48$

$$\text{Mean} = \frac{\Sigma fx}{\Sigma f} = \frac{2271}{52} = 43.7 \text{ N/mm}^2$$

$$\text{Standard deviation, } \sigma = \sqrt{\frac{\Sigma f(x - \bar{x})^2}{\Sigma f}} = \sqrt{\frac{1267.48}{52}} = \mathbf{4.94 \text{ N/mm}^2}$$

Comparison: **From example 17.6:** Interquartile range = 7 N/mm^2

Range = highest value – lowest value = 19 N/mm^2

17.7 Distribution curves

Depending on the data, a frequency distribution curve can have several shapes; some are symmetrical about a central value whereas the others are not symmetrical (or asymmetric).

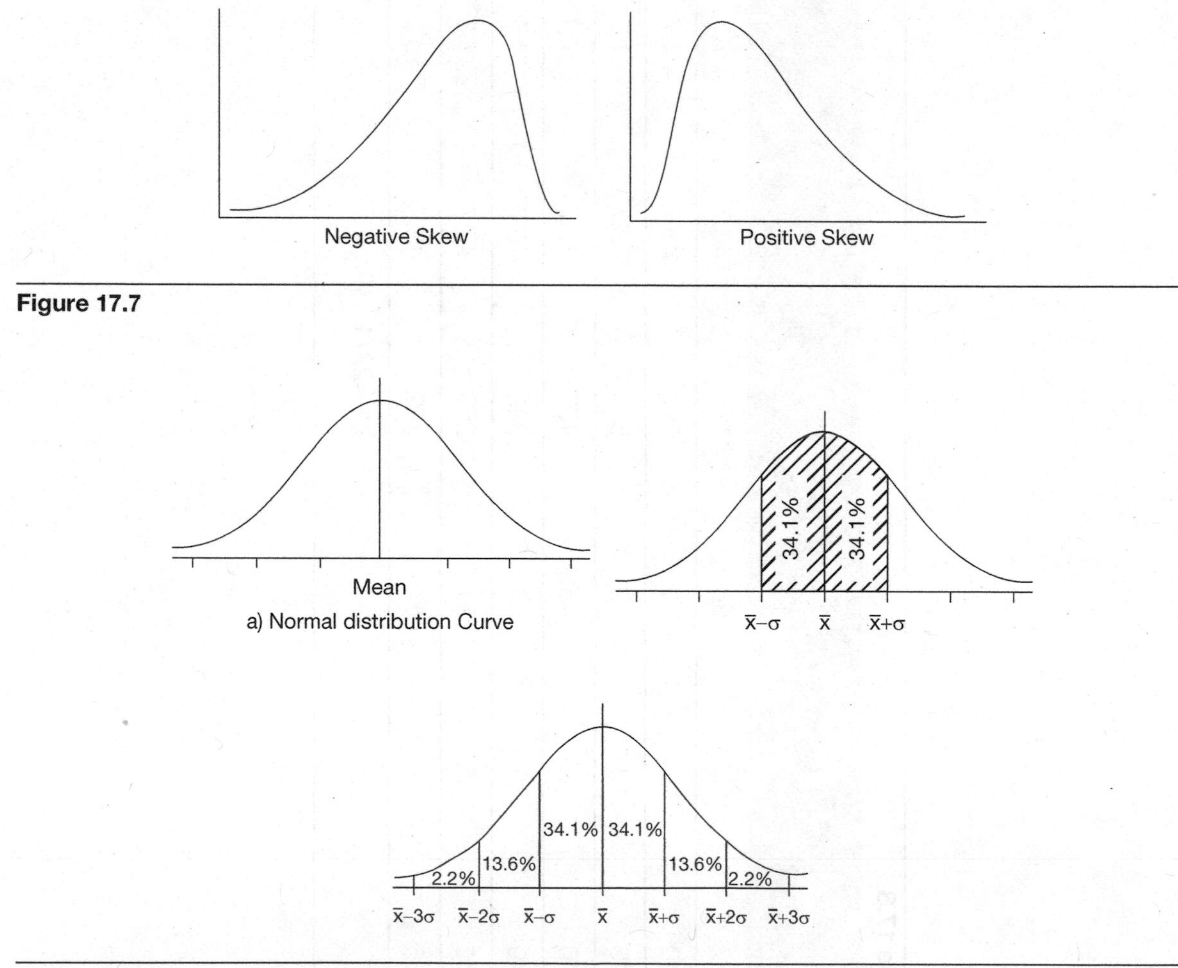

Figure 17.7

Figure 17.8

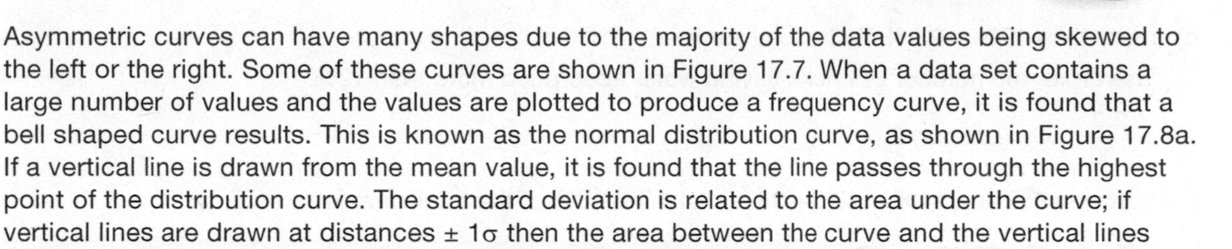

Asymmetric curves can have many shapes due to the majority of the data values being skewed to the left or the right. Some of these curves are shown in Figure 17.7. When a data set contains a large number of values and the values are plotted to produce a frequency curve, it is found that a bell shaped curve results. This is known as the normal distribution curve, as shown in Figure 17.8a. If a vertical line is drawn from the mean value, it is found that the line passes through the highest point of the distribution curve. The standard deviation is related to the area under the curve; if vertical lines are drawn at distances $\pm 1\sigma$ then the area between the curve and the vertical lines is 68.2% of the total area under the curve (Figure 17.8b). This also means that 68.2% of the total frequency also lies between these values of standard deviation. Similarly, areas at mean $\pm 2\sigma$ and mean $\pm 3\sigma$ are also shown in Figure 17.8c.

Example 17.9:

The mean water absorption of 99 bricks is 10%, and the standard deviation is 1.0%. Assuming that the results are normally distributed, determine how many bricks have water absorption:

a) less than 9%
b) more than 12%
c) between 8% and 11%

Solution:

Mean + 1 σ = 10 + 1 = 11%

Mean − 1 σ = 10 − 1 = 9%

Similarly Mean + 2 σ = 10 + 2 = 12%, and Mean − 2 σ = 10 − 2 = 8%

Mean + 3 σ = 10 + 3 = 13%, and Mean − 3 σ = 10 − 3 = 7%

These values are shown on the x-axis in Figure 17.9.

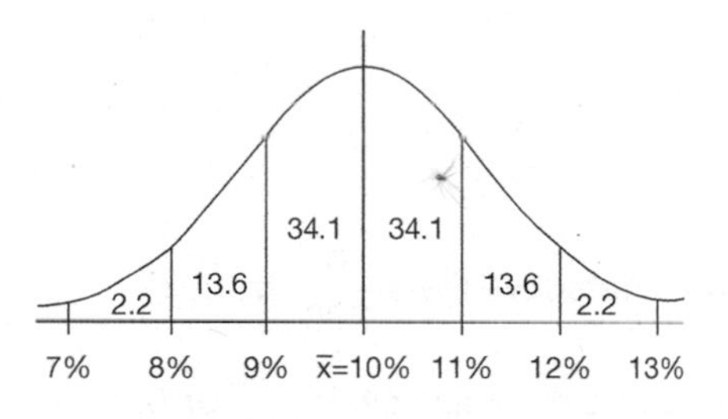

Figure 17.9

a) less than 9% = 2.2 + 13.6 = 15.8% of the total sample

Therefore, the number of bricks having water absorption less than 9%

$$= \frac{15.8}{100} \times 99 = 15.64, \text{say } \mathbf{16}$$

b) more than 12% = 2.2% of the total sample

Therefore, the number of bricks having water absorption more than 12%

$$=\frac{2.2}{100}\times 99=2.18,\text{say}\ \mathbf{2}$$

c) between 8% and 11% = 13.6 + 34.1 + 34.1 = 81.8% of the total sample

$$=\frac{81.8}{100}\times 99=80.98,\text{say}\ \mathbf{81}$$

17.8 Probability

The likelihood of an event happening is known as probability. A probability of 0 means that the event will not happen, and a probability of 1 means that the event is certain to happen. The result can be shown as a percentage, a fraction or a decimal number. When we know the probability of an event happening, we can use its value to predict the future results. Many organisations use the statistical figures from the past to predict what is likely to happen in the future and thus can plan their work in a more satisfactory manner.

$$\text{Probability}=\frac{\text{number of successful trials/outcomes}}{\text{total number of trials/outcomes}}$$

Example 17.10

There are 6 defective tiles in a sample of 200 floor tiles.

a) If 1 tile is picked at random, find the probability that it is:

i) a defective tile

ii) not a defective tile

b) If the first tile was defective and not put back into the sample, then find the probability that the second tile, picked at random, is:

i) a defective tile

ii) not a defective tile

Solution:

a) i) The probability of picking a defective tile is $\frac{6}{200}$ or 0.03

ii) The probability of not picking a defective tile is $\frac{194}{200}$ or 0.97

b) i) The probability of picking a defective tile is $\frac{5}{199}$ or 0.025

ii) The probability of not picking a defective tile is $\frac{194}{199}$ or 0.975

17.8.1 Mutually exclusive events (the OR rule)

If there are 2 mutually exclusive events A and B (i.e. either event can occur without the other event occurring), then the probability of either A or B occurring is equal to the sum of individual probabilities:

P (A or B) = P (A) + P (B)

Example 17.11

In a sample of 100 bricks there are 4 defective bricks, 1 overburnt brick and 95 good bricks. If 1 brick is picked from the sample, at random, calculate the probability that it is:

i) a defective brick
ii) a good brick
iii) a defective or overburnt brick

Solution:

i) The probability of picking a defective brick is, $\frac{4}{100}$ or 0.04

ii) The probability of picking a good brick is, $\frac{95}{100}$ or 0.95

iii) The probability of picking a defective or an overburnt brick is, $\frac{4}{100} + \frac{1}{100} = \frac{5}{100}$ or **0.05**

17.8.2 Independent events (the AND rule)

Two events, where the outcome of 1 event does not depend on the outcome of the other event, are known as independent events. If there are 2 independent events A and B, then the probability of both occurring is:

P (A and B) = P(A) × P(B)

Example 17.12

In a warehouse building 2 smoke alarms are situated very close to each other. The probability of each detecting smoke in the building is 0.85. What is the probability that the alarm will sound when there is smoke in the building?

Solution:

Let us suppose that there are alarms A and B in the building.

If N stands for 'not detected' and D for 'detected', then the possible outcomes for alarms A and B are: NN, ND, DN, DD.

The probability of a smoke alarm detecting smoke = 0.85
The probability of a smoke alarm not detecting smoke = 0.15
Probability of both alarms not detecting = NN = 0.15 × 0.15 = 0.0225
Probability of Alarm A sounding = DN = 0.85 × 0.15 = 0.1275
Probability of Alarm B sounding = ND = 0.15 × 0.85 = 0.1275
Probability of both alarms sounding = DD = 0.85 × 0.85 = 0.7225
Probability of at least 1 smoke alarm detecting smoke = 0.1275 + 0.1275 + 0.7225
= **0.9775**

Alternatively, no alarm sounding, NN = 0.0225
The probability of 1 alarm sounding = 1 − 0.0225 = 0.9775

17.8.3 Tree diagrams

The combined results of an investigation can be shown on a tree diagram when there are 2 or more independent events.

Example 17.13

A large batch of light bulbs, manufactured by Roshni Industries has 8% defective bulbs. If the quality control technician picks 3 bulbs at random, produce a tree diagram and hence find the probability of selecting, with replacement:

a) no defective bulb
b) 1 defective bulb
c) 2 defective bulbs

Solution:

Let, D = defective bulb; N = not defective bulb. The tree diagram is shown in Figure 17.10.

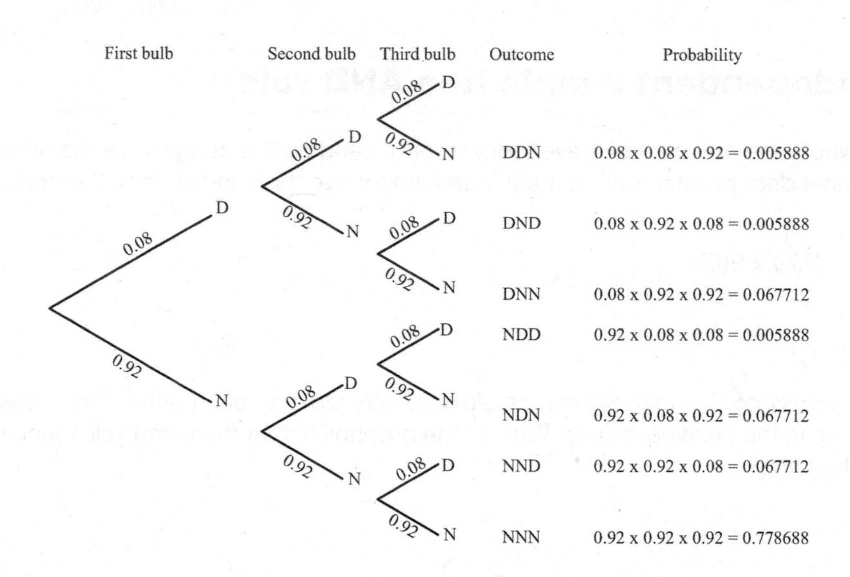

Figure 17.10 Tree diagram

a) From the tree diagram, probability of no defective bulb = NNN = **0.7787**
b) 1 defective bulb = DNN + NDN + NND = (0.067712) × 3 = **0.2031**
c) 2 defective bulbs = DDN + DND + NDD = (0.005888) × 3 = **0.0177**

17.9 Binomial distribution

The binomial distribution deals with the probability of a success or failure outcome in an experiment/investigation that is repeated several times. There are only 2 possible outcomes, for example when a coin is tossed there are only 2 possible outcomes, heads or tails. If p is the probability of getting a head and q is the probability of getting a tail, then $p + q = 1$. When a coin is tossed several times, the probabilities can be determined using a tree diagram. However, as the number of tosses increases, the difficulty of using a tree diagram increases as well. The use of binomial distribution is much easier and will give the probability of discrete events, provided:

- the number of trials is constant
- each trial or observation can result in only 2 possible outcomes, i.e. 1 of the outcomes is known as a success and the other a failure.
- the probability of success or failure is the same in each trial
- the outcome of 1 trial (success or failure) does not affect the outcome of other trials.

The probabilities of getting 0, 1, 2, 3, 4 .... successes in n trials are given by the binomial expansion of $(q + p)^n$:

$$(q+p)^n = q^n + nq^{n-1}p + \frac{n(n-1)}{2!}q^{n-2}p^2 + \frac{n(n-1)(n-2)}{3!}q^{n-3}p^3 + \ldots\ldots + p^n$$

Example 17.14

In a compression test involving concrete cubes, it is found that the 5% fail to meet the required strength. If 7 cubes are selected at random, determine the probabilities that:

a) all cubes have the required strength
b) the strength of 1 cube is less than the required strength
c) the strength of 2 cubes is less than the required strength

Solution:

The probability of a cube failing to meet the required strength, p, is 5% or 0.05.

The probability of a cube having the required strength, q, is 1 – 0.05, or 0.95

The probability of 0, 1, 2, 3, ... cubes failing to meet the required strength is given by the successive terms of $(q + p)^7$:

$$(q + p)^7 = q^7 + 7q^{7-1}p + \frac{7(7-1)}{2!}q^{7-2}p^2 + \frac{7(7-1)(7-2)}{3!}q^{7-3}p^3 + \ldots\ldots$$

$$= q^7 + 7q^6 p + 21q^5p^2 + 35q^4p^3 + \ldots\ldots$$

a) The probability of all cubes having the required strength $= q^7 = 0.95^7 =$ **0.6983**
b) The probability of 1 cube having strength less than the required $= 7q^6 p$
$= 7 \times 0.95^6 \times 0.05 =$ **0.2573**
c) The probability of 2 cubes having strength less than the required $= 21q^5 p^2$
$= 21 \times 0.95^5 \times 0.05^2 =$ **0.0406**

17.10 Poisson distribution

When the number of trials, n, is very large (> 50) and the probability of an event occurring is very small ($p < 0.1$), it is not easy and very time consuming to evaluate the terms of binomial distribution. In this situation, Poisson distribution can be used to get a good approximation to a binomial distribution, provided the following conditions are satisfied:

- the events are independent
- 2 or more events cannot occur simultaneously
- the mean number of events in a given interval is constant

The average number of events in an interval is $\bar{x}$ and is equal to the product of n and p. The Binomial and Poisson distributions are both discrete probability distributions, and in some circumstances they are very similar. The probabilities that an event will happen 0, 1, 2, ….. n times in n trials are given by the terms of the expression:

$$e^{-\bar{x}}\left[1+\frac{\bar{x}}{1!}+\frac{(\bar{x})^2}{2!}+\frac{(\bar{x})^3}{3!}+......\right]$$

$$1 = P(0) + P(1) + P(2) + P(3) + ...$$

Example 17.15

A company manufactures bolts for structural steel frames, and finds that 2% are defective. They are packed in boxes containing 100 bolts. What is the probability that each box will contain a) 0 b) 1 c) 2 defective and d) more than 2 defective bolts.

Solution:

$$\bar{x} = np = 100 \times 0.02 = 2$$

a) Probability of 0 defective $= P(0) = e^{-\bar{x}} = e^{-2} = \mathbf{0.1353}$

b) Probability of 1 defective $= P(1) = e^{-\bar{x}}\frac{\bar{x}}{1!} = e^{-\bar{x}}\frac{\bar{x}}{1!} = e^{-2}\frac{2}{1!} = \mathbf{0.2706}$

c) Probability of 2 defective $= P(2) = e^{-\bar{x}}\frac{(\bar{x})^2}{2!} = e^{-\bar{x}}\frac{(\bar{x})^2}{2\times 1} = e^{-2}\frac{2^2}{2\times 1} = \mathbf{0.2706}$

d) Probability of more than 2 defective = 1 – [P(0) + P(1) + P(2)]

= 1 – [0.1353 + 0.2706 + 0.2706]

= 1 – 0.6765 = **0.3235**

17.11 Normal distribution

When the number of trials is very large, the frequency polygon shown in Figure 17.3 will become a curve that has a characteristic bell shape, as shown in Figure 17.8a. A continuous curve cannot be drawn for a discrete variable such as number of goals scored by a team, but it can be drawn for a continuous variable such as height of people. The curve is symmetrical about the mean value of the data, and is also the highest at the mean value. The curve tends to zero as the value of x increases or decreases from the mean and can be described in terms of its mean and standard deviation. The equation of the normal curve is:

$$y = \frac{1}{\sigma\sqrt{2\pi}} e^{-(x-\bar{x})^2/2\sigma^2}$$

where x is the variable, $\bar{x}$ is the mean value and σ is the standard deviation.

The shape of the curve and the area enclosed by the curve depend on the value of standard deviation. The area under the curve can be calculated by integration, however, tables of results are available (see Table 17.9) to make the calculations much easier. To avoid the use of several tables, as the area under the curve depends on the value of σ, the curve is standardised in terms of the value $(x-\bar{x})/\sigma$, which commonly is designated by the symbol z. The total area under the curve between $z = -\infty$ and $z = +\infty$ is equal to unity, and the area between any 2 values, e.g. z_1 and z_2 represents the probability of the variable occurring between these values.

Example 17.16

The mean crushing strength of 120 concrete cubes is 24 N/mm² and the standard deviation of the results is 3 N/mm². Assuming the results are normally distributed, calculate the number of concrete cubes likely to have:

a) strength < 20 N/mm²

b) strength between 22 and 30 N/mm²

c) strength > 25 N/mm²

Solution:

Mean crushing strength, $\bar{x} = 24\ N/mm^2$. Its z value is 0 on the standardized normal curve

$$(z = \frac{x-\bar{x}}{\sigma} = \frac{24-24}{3} = 0;\ \text{here x is also} = 24)$$

a) z-value of $20 N/mm^2 = \frac{x-\bar{x}}{\sigma} = \frac{20-24}{3} = -\frac{4}{3} = -1.33$

The negative value shows that it lies to the left of z = 0, i.e. the mean value.

The area between z = 0 and z = – 1.33, from Table 17.9 is 0.4082

The area under the curve (Figure 17.11a) up to the z-value of – 1.33 represents the number of concrete cubes having strength < 20 N/mm². This is equal to:

0.5 – 0.4082 = 0.0918. Therefore the number of samples having crushing strength < 20 N/mm² is: 120 × 0.0918 = **11 samples.**

b) z-value of $22\ N/mm^2 = \frac{x-\bar{x}}{s} = \frac{22-24}{3} = -\frac{2}{3} = -0.67$

z-value of $30\ N/mm^2 = \frac{x-\bar{x}}{\sigma} = \frac{30-24}{3} = \frac{6}{3} = 2$

Using Table 17.9, z-value of – 0.67 corresponds to an area of 0.2486 between the mean value and the vertical line at – 0.67. The negative value shows that it lies to the left of z = 0, i.e. the mean value.

Table 17.9 Areas under the standard normal curve

z	0	1	2	3	4	5	6	7	8	9
0.0	0.0000	0.0040	0.0080	0.0120	0.0159	0.0199	0.0239	0.0279	0.0319	0.0359
0.I	0.0398	0.0438	0.0478	0.0517	0.0557	0.0596	0.0636	0.0678	0.0714	0.0753
0.2	0.0793	0.0832	0.0871	0.0910	0.0948	0.0987	0.1026	0.1064	0.1103	0.1141
0.3	0.1179	0.1217	0.1255	0.1293	0.1331	0.1388	0.1406	0.1443	0.1480	0.1517
0.4	0.1554	0.1591	0.1628	0.1664	0.1700	0.1736	0.1772	0.1808	0.1844	0.1879
0.5	0.1915	0.1950	0.1985	0.2019	0.2054	0.2086	0.2123	0.2157	0.2190	0.2224
0.6	0.2257	0.2291	0.2324	0.2357	0.2389	0.2422	0.2454	0.2486	0.2517	0.2549
0.7	0.2580	0.2611	0.2642	0.2673	0.2704	0.2734	0.2760	0.2794	0.2823	0.2852
0.8	0.2881	0.2910	0.2939	0.2967	0.2995	0.3023	0.3051	0.3078	0.3106	0.3133
0.9	0.3159	0.3186	0.3212	0.3238	0.3264	0.3289	0.3315	0.3340	0.3365	0.3389
1.0	0.3413	0.3438	0.3451	0.3485	0.3508	0.3531	0.3554	0.3577	0.3599	0.3621
1.1	0.3643	0.3665	0.3686	0.3708	0.3729	0.3749	0.3770	0.3790	0.3810	0.3830
1.2	0.3849	0.3869	0.3888	0.3907	0.3925	0.3944	0.3962	0.3980	0.3997	0.4015
1.3	0.4032	0.4049	0.4066	0.4082	0.4099	0.4115	0.4131	0.4147	0.4162	0.4177
1.4	0.4192	0.4207	0.4222	0.4236	0.4251	0.4265	0.4279	0.4292	0.4306	0.4319
1.5	0.4332	0.4345	0.4357	0.4370	0.4382	0.4394	0.4406	0.4418	0.4430	0.4441
1.6	0.4452	0.4463	0.4474	0.4484	0.4495	0.4505	0.4515	0.4525	0.4535	0.4545
1.7	0.4554	0.4564	0.4573	0.4582	0.4591	0.4599	0.4608	0.4616	0.4625	0.4633
1.8	0.4641	0.4649	0.4656	0.4664	0.4671	0.4678	0.4686	0.4693	0.4699	0.4706
1.9	0.4713	0.4719	0.4726	0.4732	0.4738	0.4744	0.4750	0.4756	0.4762	0.4767
2.0	0.4772	0.4778	0.4783	0.4785	0.4793	0.4798	0.4803	0.4808	0.4812	0.4817
2.1	0.4821	0.4826	0.4830	0.4834	0.4838	0.4842	0.4846	0.4850	0.4854	0.4857
2.2	0.4861	0.4864	0.4868	0.4871	0.4875	0.4878	0.4881	0.4884	0.4887	0.4890
2.3	0.4893	0.4896	0.4898	0.4901	0.4904	0.4906	0.4909	0.4911	0.4913	0.4916
2.4	0.4918	0.4920	0.4922	0.4925	0.4927	0.4929	0.4931	0.4932	0.4934	0.4936
2.5	0.4938	0.4940	0.4941	0.4943	0.4945	0.4946	0.4948	0.4949	0.4951	0.4952
2.6	0.4953	0.4955	0.4956	0.4957	0.4959	0.4960	0.4961	0.4962	0.4963	0.4964
2.7	0.4965	0.4966	0.4967	0.4968	0.4969	0.4970	0.4971	0.4972	0.4973	0.4974
2.8	0.4974	0.4975	0.4976	0.4977	0.4977	0.4978	0.4979	0.4980	0.4980	0.4981
2.9	0.4981	0.4982	0.4982	0.4983	0.4984	0.4984	0.4985	0.4985	0.4986	0.4986
3.0	0.4987	0.4987	0.4987	0.4988	0.4988	0.4989	0.4989	0.4989	0.4990	0.4990
3.1	0.4990	0.4991	0.4991	0.4991	0.4992	0.4992	0.4992	0.4992	0.4993	0.4993
3.2	0.4993	0.4993	0.4994	0.4994	0.4994	0.4994	0.4994	0.4995	0.4995	0.4995
3.3	0.4995	0.4995	0.4995	0.4996	0.4996	0.4996	0.4996	0.4996	0.4996	0.4997
3.4	0.4997	0.4997	0.4997	0.4997	0.4997	0.4997	0.4997	0.4997	0.4997	0.4998
3.5	0.4998	0.4998	0.4998	0.4998	0.4998	0.4998	0.4998	0.4998	0.4998	0.4998
3.6	0.4998	0.4998	0.4999	0.4999	0.4999	0.4999	0.4999	0.4999	0.4999	0.4999
3.7	0.4999	0.4999	0.4999	0.4999	0.4999	0.4999	0.4999	0.4999	0.4999	0.4999
3.8	0.4999	0.4999	0.4999	0.4999	0.4999	0.4999	0.4999	0.4999	0.4999	0.4999
3.9	0.5000	0.5000	0.5000	0.5000	0.5000	0.5000	0.5000	0.5000	0.5000	0.5000

30 N/mm² has a z-value of 2 standard deviations. From Table 17.9 the area under the curve is 0.4772; the positive value shows that it lies to the right of the mean value, as shown in Figure 17.11b and 17.11c.

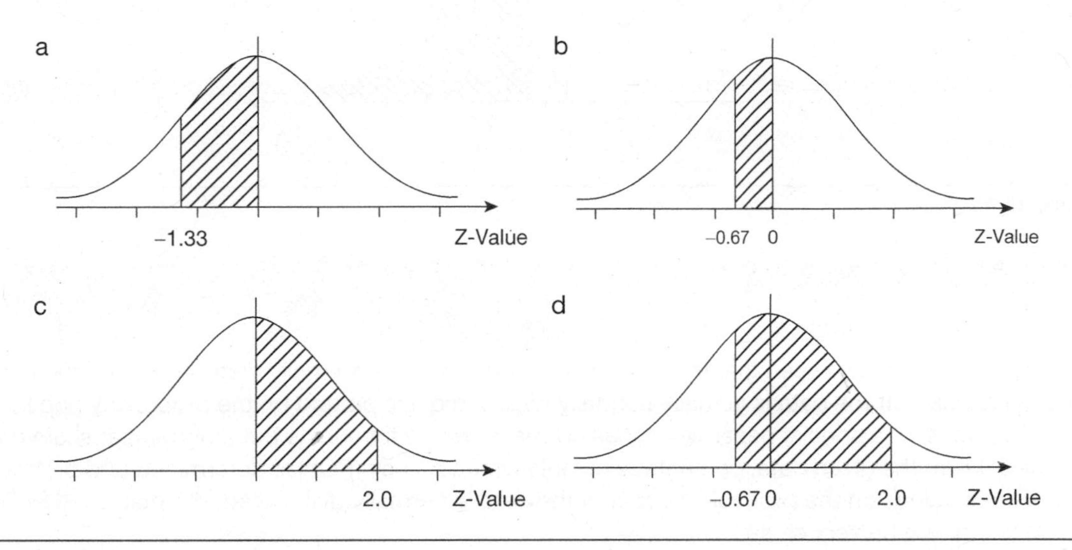

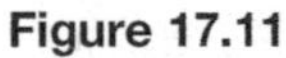
Figure 17.11

The total area under the curve is: 0.2486 + 0.4772 = 0.7258 (Figure 17.11d)

This area is directly proportional to probability. Thus the probability that the crushing strength of concrete cubes lies between 22 N/mm² and 30 N/mm² is **0.7258**.

Therefore, from 120 cubes, 120 × 0.7258 or **87** cubes are likely to have strength between 22 N/mm² and 30 N/mm².

c) z-value of 25 N/mm² $= \frac{x - \bar{x}}{s} = \frac{25 - 24}{3} = \frac{1}{3} = 0.33$

Using Table 17.9, z-value of 0.33 corresponds to an area of 0.1293 between the mean value and the vertical line at 0.33. The positive value shows that it lies to the right of z = 0, i.e. the mean value, as shown in Figure 17.12a. The shaded portion of Figure 17.12b corresponds to an area of 0.3707 (0.5 – 0.1293 = 0.3707). This area represents the probability that the crushing strength of a concrete cube is more than 25 N/mm². For 120 concrete cubes, the number of cubes having strength more than 25 N/mm² is:

120 × 0.3707 = 44.48, say **45**

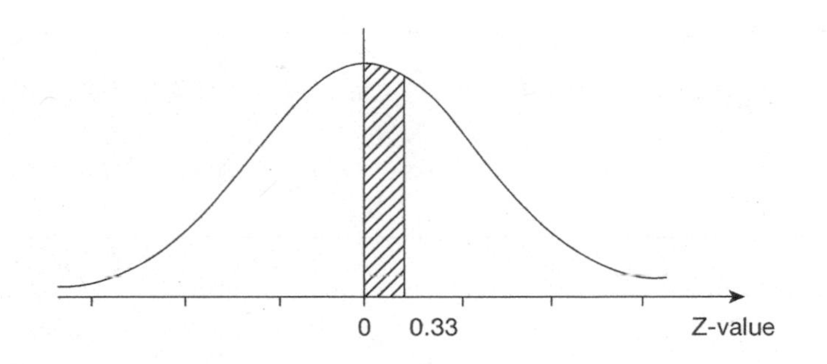

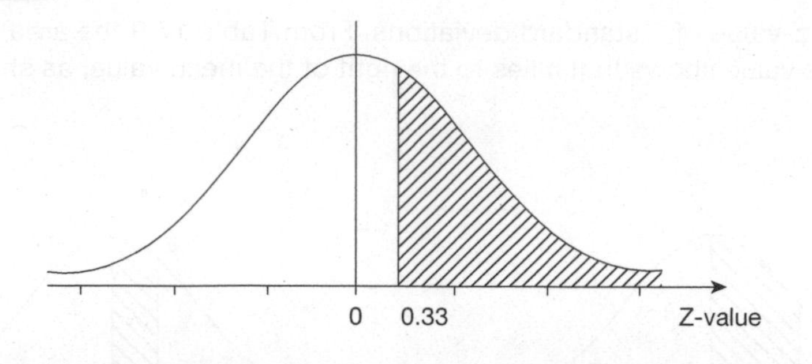

Figure 17.12

17.12 Normal distribution test

In example 17.5 we have seen that the data, which appears to be normally distributed, produces an ogive. To check that the data is actually normally distributed it is plotted on the probability paper. The probability paper is a special paper which has normal scale on the x-axis but a non-linear scale on the y-axis. From the given data, percentage cumulative frequency (PCF) is determined and plotted against the **x** values on the probability paper. If the data is normally distributed, the points either lie on a straight line or will be very close.

The mean occurs at 50% PCF

The standard deviation $= x_s - \bar{x}$, x_s is the x value at 84.13 PCF.

Example 17.17

The following data shows the compressive strength of 52 bricks. Use normal probability paper to check if the results have a normal distribution. If the data is normally distributed, use the graph to find the mean and the standard deviation.

Compressive strength (N/mm^2)	35–37	38–40	41–43	44–46	47–49	50–52	53–55
Frequency	5	10	13	10	6	5	3

Solution:

The cumulative frequency and percentage cumulative frequency are shown in Table 17.10.

Table 17.10

Class interval (Compressive strength)	Frequency	Cumulative frequency	Percentage cumulative frequency
35–37	5	5	$5\times100 \div 52 = 9.6$
38–40	10	$5 + 10 = 15$	28.8

Class interval (Compressive strength)	Frequency	Cumulative frequency	Percentage cumulative frequency
41–43	13	15 + 13 = 28	53.8
44–46	10	28 + 10 = 38	73.1
47–49	6	38 + 6 = 44	84.6
50–52	5	44 + 5 = 49	94.2
53–55	3	49 + 3 = 52	100

The percentage cumulative frequencies are plotted against the upper class boundaries and a best fit straight line drawn as illustrated in Figure 17.13. Since the points lie close to the straight line, the results may be considered to be normally distributed.

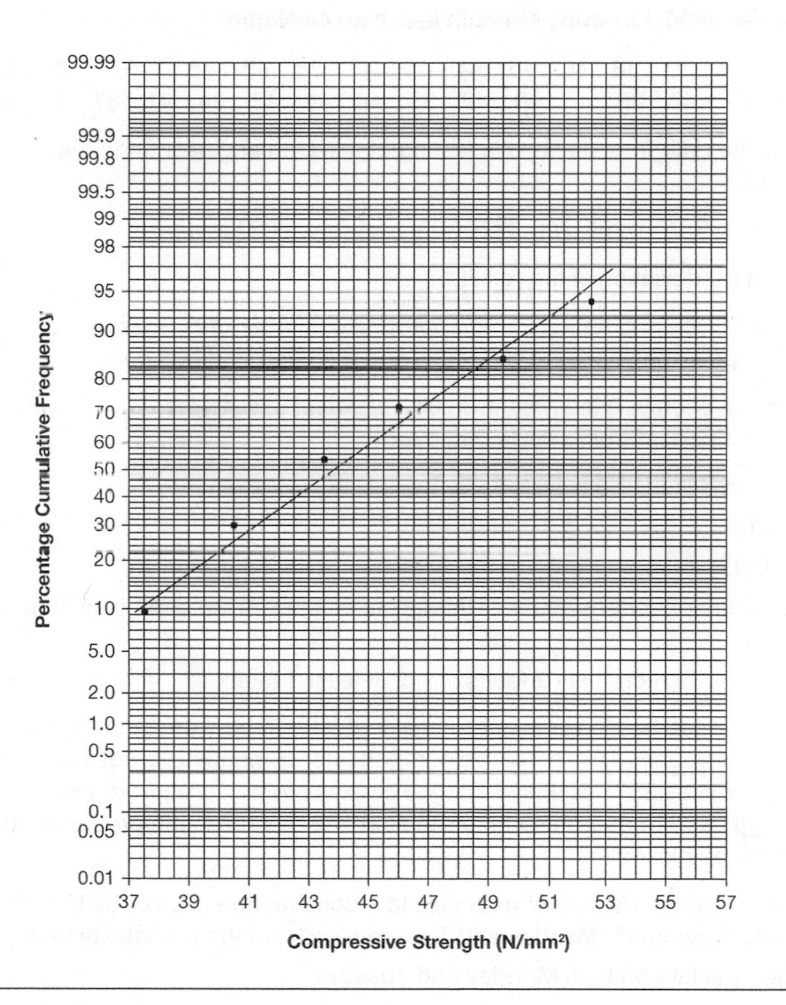

Figure 17.13

From the graph, Mean$(\bar{x})$ = **43.8N / mm²** (mean at 50% PCF)

x_s (at 84.13% PCF) = 48.9 N/mm²

Therefore, standard deviation = $x_s - \bar{x}$ = 48.9 – 43.8 = **5.1N / mm²**

Exercise 17.1

1. The moisture contents (%) of 12 samples of aggregates are shown below. Calculate the mean, mode, median and range:

   11.4, 13.0, 11.0, 12.2, 13.9, 10, 12.0, 12.2, 13.0, 12.3, 10.4 and 10.8

2. The crushing strengths (unit: N/mm²) of 40 bricks are given below. Group the data into 6 classes and find:
   a) the frequency of each class
   b) the mean crushing strength
   c) the median crushing strength from the grouped data
   d) the number of bricks having strength less than 44 N/mm²

   34 40 45 39 35 38 46 45 44 50 45 38 43 48 37 42 50 39 46 41
   44 41 51 42 47 36 47 48 49 38 44 44 43 34 41 37 42 40 42 40 N/mm²

3. On analysing the data from the tensile testing of steel, an engineer finds the values of Young's modulus to be:

   200, 198, 210, 215, 202, 195, 200, 205, 203, 212, 214, 203 kN/mm²

   Find the standard deviation of the results.

4. Group the data given in question 2 into appropriate number of classes and determine the standard deviation. Compare the answer to the range and the interquartile range.

5. A machine makes components for use in air-conditioning systems. The probabilities of the number of faults per component from a large batch are:

   Probability of zero fault, P(0 fault) = 0.82

   P(1 fault) = 0.10

   P(2 faults) = 0.05

   The remaining components have more than 2 faults. What is the probability that a component picked at random has:

   a) 1 or 2 faults b) more than 2 faults c) at least 1 fault

6. A component used in heating appliances is manufactured by 3 companies, company A, B and C. Company A produces 10% defective components, company B produces 20% defective and company C produces 25% defective. Large samples are procured from these companies, and if 1 component is picked from each sample simultaneously, what is the probability that only 1 defective component will be picked?

7. The probability that the delivery of materials to a construction site will be late on any day is $\frac{1}{5}$. Produce a tree diagram for Monday and Tuesday, and find the probability that:
   a) the delivery is late on both Monday and Tuesday
   b) the delivery is late on just 1 of the 2 days
   c) the delivery is on time on Monday and Tuesday

8. A machine making bridge components produces 3 defective components out of a batch of 75. If a sample of 5 components is chosen at random, what are the chances that the sample contains a) no defective component b) 2 defective components.
9. Sunbeam manufacturing company produces kitchen units, 10% of which are found to be defective. The quality control unit selects 20 units out of a large batch, and rejects the whole batch if 3 or more units are defective. Find the probability that the batch is rejected.
10. A company manufactures PTFE bearings for use in bridge construction, and finds that 2% are defective. They are packed in boxes containing 75 bearings. What is the probability that each box will contain a) 0 b) 1 c) 2 defective bearings
11. A factory produces components which are used in the construction of steel bridges. The mean diameter of the component is 22 mm and the standard deviation is 0.2 mm. Assuming the data to be normally distributed, calculate the proportion of the components which have a diameter:
    a) less than 21.6 mm
    b) between 21.7 mm and 22.2 mm
12. The following data shows the compressive strength of 85 bricks. Use normal probability paper to check if the results have a normal distribution. If the data is normally distributed, use the graph to find the mean and the standard deviation.

Compressive strength (N/mm^2)	32–34	35–37	38–40	41–43	44–46	47–49	50–52
Frequency	4	12	16	20	17	13	3

Answers – Exercise 17.1

1. Mean = 11.85; Median = 12.1; Mode = 12.2 and 13.0
2. b) Mean = 42.35 N/mm^2 c) Median = 42.3 N/mm^2 d) 26
3. 6.26
4. σ = 4.34 N/mm^2;   Range = 17 N/mm^2;   Interquartile range = 6.4 N/mm^2
5. a) 0.15   b) 0.03   c) 0.18
6. 0.375
7. a) 0.04   b) 0.32   c) 0.64
8. a) 0.8154   b) 0.0142
9. 0.3230
10. a) 0.2231   b) 0.3347   c) 0.251
11. a) 2%   b) 77%
12. σ = 4.5 N/mm^2;   Mean = 41.9 N/mm^2

CHAPTER 18

Computer techniques

Topics covered in this chapter:

- Calculations involving addition, subtraction, multiplication and division
- Mean, range and standard deviation
- Solution of problems involving matrices, centroids etc.

18.1 Introduction

We have at our disposal many devices like programmable calculators and computers to make mathematical calculations easier and faster than the manual techniques. There are many computer software available as well, especially in statistics, to make complex problems easier to solve. In this chapter only Microsoft Excel software will be used to solve a range of problems.

Microsoft Excel is a spreadsheet programme that lets us work with numbers and text. An Excel file is known as a workbook, and one workbook can hold several sheets, e.g. sheet 1, sheet 2, sheet 3 etc. Each sheet is divided into rows and columns and their intersections create cells which are known by a reference. A cell reference is a combination of letter/s and number, e.g. B10, AA5 etc. to signify the intersection of a column and a row. Figure 18.1 shows a spreadsheet in which the reference of the selected cell is C7.

Each cell can hold text or a number or a formula. A formula is a special way to tell Excel to perform a calculation using information present in other cells. Formulae can use not only the normal arithmetic operations like plus (+), minus (–), multiply (*) and divide (/) but also the built in functions to find square roots, average, sine of an angle etc. A brief selection of these is given in Table 18.1.

Table 18.1

Symbol/Function	Action
+	Addition
–	Subtraction
*	Multiplication
/	Division

(continued)

Table 18.1 *(continued)*

Symbol/Function	Action
=	Equal to
^	Raised to a power
Sum	The sum of the values
Average	The average (or mean) of the values

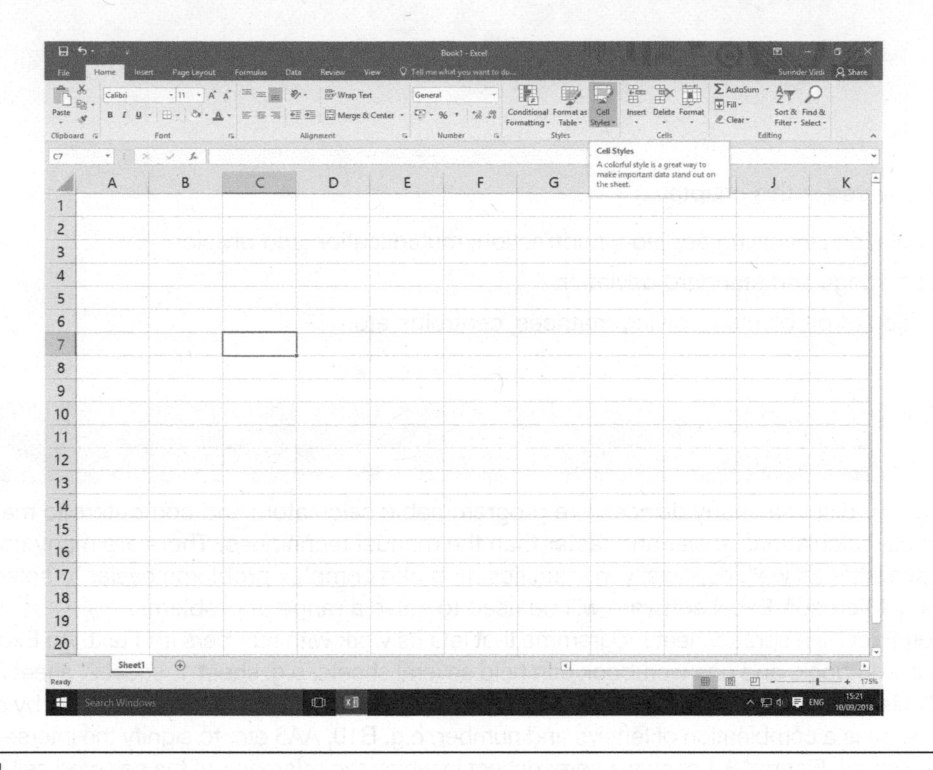

Figure 18.1

MS Excel can be used to perform calculations in mathematics, structural mechanics, construction science, land surveying and many other subjects. Examples 18.1 to 18.8 illustrate how MS Excel can be used to perform mathematical calculations.

Example 18.1

Anisha is planning to renovate her house and wants to replace the old carpets with laminated flooring in 2 rooms. If the rooms measure 5.8 m x 4.2.m and 4.5 m x 4.1 m, and 1 pack of the laminated boards covers 2.106 m², find the number of packs required. Consider wastage @ 10%.

Solution:

The cells are labelled as length, width, area etc. on Microsoft Excel spreadsheet, as shown in Figure 18.2. The length and the width of Room 1 are entered and multiplied to determine the floor area. The formula

example 18.1a - Excel

File Home Insert Page Layout Formulas Data Review View Tell me what you want to do... Surinder Virdi Share

Times New Roma 12 Wrap Text General Merge & Center Conditional Formatting Format as Table Cell Styles Insert Delete Format AutoSum Fill Clear Sort & Filter Find & Select

L8 m2

	A	B	C	D	E	F	G	H	I
1									
2									
3		Length (m)	Width (m)	Area (m^2)	Number of packs	Watage @ 10%	Number of packs required		
4									
5	ROOM 1	5.8	4.2	24.36	11.57	1.16	12.72		
6	ROOM 2	4.5	4.1	18.45	8.76	0.88	9.64		
7									
8	TOTAL						22.36		
9									
10					Number of packs required = 23				
11									
12									
13									
14									
15									
16									
17									
18									
19									
20									

Sheet1 Ready Search Windows ENG 16:50 18/09/2018

Figure 18.2a

example 18.1b - Excel

File Home Insert Page Layout Formulas Data Review View Tell me what you want to do... Surinder Virdi Share

Calibri 12 Wrap Text General Merge & Center Conditional Formatting Format as Table Cell Styles Insert Delete Format AutoSum Fill Clear Sort & Filter Find & Select

I13

	A	B	C	D	E	F	G
1							
2							
3		Length (m)	Width (m)	Area (m^2)	Number of packs	Watage @ 10%	Number of packs required
4							
5	ROOM 1	5.8	4.2	=B5*C5	=D5/2.106	=E5*0.1	=E5+F5
6	ROOM 2	4.5	4.1	=B6*C6	=D6/2.106	=E6*0.1	=E6+F6
7							
8	TOTAL						=SUM(G5:G7)
9							
10					Number of packs required		
11							
12							
13							
14							
15							
16							
17							
18							
19							
20							

Sheet1 Ready Search Windows ENG 16:52 18/09/2018

Figure 18.2b

for the floor area (= B5*C5) is entered in cell D5. It is necessary to use the equal sign before a formula otherwise the computer will not do the required calculation. The area (cell D5) is divided by 2.106 to calculate the number of packs required, as shown in cell E5. Cell F5 shows the formula to work out the wastage, 10/100 representing 10%. The total number of packs required (cell G5) is the sum of the answers in cells E5 and F5, i.e. the number of actual packs required plus the wastage.

Enter the dimensions of Room 2, and replicate all the formulae. The total number of packs is the sum of cells E6 and F6.

Figure 18.2a shows the results and Figure 18.2b shows the formulae used.

Example 18.2

5 groups of students were asked to prepare concrete cubes and test them to failure for determining the crushing strength. The test results were:

Group 1: 50, 52, 53, 53, 55, 57, 60, 54, 54, 55 N/mm^2

Group 2: 50, 52, 53, 54, 55, 58, 61, 54, 54, 55 N/mm^2

Group 3: 49, 52, 52, 53, 55, 58, 60, 55, 54, 56 N/mm^2

Group 4: 50, 52, 53, 60, 55, 59, 62, 54, 54, 55 N/mm^2

Group 5: 48, 52, 52, 53, 55, 57, 58, 54, 53, 55 N/mm^2

Find the mean and the range of each group's data.

Solution:

The data of all the groups are entered as shown in Figure 18.3a. To find the mean, the formula is: =average (first cell:last cell). For Group 1, this is: =average(B5:K5). The formula is entered without any spaces. The range of a set of data is equal to the maximum value minus the minimum value. Max(B5:K5) means the maximum value from the cells ranging between B5 and K5. Similarly, Min(B5:K5) means the minimum value from the cells ranging between B5 and K5. The formula to find the range is entered in cell M5 and replicated in cells M6, M7, M8 and M9.

The results are shown in Figure 18.3a and the formulae used in Figure 18.3b.

Example 18.3

26 bricks were tested to determine their compressive strength in N/mm^2. The results are:

49	50	55	54	51	52	56	55	53	54	54	53
60	55	53	58	61	56	57	52	54	57	55	58
56	59										

Arrange the data into 5 groups and calculate the mean compressive strength.

Solution:

The data have been arranged into 5 groups as shown in Figure 18.4a. The class mid-points are entered in cells C6 to C10. As the mean of grouped data is determined by the formula:

	A	B	C	D	E	F	G	H	I	J	K	L	M	N
1														
2														
3	Group	Crushing Strength										Mean	Range	
4														
5	Group 1	50	52	53	53	55	57	60	54	54	55	54.3	10.0	
6	Group 2	50	52	53	54	55	58	61	54	54	55	54.6	11.0	
7	Group 3	49	52	52	53	55	58	60	55	54	56	54.4	11.0	
8	Group 4	50	52	53	60	55	59	62	54	54	55	55.4	12.0	
9	Group 5	48	52	52	53	55	57	58	54	53	55	53.7	10.0	

Figure 18.3a

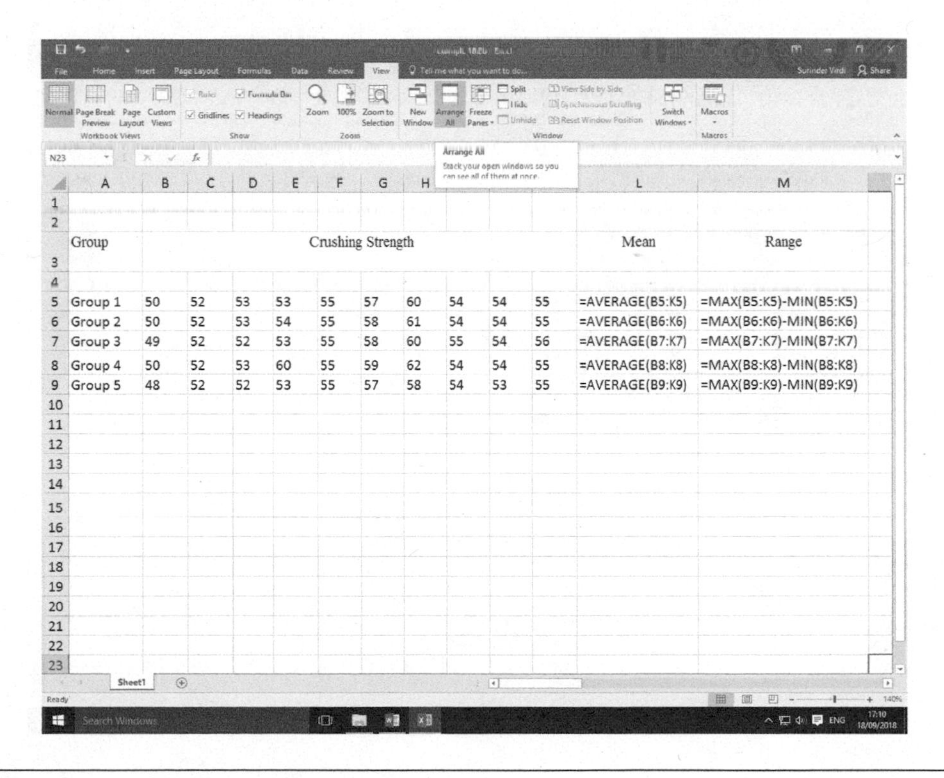

	A	B	C	D	E	F	G	H	I	J	K	L	M
3	Group					Crushing Strength						Mean	Range
5	Group 1	50	52	53	53	55	57	60	54	54	55	=AVERAGE(B5:K5)	=MAX(B5:K5)-MIN(B5:K5)
6	Group 2	50	52	53	54	55	58	61	54	54	55	=AVERAGE(B6:K6)	=MAX(B6:K6)-MIN(B6:K6)
7	Group 3	49	52	52	53	55	58	60	55	54	56	=AVERAGE(B7:K7)	=MAX(B7:K7)-MIN(B7:K7)
8	Group 4	50	52	53	60	55	59	62	54	54	55	=AVERAGE(B8:K8)	=MAX(B8:K8)-MIN(B8:K8)
9	Group 5	48	52	52	53	55	57	58	54	53	55	=AVERAGE(B9:K9)	=MAX(B9:K9)-MIN(B9:K9)

Figure 18.3b

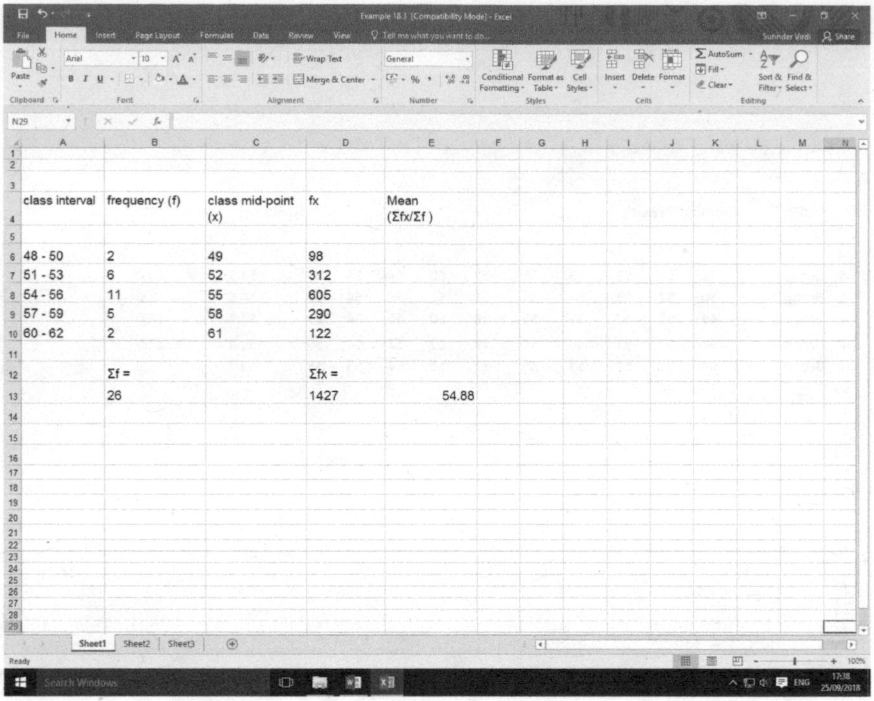

Example 18.3 [Compatibility Mode] - Excel
File Home Insert Page Layout Formulas Data Review View Tell me what you want to do... Sunnder Virdi Share
Paste Arial 10 Wrap Text Merge & Center General % Conditional Formatting Format as Table Cell Styles Insert Delete Format AutoSum Fill Clear Sort & Filter Find & Select
Clipboard Font Alignment Number Styles Cells Editing
N29

	A	B	C	D	E
4	class interval	frequency (f)	class mid-point (x)	fx	Mean (Σfx/Σf)
6	48 - 50	2	49	98	
7	51 - 53	6	52	312	
8	54 - 56	11	55	605	
9	57 - 59	5	58	290	
10	60 - 62	2	61	122	
12		Σf =		Σfx =	
13		26		1427	54.88

Sheet1 Sheet2 Sheet3
Ready 100%
Search Windows ENG 17:38 25/09/2018

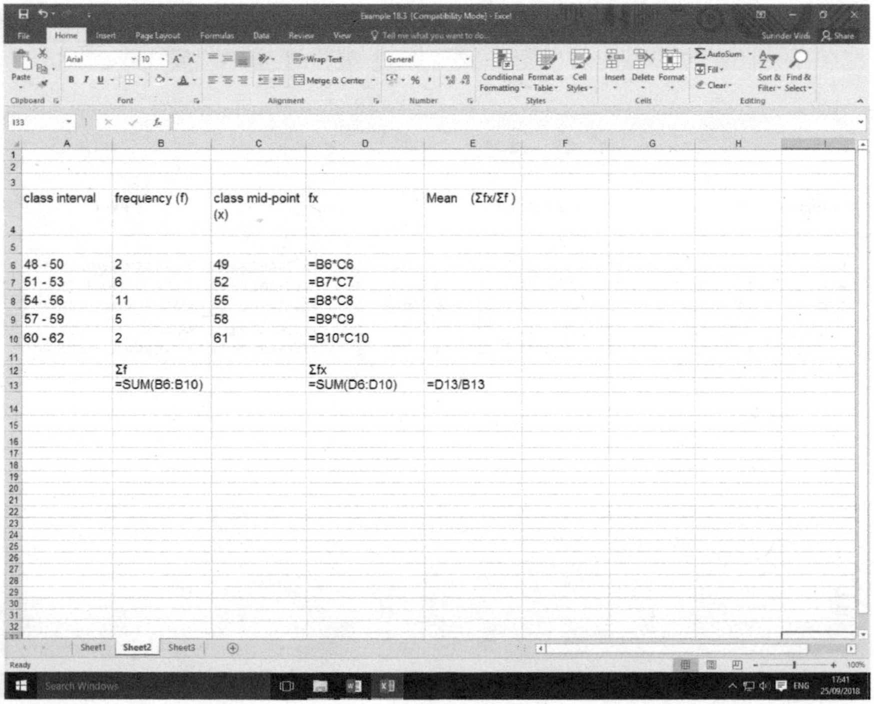

Figure 18.4a

Example 18.3 [Compatibility Mode] - Excel
File Home Insert Page Layout Formulas Data Review View Tell me what you want to do... Sunnder Virdi Share
Paste Arial 10 Wrap Text Merge & Center General % Conditional Formatting Format as Table Cell Styles Insert Delete Format AutoSum Fill Clear Sort & Filter Find & Select
Clipboard Font Alignment Number Styles Cells Editing
I33

	A	B	C	D	E
4	class interval	frequency (f)	class mid-point (x)	fx	Mean (Σfx/Σf)
6	48 - 50	2	49	=B6*C6	
7	51 - 53	6	52	=B7*C7	
8	54 - 56	11	55	=B8*C8	
9	57 - 59	5	58	=B9*C9	
10	60 - 62	2	61	=B10*C10	
12		Σf		Σfx	
13		=SUM(B6:B10)		=SUM(D6:D10)	=D13/B13

Sheet1 Sheet2 Sheet3
Ready 100%
Search Windows ENG 17:41 25/09/2018

Figure 18.4b

Mean $= \frac{\Sigma fx}{\Sigma f}$, the next step is to find the product of frequency (f) and class mid-point (x).

The formula to determine fx is entered in cell D6 and replicated for the other cells. The sum of the frequencies (Σ f) and the sum of fx (Σ fx) is found by entering the formulae shown in Figure 18.4b. Finally, the mean is shown in cell E13.

Example 18.4

10 bricks were tested to determine their crushing strength; the results are:

40, 35, 31, 32, 36, 38, 38, 40, 34, 36 N/mm^2

Find: a) the median, b) the standard deviation of the data.

Solution:

The data are entered as shown in Figure 18.5a.

The formula for determining the median of the data is: =median(B5:K5), and it has been entered in cell M5.

The formula for determining the standard deviation of the data is: =stdev(B5:K5), and it has been entered in cell N5.

The results are shown in Figure 18.5a and the formulae used in Figure 18.5b.

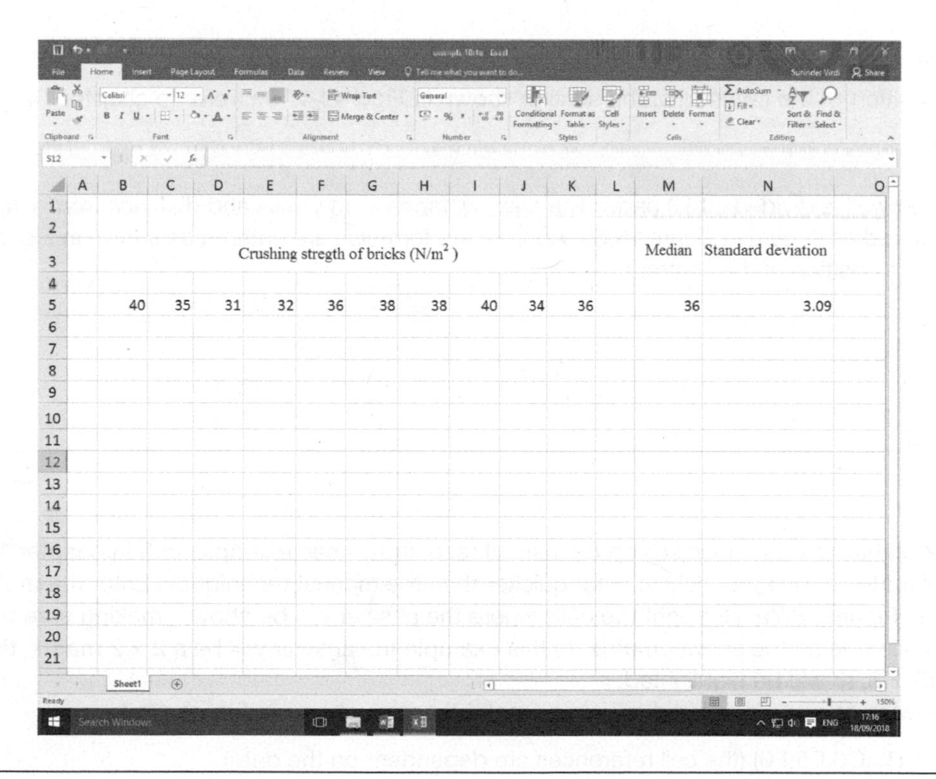

Figure 18.5a

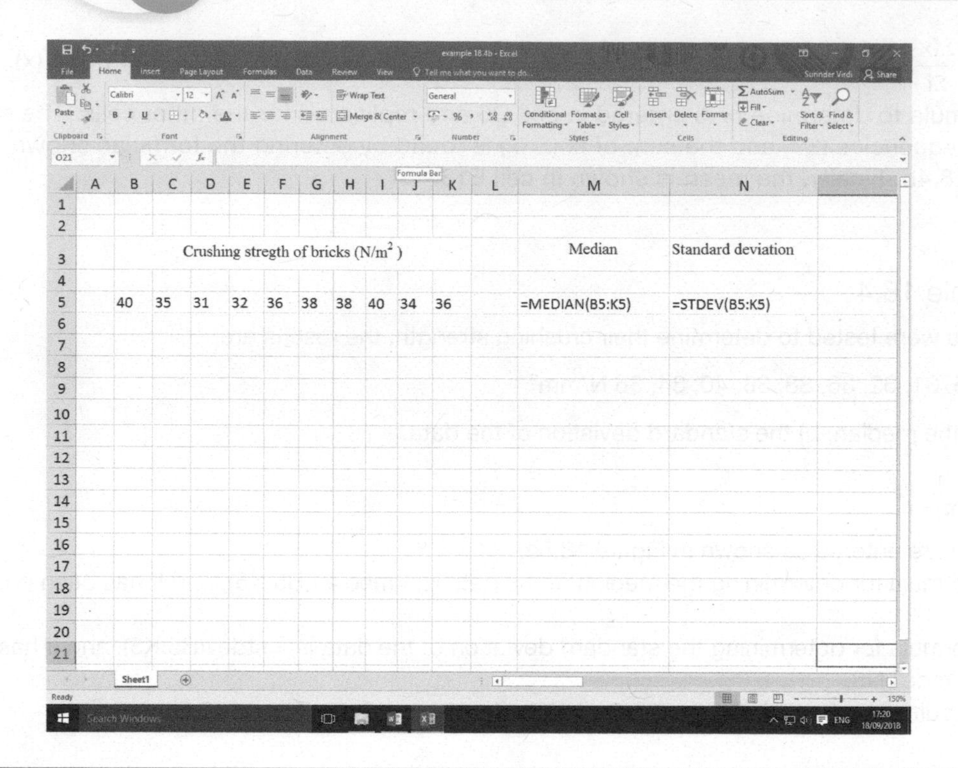

Figure 18.5b

Example 18.5

Find the position of the centroid of the section shown in Figure 14.19a (refer to chapter 14).

Solution:

The retaining wall is divided into 3 parts. The area, distance from y-axis and distance from x-axis for each part are entered as shown in Figure 18.6a. Appropriate formulae are entered as shown in Figure 18.6b to determine the position of the centroid.

Example 18.6

Evaluate $\begin{pmatrix} 2 & 4 \\ 3 & 5 \end{pmatrix} \times \begin{pmatrix} 6 & 2 \\ 0 & 5 \end{pmatrix}$

Solution:

This question has already been solved by manual technique – see example 15.3 in chapter 15. The solution of matrices by spreadsheet is far quicker than the manual techniques. Enter the matrices as shown in Figure 18.7a. Highlight the cells where the answer will be shown, making sure that their dimension is same as the answer matrix. In this example the answer will be a 2 × 2 matrix, therefore cells H5, I5, H6, I6 will be highlighted.

The following formula is entered only once, not 4 times as shown in Figure 18.7b.

=MMULT(B5:C6,E5:F6) (the cell references are dependent on the data)

After entering the formula press CTRL+SHIFT+ENTER keys to get the answer.

Example18.5a (Compatibility Mode) - Excel

File Home Insert Page Layout Formulas Data Review View Tell me what you want to do... Surinder Virdi Share

Paste Arial 10 A A Wrap Text General Conditional Formatting Format as Table Cell Styles Insert Delete Format AutoSum Fill Clear Sort & Filter Find & Select

Clipboard Font Alignment Number Styles Cells Editing

M21

	A	B	C	D	E	F	G	H	I	J	K	L
1												
2												
3	Part	Area (A)	x	y	Ax	Ay	$\bar{x}$	$\bar{y}$				
4		m^2	m	m			m	m				
5												
6	1	2	0.5	1	1	2						
7	2	4	1.5	2	6	8						
8	3	6	2.5	3	15	18						
9	Total	12			22	28						
10							**1.833**	**2.333**				
11												
12												
13												
14												
15												
16												
17												
18												
19												
20												
21												

Sheet1 Sheet2 Sheet3

Ready 150%

Search Windows ENG 17:28 18/09/2018

Figure 18.6a

Example18.5b (Compatibility Mode) - Excel

File Home Insert Page Layout Formulas Data Review View Tell me what you want to do... Surinder Virdi Share

Paste Arial 10 A A Wrap Text General Conditional Formatting Format as Table Cell Styles Insert Delete Format AutoSum Fill Clear Sort & Filter Find & Select

I18

	A	B	C	D	E	F	G	H
1								
2								
3	Part	Area (A)	x	y	Ax	Ay	$\bar{x}$	$\bar{y}$
4		m^2	m	m			m	m
5								
6	1	2	0.5	1	=B6*C6	=B6*D6		
7	2	4	1.5	2	=B7*C7	=B7*D7		
8	3	6	2.5	3	=B8*C8	=B8*D8		
9	Total	=SUM(B6:B8)			=SUM(E6:E8)	=SUM(F6:F8)		
10							**=E9/B9**	**=F9/B9**
11								
12								
13								
14								
15								
16								
17								
18								

Sheet1 Sheet2 Sheet3

Ready 169%

Search Windows ENG 17:34 18/09/2018

Figure 18.6b

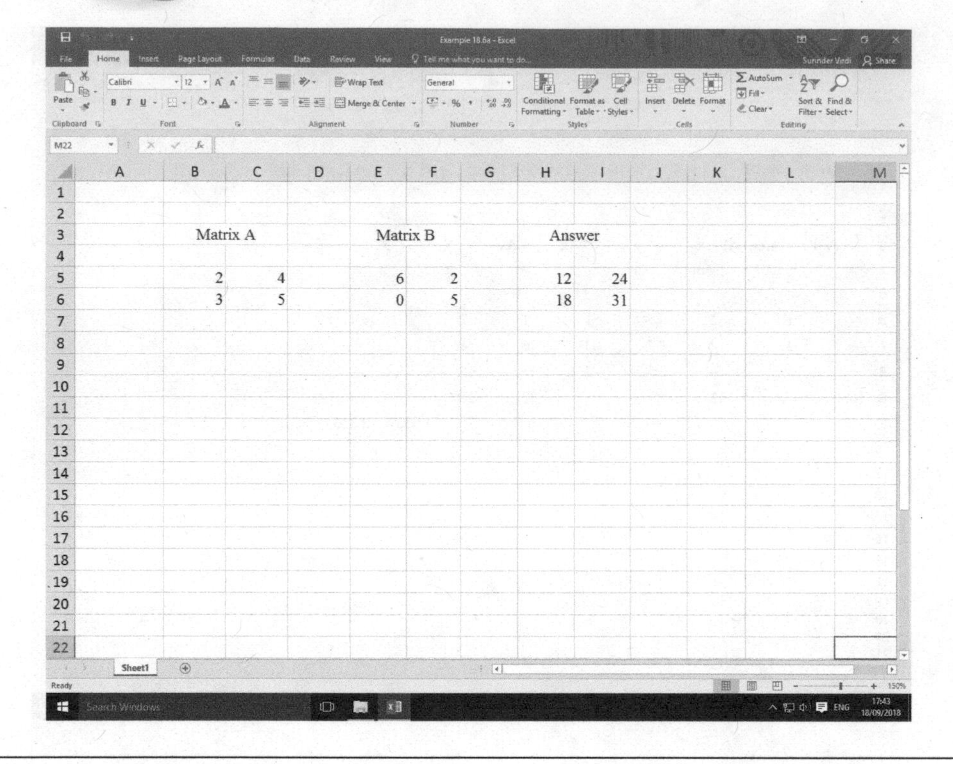

Figure 18.7a

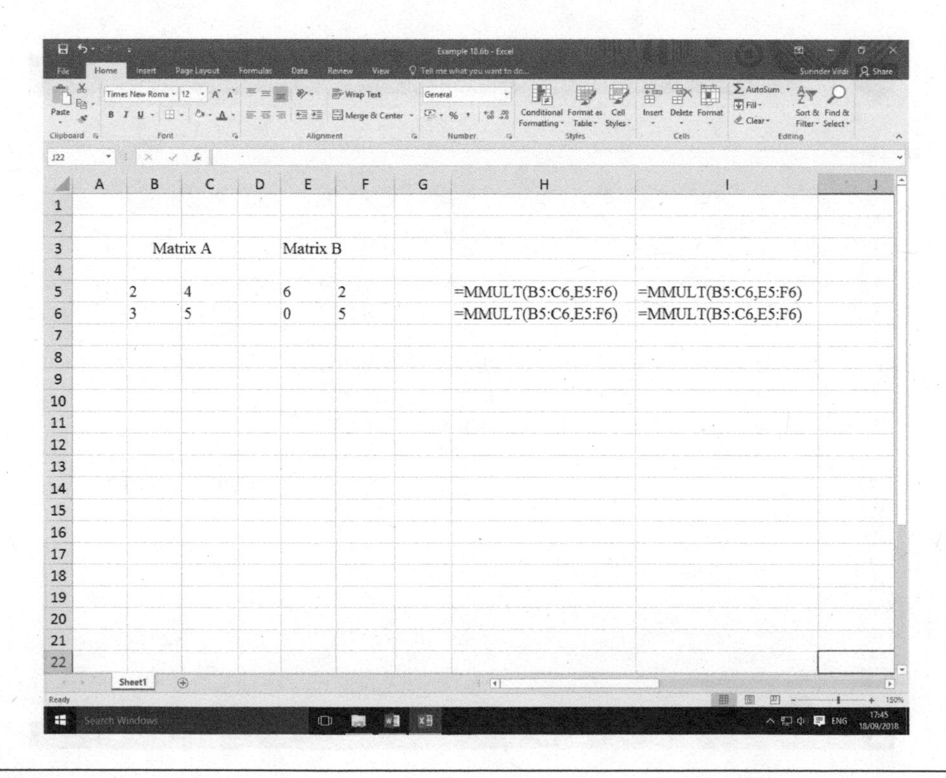

Figure 18.7b

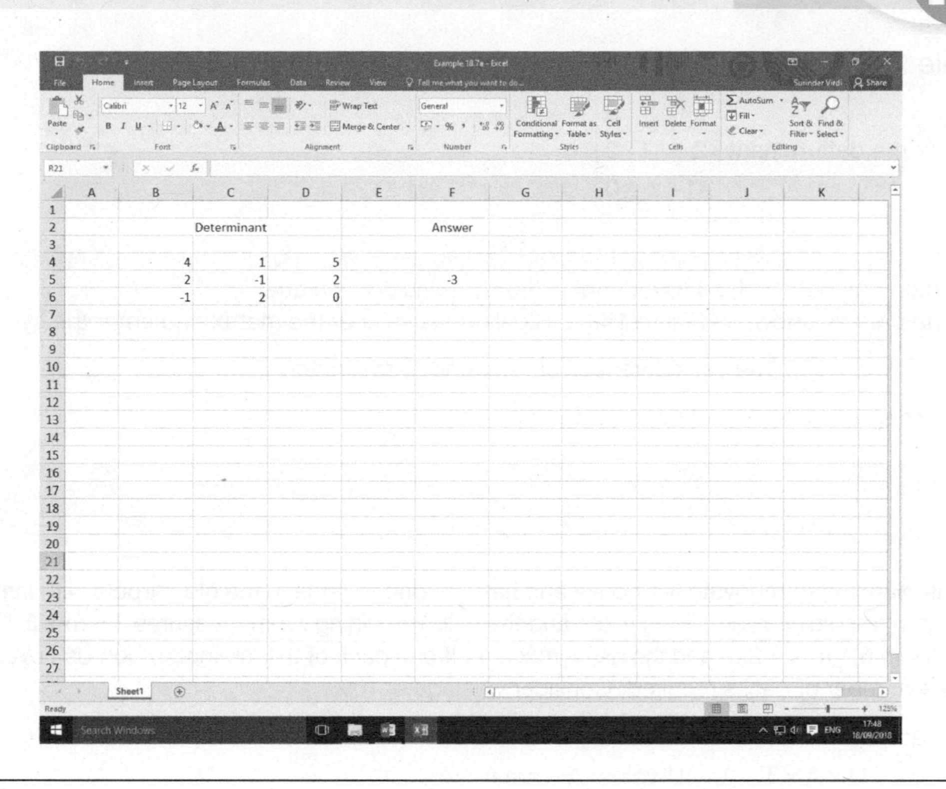

Figure 18.8a

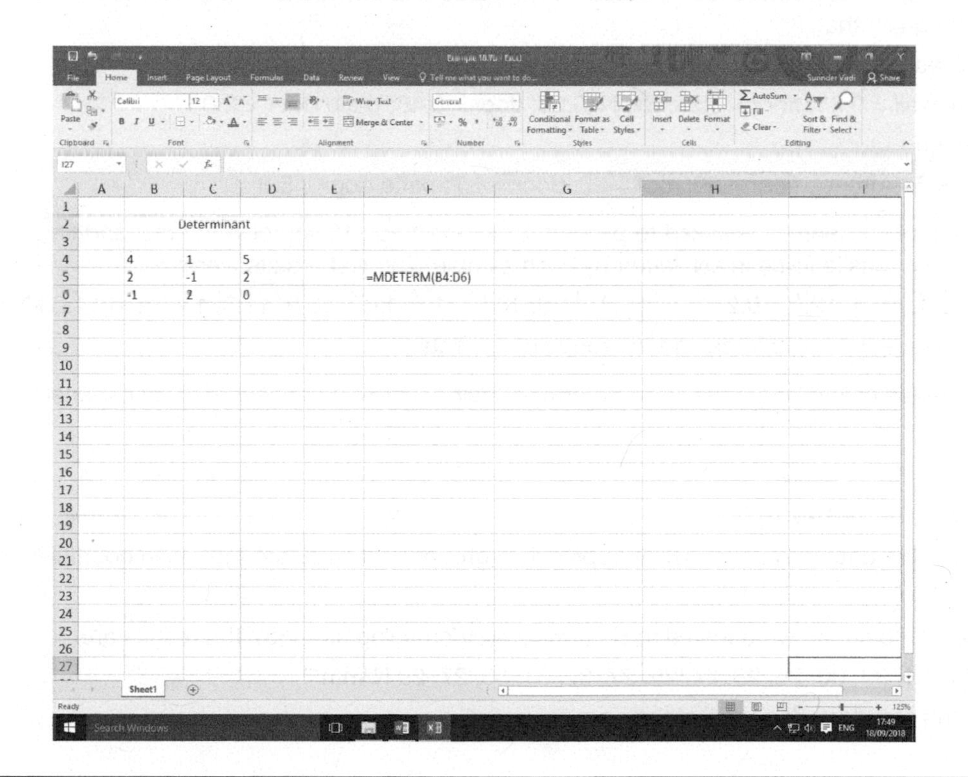

Figure 18.8b

Example 18.7

Evaluate the determinant $\begin{vmatrix} 4 & 1 & 5 \\ 2 & -1 & 2 \\ -1 & 2 & 0 \end{vmatrix}$

Solution:

This question has been solved as example 15.6 by manual technique.

Enter the data as shown in Figure 18.8a. Highlight a cell near the matrix and enter the syntax:

=MDETERM(B4:D6). The cell references correspond to the matrix.

Figure 18.8b shows how to enter the above formula.

Exercise 18.1

1. Jane is planning to renovate her house and has decided to replace the old carpets with laminated flooring in the dining room, living room and the hall. The dining room measures 3.5 m x 3.5 m, the living room 6.1 m x 4.2 m and the hall 4 m x 2 m. If one pack of the laminated boards covers 2.106 m^2, find the number of packs required. Consider wastage @ 10%.
2. The heat loss from a room is given by:

   Heat loss = U x A x T (U = U-value; A = area)

   If the temperature difference (T) between the inside and outside air is 20 °C, find the heat loss from a room given that:

   U-values (W/m^2 °C)

   Cavity wall = 0.25; floor = 0.25; roof = 0.25; door = 0.46; patio door = 2.0

   Areas (m^2)

   Walls = 54 (gross); floor = 20; roof = 20; door = 1.7; patio door = 5.0
3. A group of students was asked to perform tensile tests on 15 samples of steel and find the maximum tensile force in kN, taken by each sample. The test results were:

   9.9, 12.1, 10.0, 12.0, 10.2, 11.9, 10.3, 11.6, 10.5, 11.5, 11.3, 10.9, 11.3, 11.6 and 11.0 kN.

   Find the mean tensile force and the range of the data.
4. 25 samples of PVC were tested to determine their coefficient of linear thermal expansion. The results were:

   63 70 66 70 74 71 67 77 78 72 66 76 64 69 73 82 69 76 70 65 71 72 74 78 73 x10^{-6}/°C

   Arrange the data into a number of appropriate groups and calculate the mean coefficient of thermal expansion.
5. 15 concrete cubes were tested to determine their crushing strength; the results are:

   25, 27, 31, 32, 30, 26, 29, 24, 25, 27, 32, 31, 30, 27, 29 N/mm^2

   Find the standard deviation of the data.

6. Evaluate $\begin{pmatrix} 3 & 2 \\ 6 & 1 \end{pmatrix} \times \begin{pmatrix} 5 & 3 \\ 6 & 0 \end{pmatrix}$

7. Evaluate the determinant $\begin{vmatrix} -2 & 3 & 0 \\ -3 & 1 & 1 \\ -2 & 3 & 4 \end{vmatrix}$

8. Refer to Figure 14.19b (chapter 14) that shows a retaining wall. Find the position of its centroid.

Answers – Exercise 18.1

1. 24 packs
2. 685.64 W
3. Mean = 11.07 N; Range = 2.2 N
4. 71.4×10^{-6}/°C
5. 2.66
6. $\begin{pmatrix} 27 & 9 \\ 36 & 18 \end{pmatrix}$
7. 28
8. $\bar{x} = 1.826\text{m}; \bar{y} = 2.522\text{m}$

End of unit assignment

1. Transpose $\frac{w}{G} = w - \frac{R}{1000}$ to make w the subject.
2. The area of steel (A) in a reinforced concrete beam is given by $A = \frac{M}{tjd}$. Find the approximate percentage change in the area of steel if M increases by 3% and t decreases by 2%.
3.
   a) Solve simultaneously: $5y = 0.5 - 4x$ and $2x = 4 - 4y$
   b) The perimeter of a rectangular plot of land is 84 m and the length of its diagonal is 30 m. Calculate the dimensions of the plot.
4. A pitched roof is 12 m long, 4 m high and has a span of 14 m. Calculate:
   a) Pitch of the roof
   b) True lengths of the common rafters
   c) Surface area of the roof
5. A surveyor wants to find the distance between 2 buildings A and B. However, the distance cannot be calculated directly as there is a small lake between the 2 buildings. The surveyor sets up a station at point C and measures angle ACB to be 120°. If distances AC and CB are 220 m and 260 m respectively, find the distance between the two buildings.
6. Show that $\frac{\sec A}{\tan A + \cot A} = \sin A$
7. An oscillating mechanism in a machine has a maximum displacement of 0.5 m and a frequency of 40 Hz. The displacement at time $t = 0$ is 20 cm. Express the displacement in the general form: displacement $= A \sin(\omega t \pm \alpha)$
8. Solve: a) $\log(x^2 - 1) = \log(3x - 3)$
9. The city council of a town needs to predict the future population of the city so that the road network and other facilities could be planned. If the population increases at a rate of 1.6% each year, what will be the town's population after 25 years. The current population is 100 000 and assume that it increases continuously.
10. Solve the following equation:

    $2.4 \cosh x + 5 \sinh x = 7.5$
11. Differentiate: a) $y = (z^2 + 2z - 2) \sin z$

    b) $y = \frac{\cos x}{x}$
12. A steel plate, 700 mm wide, is used to manufacture a hollow beam having a width of b mm and depth of h mm in cross-section. If the strength of this beam is proportional to bh^3, find the dimensions of the strongest beam.
13. Evaluate $\int_1^2 \frac{2\log_e x}{x} dx$

14. A retaining wall is to be designed to retain a 6 m deep soil deposit. If the unit weight of the retained soil is 18.0 kN/m^3 and the coefficient of active earth pressure is 0.33, calculate the force on the wall due to the soil deposit at a depth of 4 m.
15. Find the position of the centroid of the area shown in Figure Asn1.

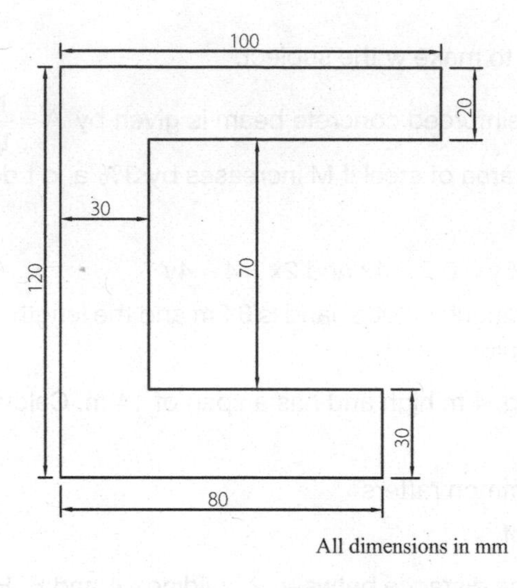

Figure Asn1

16. Determine I_{XX} and I_{YY} of the channel section shown in Figure Asn2.

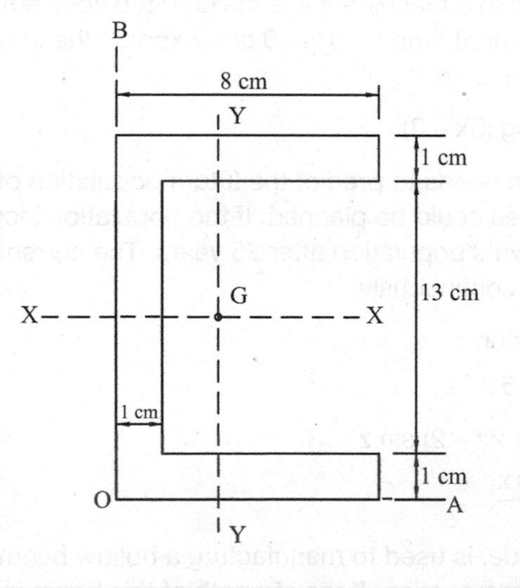

Figure Asn2

17. Use matrices to solve the simultaneous equations, $2x + 3y = 4$ and $x - y = -3$

18. Use a graphical method to determine the magnitude and direction of the resultant of the forces shown in Figure Asn3.

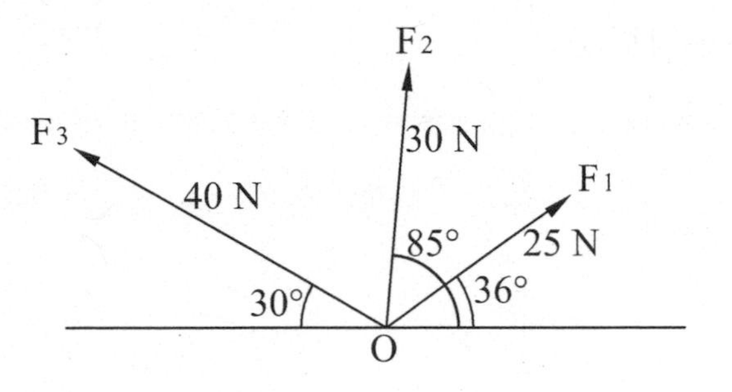

Figure Asn3

19. The data from a tensile test was analysed to determine the values of Young's modulus, which are given below:

    200, 198, 210, 215, 202, 195, 200, 205, 203, 212, 214, 203 kN/mm^2

    Find the standard deviation of the results.

20. A factory produces components which are used in the construction of bridges. The mean diameter of the component is 22 mm and the standard deviation is 0.2 mm. Assuming the data to be normally distributed, calculate the proportion of the components which have a diameter between 21.7 mm and 22.2 mm

Answers

1. $w = \frac{R}{1000} \times \frac{G}{G-1}$
2. +5%
3. x = – 3; y = 2.5   Length = 24 m; width = 18 m
4. a) 29.745°   b) 8.062 m   c) 193.488 m^2
5. 416.173 m
7. Displacement = 0.5 sin(80 πt + 0.412) m
8. x = 2 or x = 1
9. 149182
10. 0.7828
11. a) $(2z + 2)\sin z + (z^2 + 2z - 2)\cos z$ b) $\frac{-x\sin x - \cos x}{x^2}$
12. b = 87.5 mm, h = 262.5 mm
13. 0.48
14. 47.52 kN

15. $\bar{x} = 35.0\,mm, \bar{y} = 60.38\,mm$
16. $I_{xx} = 968.42$ cm, $I_{yy} = 174.43\ cm^4$
17. $x = -1, y = 2$
18. 65.5 N acts at 65° to 25 N force
19. 6.26
20. 77%

APPENDIX 1

Bending moment and shear force

1. Moment of a force

A force or a system of forces causes deflection in beams and other elements of a building. The deflection of a beam is due to the turning or rotational effect of a force; this turning effect is known as the **moment of a force** or **bending moment**:

Moment of a force = Force × distance

Figure A1.1 shows 2 cantilevers (beams with 1 support) each acted upon by force F. The force will cause the rotation of the cantilevers about their supports.

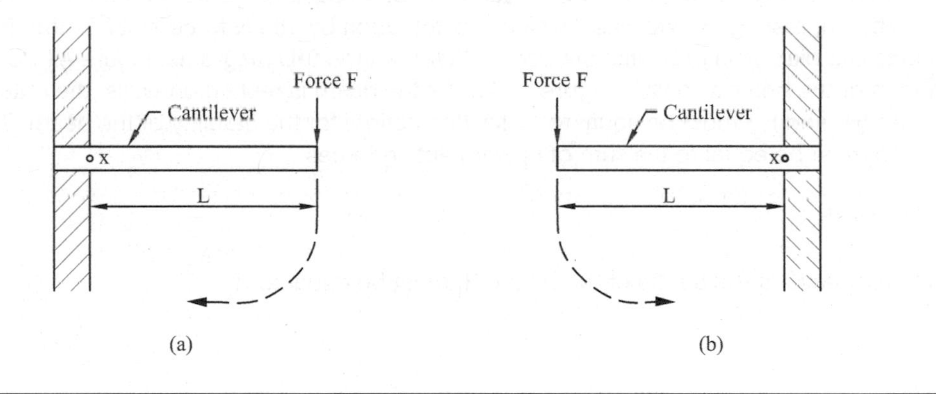

Figure A1.1

The moment of force F, about point x = F × L
where L is the perpendicular distance between the force and point x
In Figure A1.1a, the effect of the force is to cause clockwise rotation of the cantilever as shown by the arrow. The moment in this case is known as clockwise moment. The cantilever shown in Figure 11.1b will be subjected to anti-clockwise rotation due to the action of force F, and hence the moment is anti-clockwise.

2. Sign convention

Clockwise (CW) moments are considered positive.

Anti-clockwise (ACW) moments are considered negative.

3. Laws of equilibrium

Building elements such as beams, slabs, walls and columns are acted upon by a number of external forces. An element is in equilibrium if it is not disturbed from its state of rest by the external forces. There are 3 conditions for maintaining the equilibrium of a body:

a) the sum of vertical forces is 0, i.e. $\Sigma V = 0$

b) the sum of horizontal forces is 0, i.e. $\Sigma H = 0$

c) the sum of moments about any point is 0, i.e. $\Sigma M = 0$

4. Beam reactions

Newton's third law of motion states that: to every action there is an equal and opposite reaction.

In the case of a beam the action is due to the dead, imposed and other loads acting on it. Since beams are supported on walls or columns, the reactions are provided by these supports.

If the loading is symmetrical, then the reactions must be equal. Figure A1.2a shows a simply supported beam, resting on two walls, which is acted upon by 16 kN force at its centre. For structural calculations line diagrams of beams are used rather than the 3-D diagrams. Figure A1.2b shows the line diagram of the beam shown in Figure A1.2a. As the beam is resting on walls, the reactions offered by the walls (R_1 and R_2) must be equal to 16 kN (the action) for the stability of the beam. The downward acting force must be equal to the sum of upward acting forces.

$$16 \text{ kN} = R_1 + R_2$$

As 16 kN force acts at the centre of the beam, R_1 must be equal to R_2.

$$\text{or } R_1 = R_2 = 8 \text{ kN}$$

The beam shown in Figure A1.3 is not symmetrically loaded. In this case the magnitude of the reactions will not be the same and cannot be calculated as easily as done before. However a simple method, shown in example 1, may be used.

Example 1

Figure A1.3 shows the line diagram of a beam and the force acting on it. Calculate reactions R_1 and R_2.

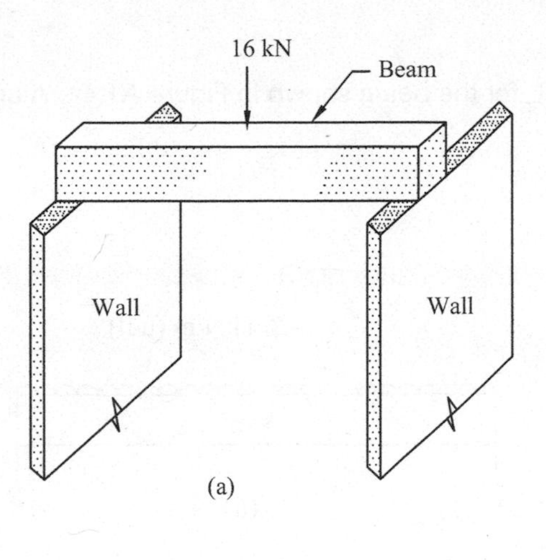

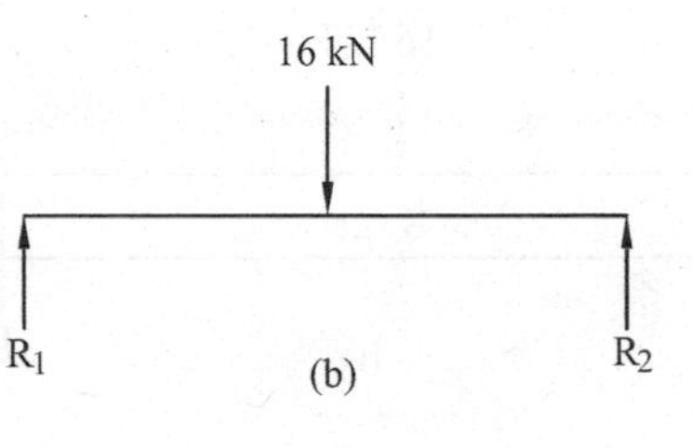

Figure A1.2

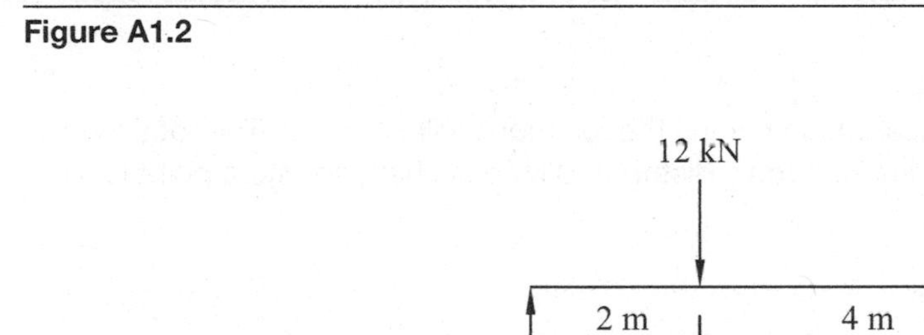

Figure A1.3

Solution:

As the loading is not symmetrical, R_1 and R_2 will be unequal, but can be determined as shown below:

Calculation of reaction R_1: $$R_1 = \frac{12\text{kN} \times 4\text{m}}{6\text{m}} = \mathbf{8\,kN}$$

Calculation of reaction R_2: $$R_2 = \frac{12\text{kN} \times 2\text{m}}{6\text{m}} = \mathbf{4\,kN}$$

Check: $R_1 + R_2 = 8 + 4 = 12$ kN (this is equal to the force acting on the beam)

Example 2

Calculate reactions R_1 and R_2 for the beam shown in Figure A1.4a. A uniformly distributed load (udl) of 2 kN/m acts on the beam.

Solution:

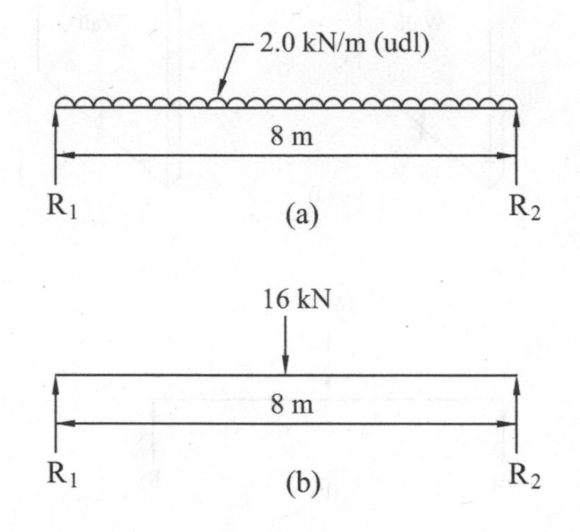

Figure A1.4

In this example the loading is symmetrical which means the reactions will be equal. The total load acting on the beam is 2 kN/m × 8 m = 16 kN (uniformly distributed load is changed into a point load)

$$\text{Therefore } R_1 = R_2 = \frac{16}{2} = 8\,\text{kN}$$

(Note: The total load (point load) is assumed to act at the centre of the distance over which the udl acts, as shown in Figure A1.4b)

5. Shear force (SF)

Shear force causes or tends to cause the horizontal or vertical movement of a part of the material against the rest. The shear force at any point on a beam is defined as the algebraic sum of all the forces acting on one side of the beam. Figure A1.5 shows the sections of a typical beam where failure due to vertical shear may occur.

Shear force at point A = algebraic sum of all forces to the left of point A = Reaction R_1

Shear force at point B = algebraic sum of all forces to the left of point B = R_1 – F

Sign convention: The use of sign convention is important in drawing the shear force diagrams. Shear force is considered positive if the forces act 'up on the left and down on the right'. This can also be represented as **↑↓**.

Shear force is negative if the forces act down on the left and up on the right **(↓↑)**.

For producing a shear force diagram, a base line is drawn first and lines are drawn up for upward acting forces, down for downward acting forces.

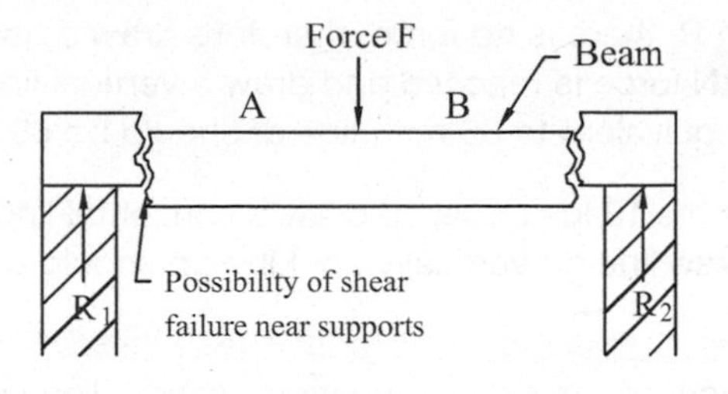

Figure A1.5

Example A1.3

Draw the shear force diagram for the beam shown in Figure A1.3.

Solution:

Before producing the shear force (SF) diagram we need to calculate reactions R_1 and R_2. This is already done: $R_1 = 8$ kN and $R_2 = 4$ kN.

To draw the SF diagram, draw up for upward acting forces and down for downward acting forces. The process is explained below:

1) Select a suitable scale, e.g. 1 kN = 5 mm (see Figure A1.6).
2) As shown in Figure A1.6b, draw base line ab; although its length is not important, usually it is drawn below the beam. As 8 kN force (R_1) acts upwards at A, draw line ac = 40 mm

   (8 kN = 8 × 5 mm = 40 mm) vertically up from the base line.

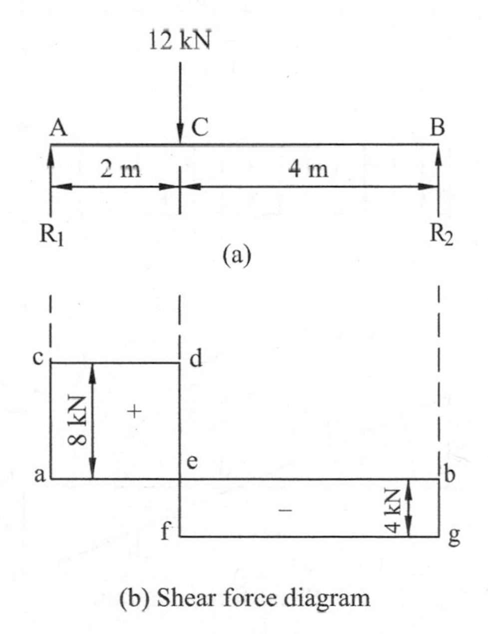

Figure A1.6

3) Immediately after reaction R_1 there is no force, therefore draw a horizontal line up to the 12 kN force. Stop when the 12 kN force is reached and draw a vertical line downwards to represent the 12 kN force. As 12 kN is equivalent to 60 mm, line df should be 60 mm long.

4) There is no force just after the 12 kN force, so draw a horizontal line up to reaction R_2. As R_2 acts in the upward direction, draw line gb vertically up. Line gb should be 20 mm long to represent reaction R_2.

5) The sign is decided by following the sign convention already discussed

Example A1.4

Draw the shear force diagram for the beam shown in Figure A1.4.

Solution:

1) The beam reactions have already been determined: $R_1 = R_2 = 8$ kN.

2) Use a convenient scale, say 1 kN = 3 mm.

3) As a udl is similar to closely spaced point loads (Figure A1.7b), it is necessary to convert it to a point load for calculating the shear force at a point. Draw horizontal line ab as the base line (Figure A1.7c). As reaction R_1 acts upwards, draw vertical line af up from point a. Measure af = 24 mm to represent R_1 (8 kN × 3 mm = 24 mm) and hence the shear force at A.

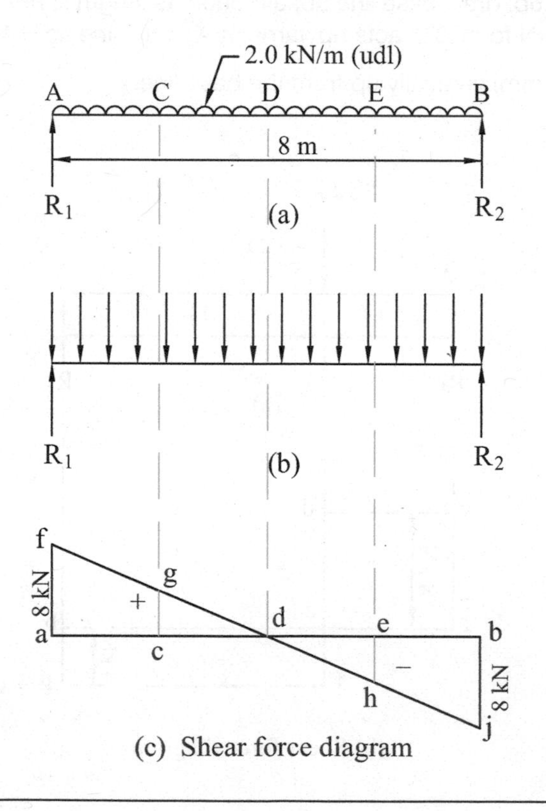

Figure A1.7

4) SF at point C = net vertical force to the left of C (AC = CD = 2 m)

= R_1 – downward acting force between A and C

= 8 kN – (2 kN/m × 2 m) = 4 kN.

5) Mark point g vertically below C to represent the SF at C. Distance gc should be 12 mm.

6) Shear force at D = net vertical force to the left of point D

= R1 – downward acting force between A and D

= 8 kN – (2 kN/m × 4 m) = 0 kN

Mark point d on the base line, vertically below D.

7) Shear force at E = net vertical force to the left of point E

= R_1 – downward acting force between A and E

= 8 kN – (2 kN/m × 6 m) = – 4 kN.

As the SF is negative, this part of the SF diagram will be drawn below the base line. Mark point h so that eh = 12 mm (4 kN × 3 mm = 12 mm)

9) Repeat the procedure to complete the SF. Join points f, g, d, h and j, to produce a straight line.

6. Bending Moment (BM)

The bending moment is a measure of the amount of bending at a point on the beam. It is calculated by taking algebraic sum of moments about the point under consideration. Depending on the type of the beam the forces can cause either sagging or hogging of the beam. In producing the BM diagrams:

i) sagging moment is considered positive
ii) hogging moment is considered negative

Also, as described before, **clockwise (CW)** moments are taken **positive; anti-clockwise (ACW)** moments are taken **negative.**

The structural design of a beam is based on the maximum bending moment which can be determined from the BM diagram.

Example A1.5

Draw bending moment diagram for the beam shown in Figure A1.3.

Solution:

Before the BM diagram can be drawn we need to calculate the bending moment at various points i.e. at points A, C and B. To determine BM at any point take moments of all forces to its left and find their algebraic sum.

1) Determine reactions R_1 and R_2, as explained earlier. R_1 = 8 kN and R_2 = 4 kN.

2) **BM at A**: To calculate the bending moment at point A, we need to take moments of all forces to its left. As there is no force to the left of point A, the BM at A is zero. Although reaction R_1 acts at A, the moment produced is 0:

Moment about A (M_A) due to R_1 = Force R_1 × distance between R_1 and A

= 8 kN × 0 = 0 kNm

3) **BM at C**: Consider all forces to the left of point C and take the algebraic sum of their moments about C. There is only R_1 that acts on the left of C, and it produces a clockwise moment (positive):

Moment about C (M_C) due to R_1 = Force R_1 × perpendicular distance between R_1 and point C

= 8 kN × 2 m = 16 kNm

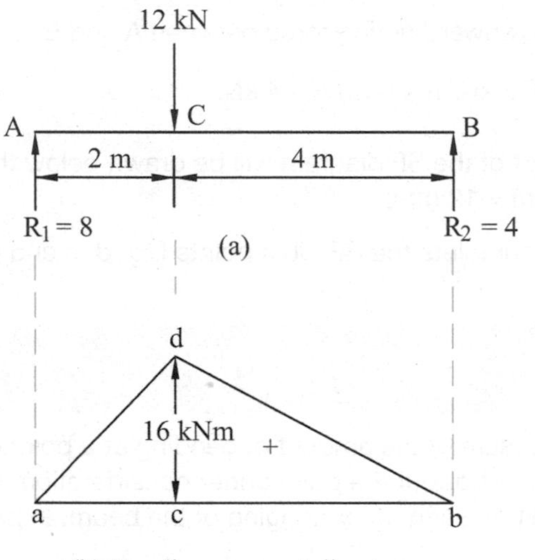

(b) Bending moment diagram

Figure A1.8

4) **BM at B**: Consider all forces to the left of point B. R_1 and 12 kN act to the left of B; they produce clockwise and anti-clockwise moments, respectively.

Clockwise moment about B = R_1 × perpendicular distance between R_1 and point B

= 8 kN × 6 m = 48 kNm (CW moment is positive)

Anti-clockwise moment about B = 12 kN × perpendicular distance between 12 kN force and B

= 12 kN × 4 m = – 48 kNm (ACW moment is considered negative)

R_2 will produce 0 moment about B as its distance from B is 0.

B.M. at B (M_B) = algebraic sum of moments about B

= CW moment + ACW moment

= 40 + (–40) = 0 kNm.

5) Select a suitable scale, say 1 kNm = 2 mm

   Bending moment at C = 16 kNm; this is equivalent to 32 mm (16 kN × 2 mm = 32 mm)

Draw base line acb and mark point d so that cd = 32 mm. Complete the BM diagram as shown in Figure A1.8b. The bending moment is positive as the force will cause sagging of the beam.

Example A1.6

Draw the bending moment diagram for an 8 m long beam carrying a udl of 2 kN/m.

Solution:

The beam is shown in Figure A1.9a. Distances AC = CD = DE = EB = 2 m

1) Calculate reactions R_1 and R_2 as explained earlier

   $R_1 = R_2 = 8$ kN

2) **BM at A:** As explained in the previous example, the bending moment at A is 0 as there is no force acting to the left of point A.

3) **BM at C:** Take moments about point C, of all forces acting to its left, and find their algebraic sum. Before the moments can be taken the udl needs to be converted to a point load. The udl acting to the left of C is equivalent to a point load of:

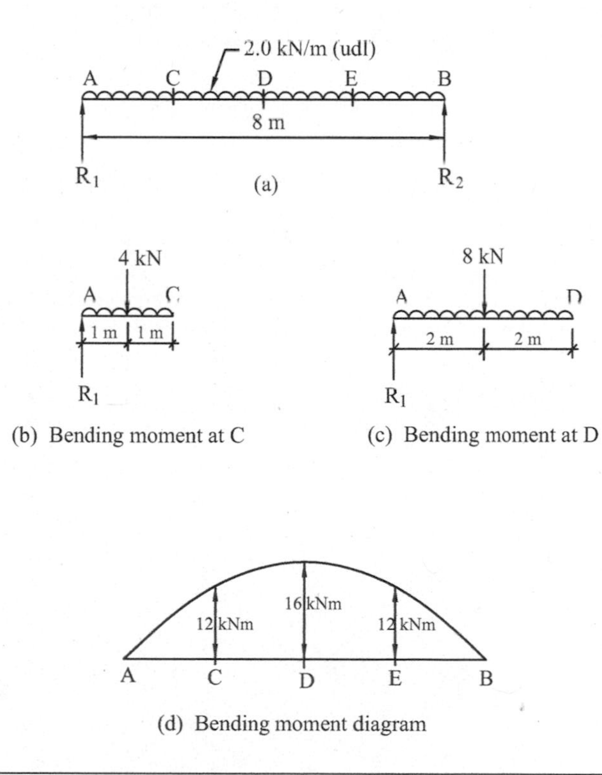

Figure A1.9

2 kN/m × 2 m = 4 kN

The force of 4 kN is supposed to act at the mid-point of AC, as shown in Figure A1.9b.

B.M. at C = R_1 × distance AC – 4 kN × 1 m

= 8 kN × 2 m – 4 kN × 1 m = 12 kNm

4) **BM at D:** Repeat step 3 to determine the bending moment at point D. The udl acting to the left of D (between A and D) is equivalent to a point load of 8 kN (2 kN/m × 4 m = 8 kN). This is also assumed to act at the mid-point of AD.

BM at D = R_1 × distance AD – 8 kN × 2 m

= 8 kN × 4 m – 8 kN × 2 m = 16 kNm

Similarly BM at other points may also be calculated.
The BM diagram is shown in Figure A1.9d.

References/further reading

1. Virdi, S., and Waters, R. (2017). *Construction Science and Materials*. Oxford: Wiley-Blackwell.
2. Durka, F., Al Nageim, H., Morgan, W. and Williams, D. (2002). *Structural Mechanics*. Harlow: Prentice Hall.

APPENDIX 2

Frame analysis

1. Triangle of forces

A force can be represented in both magnitude and direction, therefore, it is a vector quantity. Units of Newton (N), kiloNewton (kN), or MegaNewton (MN) are used to show the magnitude and the direction shown by using short lines with arrowheads.

Forces meeting at a point are known as **concurrent forces.** Forces whose line of action lies in the same plane are called **coplanar forces**. If a body is in equilibrium under the action of 3 concurrent, coplanar forces, then these forces can be represented by a triangle, with its sides drawn parallel to the direction of the forces. This is known as the law of the **triangle of forces** and can be used to determine the unknown magnitude and/or direction of 1 or 2 forces.

If the force system consists of more than 3 forces, then the law of polygon of forces can be used which is similar to the law of the triangle of forces.

If a member is subjected to compression or tension, then the internal resistance of the material acts in the opposite direction as shown in Figure A2.1. In all problems where we use the triangle of forces or the polygon of forces, we use the direction of internal resistance to show compression or tension.

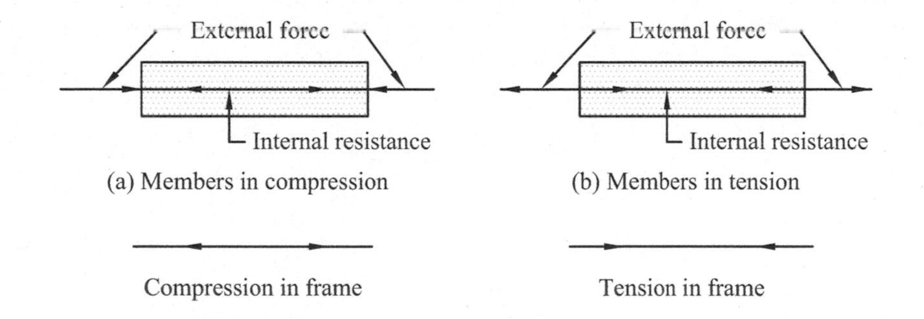

Figure A2.1

11.4.1 Bow's Notation

Bow's notation is used to identify forces in frames and force systems. It involves marking the spaces in the space diagram with capital letters, as shown in Figure A2.2. Any force and any member of the frame can be identified by letters, on either side. For example, reaction R_1 is identified as AB as it has spaces A and B on either side. Similarly 5 kN force is identified as BC.

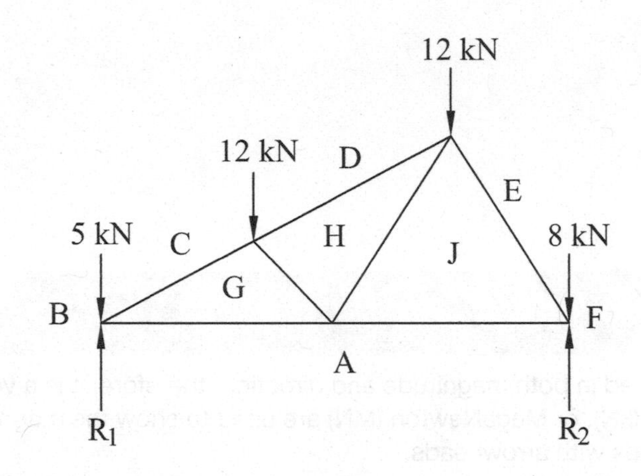

Figure A2.2

It is usual practice to start from the left hand support, and move in the clockwise direction at each joint (node). Although it is not necessary, the first few letters are used exclusively for reactions and external forces.

References/further reading

Virdi, S., and Waters, R. (2017). *Construction Science and Materials*. Oxford: Wiley-Blackwell.

APPENDIX 3

Solutions – Exercise 1.1 to Exercise 18.1

Solutions – Exercise 1.1

1. a) $6 + 5 \times 4 - 20 + 2 \times 3 = 6 + 20 - 20 + 6 = \mathbf{12}$

   b) $5 - 6 \times 1.5 - 3 \times 2.5 + 5 \times 6 \times 1.5 = 5 - 9 - 7.5 + 45 = \mathbf{33.5}$

   The solutions for questions 2, 3 and 4 are very straightforward

5. a) 620 – 380 – 120 can be approximated to 600 – 400 – 100 = **100**

   b) $\frac{45 \times 20}{50}$ can be approximated to $\frac{50 \times 20}{50} = \mathbf{20}$

6. $$\frac{39 \times 89 \times 143}{43 \times 108} = \frac{50 \times 100 \times 100}{50 \times 100} = \mathbf{100}\ (\textbf{estimated})$$

   Accurate answer = 106.88 (2 d.p.)

   Absolute error = 106.88 – 100 = **6.88**

   Relative error $= \frac{6.88}{100} \times 100 = \mathbf{6.88\%}$

7. Simplify/solve

   a) $4 \times 4^2 \times 3^4 \times 3^2 = 4^3 \times 3^6 = 64 \times 729 = \mathbf{46656}$

   b) $\frac{x^6 \times x^2 \times x^2}{x^3 \times x \times x^{-2}} = \frac{x^{10}}{x^2} = x^{10-2} = \mathbf{x^8}$

   c) $(2^2)^3 + (2x^2)^2 = 2^2 x^3 + 2^2 x^2 x^2 = 2^6 + 4x^4$

   $= \mathbf{64 + 4x^4}$

   d) $\frac{m^6 \times m^4 \times m^2}{m^3 \times m^5 \times m^4} = \frac{m^{12}}{m^{12}} = m^{12-2} = m^0 = \mathbf{1}$

Solutions – Exercise 2.1

1. i) $1.5x\,y^2 \times 6x\,y\,z = 1.5 \times 6 \times x \times x \times y^2 \times y \times z$

   $= \mathbf{9x^2\,y^3\,z}$

   ii) $2a^2 b^2 c \times 4a\,b^2 c^3 = 2 \times 4 \times a^2 \times a \times b^2 \times b^2 \times c \times c^3$

   $= \mathbf{8a^3\,b^4\,c^4}$

2. i) $4x^6y^4z^3 \div 2x^4y^2z^2 = \dfrac{4 \times x \times x \times x \times x \times x \times x \times y \times y \times y \times y \times z \times z \times z}{2 \times x \times x \times x \times x \times y \times y \times z \times z}$

$= \mathbf{2x^2\, y^2\, z}$

ii) $2x^8y^3z^3 \div x^4y^3z^{-2} = \dfrac{2 \times x^8 \times y^3 \times z^3}{x^4 \times y^3 \times z^{-2}}$

$= \dfrac{2 \times x^8 \times y^3 \times z^3 \times z^2}{x^4 \times y^3} = \mathbf{2x^4\, z^5}$

3. i) $3 + (2a + 5b - 20) = 3 + 2a + 5b - 20 = \mathbf{2a + 5b - 17}$

ii) $6 - (3a - 2b - 10) = 6 - 3a + 2b + 10 = \mathbf{16 - 3a + 2b}$

iii) $2(2a + 2.5b) = 2 \times 2a + 2 \times 2.5b = \mathbf{4a + 5b}$

iv) $1.5(2a - 3b - 5) = 1.5 \times 2a - 1.5 \times 3b - 1.5 \times 5$

$= \mathbf{3a - 4.5b - 7.5}$

4. i) $5xz - 15yz = \mathbf{5z(x - 3y)}$

ii) $4xy^2 + 8x^2y = \mathbf{4xy(y + 2x)}$

iii) $x^3 + 3x^2 + x = \mathbf{x(x^2 + 3x + 1)}$

5. i) $3x - 3 = x + 7$, $3x - x = 7 + 3$

$2x = 10$, or $\mathbf{x} = \dfrac{10}{2} = \mathbf{5}$

ii) $5x + 2 = 12$, $5x = 12 - 2 = 10$, $\mathbf{x = 2}$

iii) $\dfrac{2x}{3} = 3.2 - x$, $2x = 3 \times 3.2 - 3x$,

$2x + 3x = 9.6$, $\mathbf{x} = \dfrac{9.6}{5} = \mathbf{1.92}$

6. i) $2(x + 4) + 4(x + 1) = 6(2x + \dfrac{1}{2})$

$2x + 8 + 4x + 4 = 12x + 3$

$2x - 12x + 4x = 3 - 4 - 8$, $-6x = -9$, $\mathbf{x} = \dfrac{-9}{-6} = \mathbf{1.5}$

ii) $\dfrac{6 - A}{2} + \dfrac{A - 5}{3} = \dfrac{5}{4}$

Multiply both sides by 12 to simplify the fractions:

$12 \times \dfrac{6 - A}{2} + 12 \times \dfrac{A - 5}{3} = 12 \times \dfrac{5}{4}$

$6(6 - A) + 4(A - 5) = 15$, $36 - 6A + 4A - 20 = 15$

$-6A + 4A = 15 - 36 + 20$, $-2A = -1$, Hence $\mathbf{A = 0.5}$

iii) $\dfrac{3Y}{10} = \dfrac{Y}{3} - \dfrac{5}{6}$

Multiply both sides of the equation by 30

$30 \times \dfrac{3Y}{10} = 30 \times \dfrac{Y}{3} - 30 \times \dfrac{5}{6}$

$9Y = 10Y - 25$, $9Y - 10Y = -25$, $\mathbf{Y = 25}$

7. Let x (kg) be the quantity of cement

Quantity of sand = $1.5x$

Quantity of gravel = $1.5x + 200$

Quantity of water = $\dfrac{50}{100} \times x = 0.5x$

$x + 1.5x + 1.5x + 200 + 0.5x = 1100$

$4.5x = 1100 - 200 = 900,\ x = \dfrac{900}{4.5} = 200$ kg

Cement = 200 kg; **Sand = 300 kg;** **Gravel = 500 kg;** **Water = 100 kg**

8. Let x cm be the length of the rectangle
   Width of the rectangle = x – 4 cm
   Perimeter = 2(length + width) = 32
   $2(x + x - 4) = 32,\quad 4x - 8 = 32,\quad 4x = 40,\quad$ **x = 10**
   Length = 10 cm, Width = 10 – 4 = 6 cm
9. Let x cm be the width of the rectangle
   Length of the rectangle = 1.5x
   $x + 1.5x + x + 1.5x = 40,\quad 5x = 40,\quad$ **x = 8 cm**
   Width = 8 cm, **Length = 1.5 × 8 = 12 cm**
10. a) Potential energy = mgh = mass × m/sec² ×height

    $\left[\text{Potential energy}\right] = M\dfrac{L}{T^2}\ L = ML^2T^{-2}$

    b) Work = force × distance

    $[\text{Work}] = MLT^{-2} \times L = ML^2T^{-2}$

    c) $\text{Power} = \dfrac{\text{Work}}{\text{Time}}$

    $\left[\text{Power}\right] = \dfrac{ML^2T^{-2}}{T} = ML^2T^{-3}$
11. a) $v^2 = u^2 + 2as$

    Dimensions of 'u' and 'v' are LT^{-1}, dimensions of 'a' are: LT^{-2}, and dimension of 's' is L.

    $[LT^{-1}]^2 = [LT^{-1}]^2 + [2 \times LT^{-2} \times L]$

    $[L^2T^{-2}] = [L^2T^{-2}] + [2\ L^2T^{-2}] = [3\ L^2T^{-2}]$

    We can ignore 3 from the RHS as it has no dimension.

    As LHS = RHS, the formula is dimensionally correct

    b) $v = u + at^2$

    Replace the symbols by their dimensions:

    $[LT^{-1}] = [LT^{-1}] + [LT^{-2} \times T^2]$

    $= [LT^{-1}] + [L]$

    Here the LHS is not equal to RHS (LHS ≠ RHS)

    The formula is dimensionally **incorrect.**

    c) $s = ut + \dfrac{1}{2}at^2$

    $[L] = [LT^{-1} \times T] + [0.5\ LT^{-2} \times T^2]$

    $= [L] + [0.5L] = [1.5\ L]$

    Ignore 1.5; LHS = RHS, therefore the formula is dimensionally **correct.**
12. $\Delta L = \alpha L \Delta T$; therefore $\alpha = \dfrac{\Delta L}{L\Delta T}$

    $= \Delta L \Delta T^{-1} L^{-1}$

$[\alpha] = [L][T^{-1}][L^{-1}] = [L^{1-1}][T^{-1}]$

$= [T^{-1}]$

The dimension of α is: T^{-1}

13. The dimensions of Q (heat energy) are the same as other forms of energy, i.e. ML^2T^{-2}. The dimensions of A, $(\theta_2 - \theta_1)$, t, and d are L^2, K, T and L respectively.

Transposing the given formula, $\lambda = \frac{Qd}{A(\theta_2 - \theta_1)t} = QdA^{-1}(\theta_2 - \theta_1)^{-1}t^{-1}$

$[\lambda] = [ML^2T^{-2}]\,[L]\,[L^2]^{-1}\,[K]^{-1}\,[T]^{-1}$

$= [ML^2T^{-2}]\,[L]\,[L^{-2}]\,[K^{-1}]\,[T^{-1}]$

$= [ML^{2+1-2}T^{-2-1}K^{-1}] = [MLT^{-3}K^{-1}]$

14. Centripetal force, $F \propto m^a\,v^b\,r^c$

$F = k\,m^a\,v^b\,r^c$ (k = constant)

Leaving the constant and replacing the other symbols with their dimensions:

$MLT^{-2} = [M]^a\,[LT^{-1}]^b\,[L]^c$

$= M^a\,L^{b+c}\,T^{-1b}$

Comparing the two sides of the equation, a = 1; b = 2 and c = -1

Finally, $k\,m^1\,v^2\,r^{-1}$ or $F = \frac{kmv^2}{r}$

Solutions – Exercise 2.2

1. Let T_{12} be the twelfth term, therefore $T_{12} = a + (n - 1)d$

n = 12, a = 1 (first term of the series),

d (the common difference) = 3.5 – 1 = 2.5

$T_{12} = 1 + (12 - 1)2.5$

$= 1 + (11)2.5 = 28.5$

The sum of first 12 terms is given by, $S_{12} = \frac{n}{2}(2a + (n-1)d)$

$= \frac{12}{2}(2 \times 1 + (12-1)2.5)$

$= 6\,(2 + 27.5) = 177$

2. Let T_5 be the fifth term and T_{12} be the twelfth term

$T_5 = 28 = a + (5 - 1)d$, therefore $a + 4d = 28$ (1)

$T_{12} = 63 = a + (12 - 1)d$, therefore $a + 11d = 63$ (2)

Solving the above equations simultaneously gives, **d = 5**

Put d = 5 in either of the above equations:

a + 4d = 28, Hence a (first term) = 28 – 20 = **8**

3. The first term, a = 52 dB; n = 6

$T_6 = 52 + (6 - 1)d$

$90 - 52 = 5d$, therefore $d = \frac{90-52}{5} = \frac{38}{5} = 7.6$

Hence the noise levels are: 52, 59.6, 67.2, 74.8, 82.4 and 90.0 dB

4. a) $a = 4000$; $d = 50$; $n = 8$

$T_8 = 4000 + (8 - 1)50$

$= 4000 + 350 = 4350$ doors

b) $S_{20} = \frac{n}{2}(2a + (n-1)d)$

$= \frac{20}{2}(2 \times 4000 + (20-1)50)$

$= 10\,(8000 + 950) = 89500$ doors

c) Increase in production = 25%

Production after increase $= 4000 + \frac{4000 \times 25}{100} = 5000$

$a + (n - 1)d = 5000$

$4000 + (n - 1)50 = 5000$

$50n = 5000 - 4000 + 50$; hence $n = 21$ years

5. a) $S_{10} = \frac{10}{2}(2a + (10-1)d) = 137500$

$5(2a + 9d) = 137500$

$10a + 45d = 137500$ (1)

$S_{18} = \frac{18}{2}(2a + (18-1)d) = 427500$

$9(2a + 17d) = 427500$

$18a + 153d = 427500$ (2)

Solving equations (1) and (2) we have, $a = 2500$

Thus the production after the first year = 2500 floor units

b) Put $a = 2500$ in equation (1)

$10 \times 2500 + 45d = 137500$

$45d = 137500 - 25000$, or $d = 2500$

Therefore, the yearly increase in production is 2500 units

c) $S_{40} = \frac{40}{2}(2a + (40 - 1)d) = 2\,050\,000$

$S_{40} = 20(2 \times 2500 + 39 \times 2500)$

$= 20(102500) = 2\,050\,000$

Total production in 40 years is 2 050 000 units

6. The first term, $a = 2.2$

Common ratio, $r = \frac{4.4}{2.2} = 2$

The seventh term $= ar^{7-1} = ar^6$

$= 2.2 \times 2^6 = 140.8$

7. The first term, $a = 1$

Common ratio, $r = \frac{4}{1} = 4$

Sum of 8 terms, $S_8 = \frac{a(1-r^8)}{(1-r)}$

$$=\frac{1(1-4^8)}{(1-4)}=\frac{1(1-65536)}{-3}=21845$$

8. The third term = $ar^{3-1} = ar^2 = 22.5$ (1)

   The sixth term = $ar^{6-1} = ar^5 = 607.5$ (2)

   Divide equation (2) by (1), $\frac{ar^5}{ar^2}=\frac{607.5}{22.5}$

   $r^3 = 27$, hence $r = 3$

   From equation (1), $ar^2 = 22.5$, therefore, $a=\frac{22.5}{r^2}=\frac{22.5}{9}=2.5$

   The eleventh term = $ar^{10} = 2.5 \times 3^{10} = 147622.5$

9. a) Second blow of the hammer = $ar^{2-1} = ar = 18.4$ (1)

   eleventh blow of the hammer = $ar^{11-1} = ar^{10} = 2.47$ (2)

   Divide equation (2) by (1), $\frac{ar^{10}}{ar}=\frac{2.47}{18.4}$

   $r^9 = 0.13424$, hence $r = (0.13424)^{\frac{1}{9}} = 0.8$

   From equation (1), $a=\frac{18.4}{r}=\frac{18.4}{0.8}=23$

   Penetration due to the seventh blow = $ar^6 = 23 \times (0.8)^6 = 6.03$ mm

   b) Total penetration after 15 blows = $S_{15}=\frac{a(1-r^{15})}{(1-r)}$

   $$=\frac{23(1-0.8^{15})}{(1-0.8)}=\frac{23(1-0.03518)}{0.2}=110.95\text{ mm}$$

10. First year rent, $a = 6000$

    Second year rent, $ar = 6000+\left(6000\times\frac{10}{100}\right)=6600$

    Common ratio, $r=\frac{6600}{6000}=1.1$

    Rent in the tenth year = $ar^9 = 6000 \times (1.1)^9 =$ £14 147.69

    Total rent in 10 years = $S_{10}=\frac{a(1-r^{10})}{(1-r)}$

    $$=\frac{6000(1-1.1^{10})}{(1-1.1)}$$

    $$=\frac{6000(1-1.1^{10})}{(1-1.1)}=\frac{6000(-1.59374)}{-0.1}$$

    = **£95 624.40**

Solutions – Exercise 3.1

1. $f = c + d - e$ or $\mathbf{e} = \mathbf{c} + \mathbf{d} - \mathbf{f}$
2. $mx + c = y$ or $mx + c = y$

   $mx = y - c$ $\quad \mathbf{x}=\frac{y-c}{m}$

3. a) $k = \dfrac{Qd}{AT}$ $\quad Qd = kAT$

$\mathbf{Q} = \dfrac{kAT}{d}$

b) $\mathbf{Q} = \dfrac{0.8 \times 10 \times 40}{0.5} = \mathbf{6400}$

4. $P = \dfrac{V^2}{R}$, $\quad V^2 = PR, \quad \mathbf{V} = \sqrt{PR}$

5. a) $A = 2\pi r^2$, $\quad 2\pi r^2 = A$, $\quad r^2 = \dfrac{A}{2\pi}$, $\quad r = \sqrt{\dfrac{A}{2\pi}}$

b) $\mathbf{r} = \sqrt{\dfrac{8000}{2\pi}} = \mathbf{35.68\ cm}$

6. a) $A = \dfrac{1}{2}(a+b) \times d$, $\quad (a+b)d = 2A$

$\mathbf{d} = \dfrac{2A}{a+b}$

b) $(a+b)d = 2A$, $\quad a+b = \dfrac{2A}{d}$, $\quad \mathbf{b} = \dfrac{2A}{d} - \mathbf{a}$

7. a) Square both sides, $v^2 = c^2 \times RS$, $\quad \mathbf{R} = \dfrac{v^2}{c^2 S}$

b) $\mathbf{R} = \dfrac{(2.5)^2}{50^2 \times 0.02} = \mathbf{0.125}$

8. $\dfrac{e}{1+e} = n$, $\quad e = n(1+e)$, $\quad e = n + ne$, $\quad e - ne = n$

$e(1-n) = n \quad \mathbf{e} = \dfrac{n}{1-n}$

9. $\dfrac{5wL^3}{384EI} = d$, $\quad 5wL^3 = 384\ dEI$, $\quad L^3 = \dfrac{384\ dEI}{5w}$

$\mathbf{L} = \sqrt[3]{\dfrac{384\ dEI}{5w}}$

10. a) $\dfrac{1}{n} R^{\frac{2}{3}} S^{\frac{1}{2}} = v \quad R^{\frac{2}{3}} S^{\frac{1}{2}} = vn \quad R^{\frac{2}{3}} = \dfrac{vn}{S^{\frac{1}{2}}}$

$R = \left(\dfrac{vn}{S^{\frac{1}{2}}} \right)^{\frac{3}{2}}$

b) $R = \left(\dfrac{2.75 \times 0.01}{0.02^{\frac{1}{2}}} \right)^{\frac{3}{2}} = (0.1945)^{\frac{3}{2}}$

$\mathbf{R = 0.0857}$

11. $B(T + A(S - T)) = U$

$B(T + AS - AT) = U$

$BT + BAS - BAT = U$

$T(B - BA) = U - BAS$

$$\mathbf{T} = \frac{U\text{-}BAS}{B\text{-}BA}$$

12. $D = \frac{(G+e)d}{1+e}$

$D + De = Gd + ed$

$De - ed = Gd - D$

$e(D - d) = Gd - D, \qquad \mathbf{e} = \frac{Gd - D}{D - d}$

13. a) $A_2 = \frac{A_1}{1 - \frac{h}{H}}$

$A_2\left(1 - \frac{h}{H}\right) = A_1, \qquad 1 - \frac{h}{H} = \frac{A_1}{A_2}$

$1 - \frac{A_1}{A_2} = \frac{h}{H}, \qquad H\left(1 - \frac{A_1}{A_2}\right) = h$

or $\mathbf{h} = H\left(1 - \frac{A_1}{A_2}\right)$

b) $h = 5\left(1 - \frac{5}{10}\right) = 5 \times 0.5 = \mathbf{2.5}$

14. $w - \frac{R}{1000} = \frac{w}{G} \qquad w - \frac{w}{G} = \frac{R}{1000}$

$\left(\frac{wG - w}{G}\right) = \frac{R}{1000} \qquad w\left(\frac{G - 1}{G}\right) = \frac{R}{1000}$

$\mathbf{w} = \frac{R}{1000} \times \frac{G}{G\text{-}1}$

15. a) $C \times \log_{10}\left(\frac{n_2}{n_1}\right) = w_1 - w_2$

$\log_{10}\left(\frac{n_2}{n_1}\right) = \frac{w_1 - w_2}{C}$

$\text{antilog}\left[\log_{10}\left(\frac{n_2}{n_1}\right)\right] = \text{antilog}\left(\frac{w_1 - w_2}{C}\right)$

$\left(\frac{n_2}{n_1}\right) = \text{antilog}\left(\frac{w_1 - w_2}{C}\right)$

$\mathbf{n_2} = n_1 \times \text{antilog}\left(\frac{w_1 - w_2}{C}\right)$

b) $n_2 = n_1 \times \text{antilog}\left(\frac{25 - 24}{4.34}\right)$

$= 20 \times \text{antilog}\ (0.2304) = 20 \times 1.7 = 34$

16. $I_2 = 1 \times 10^{-12} \times \text{antilog}\left(\frac{97}{10}\right)$

$= 1 \times 10^{-12} \times \text{antilog } 9.7$

$= \mathbf{5.01 \times 10^{-3}}$

Solutions – Exercise 3.2

1. Expansion of $(2m - 3n)^4$

   This question can be written as expansion of $(2m + (-3n))^4$

   Comparing $(2m + (-3n))^4$ with $(a + x)^4$ shows that $a = 2m$, and $x = -3n$

   From Pascal's triangle, $(a + x)^4 = a^4 + 4a^3 x + 6a^2 x^2 + 4ax^3 + x^4$

   $\therefore (2m + (-3n))^4 = (2m)^4 + 4(2m)^3(-3n) + 6(2m)^2(-3n)^2 + 4(2m)(-3n)^3 + (-3n)^4$

   $= \mathbf{16m^4 - 96m^3 n + 216m^2 n^2 - 216mn^3 + 81n^4}$

2. Expansion of $(3 + x)^3$

   $$(3+x)^3 = (3)^3 + 3(3)^{3-1}(x) + \frac{3(3-1)}{2!}(3)^{3-2}(x)^2 + \frac{3(3-1)(3-2)}{3!}(3)^{3-3}(x)^3$$

   $= 27 + 3(3)^2 (x) + 3(3)^1 (x)^2 + (3)^0 (x)^3$

   $= \mathbf{27 + 27x + 9x^2 + x^3}$

3. Expansion of $(2p + q)^5$

   $$(2p+q)^5 = (2p)^5 + 5(2p)^{5-1}(q) + \frac{5(5-1)}{2!}(2p)^{5-2}(q)^2 + \frac{5(5-1)(5-2)}{3!}(2p)^{5-3}(q)^3 +$$

   $$\frac{5(5-1)(5-2)(5-3)}{4!}(2p)^{5-4}(q)^4 + \frac{5(5-1)(5-2)(5-3)(5-4)}{5!}(2p)^{5-5}(q)^5$$

   $= (2p)^5 + 5(2p)^4 (q) + 10(2p)^3 (q)^2 + 10(2p)^2 (q)^3 + 5(2p)^1 (q)^4 + (2p)^0 (q)^5$

   $= \mathbf{32p^5 + 80p^4 q + 80p^3 q^2 + 40p^2 q^3 + 10pq^4 + q^5}$

4. Without expanding $(3 + 2x)^8$ determine the fifth term

   Here $n = 8$; $a = 3$; $x = 2x$; $r = 5$ and $r - 1 = 4$

   Substituting these values, the fifth term of $(3 + 2x)^8$ is:

   $$\frac{8(8\quad 1)(8\quad 2)(8\quad 3)}{4!}(3)^{8\ (5\ 1)}(2x)^{5\quad 1}$$

   $= 70 \times (3)^4 (2x)^4 = \mathbf{90\ 720\ x^4}$

5. Without expanding $(2 - 3x)^7$ determine the fourth term

   $$\text{Fourth term} = \frac{n(n-1)(n-2)}{3!} a^{n-(r-1)} x^{r-1}$$

   Here $n = 7$; $a = 2$; $x = -3x$; $r = 4$ and $r - 1 = 3$

   $$\text{Fourth term} = \frac{7(7-1)(7-2)}{3\times2\times1}(2)^{7-(4-1)}(-3x)^{4-1}$$

   $= \mathbf{-15\ 120\ x^3}$

6. Expansion of $\frac{1}{\sqrt[3]{1+2x}}$ to 4 terms

   $$\frac{1}{\sqrt[3]{1+2x}} = \frac{1}{(1+2x)^{1/3}} = (1+2x)^{-1/3}$$

   $$\text{Expansion of } (1+2x)^{-1/3} = 1 + \left(-\frac{1}{3}\right)(1)^{(-1/3-1)}(2x) + \frac{(-1/3)(-1/3-1)}{2!}(1)^{(-1/3-2)}$$

   $$(2x)^2 + \frac{(-1/3)(-1/3-1)(-1/3-2)}{3!}(1)^{-1/3-3}(2x)^3$$

$$= 1 - \frac{2}{3}x + \frac{8}{9}x^2 - \frac{112}{81}x^3$$

This is true provided 2x < 1 numerically

$$\text{or } x < \frac{1}{2}; \qquad \text{i.e.} -\frac{1}{2} < x < \frac{1}{2}$$

7. Expansion of $(3 - 2x)^{3/2}$

$(3 - 2x)^{3/2} = (3 + (-2x))^{3/2}$

$$(3+(-2x))^{3/2} = (3)^{3/2} + \frac{3}{2}(3)^{3/2-1}(-2x) + \frac{(3/2)(3/2-1)}{2!}(3)^{3/2-2}(-2x)^2 + \frac{(3/2)(3/2-1)(3/2-2)}{3!}(3)^{3/2-3}(-2x)^3$$

$$= (3)^{3/2} - 3(3)^{1/2}(x) + \frac{3\times1}{4\times(2\times1)}(3)^{-1/2}(4x^2) + \frac{3\times1\times-1}{8(3\times2\times1)}(3)^{-3/2}(-8x^3)$$

$$= (\mathbf{3})^{3/2} - (\mathbf{3})^{3/2}x + \frac{3^{1/2}}{2}x^2 + \frac{3^{-3/2}}{2}x^3$$

This is true provided 2x < 3 numerically

$$\text{or } x < \frac{3}{2}; \qquad \text{i.e.} -1.5 < x < 1.5$$

8. Let Q' be the new quantity.

$$Q' = \frac{k[m(1-\frac{2}{100})]^2\, t(1+\frac{3}{100})}{L(1+\frac{2}{100})}$$

$$= \frac{km^2 t(1-\frac{2}{100})^2(1+\frac{3}{100})}{L \qquad (1+\frac{2}{100})}$$

$$= \frac{km^2 t}{L}\left(1-\frac{2}{100}\right)^2\left(1+\frac{3}{100}\right)^1\left(1+\frac{2}{100}\right)^{-1}$$

$$= \frac{km^2 t}{L}\left(1-\frac{4}{100}+\frac{3}{100}+\frac{2\times(-1)}{100}\right)$$

$$= \frac{km^2 t}{L}\left(1-\frac{4}{100}+\frac{3}{100}-\frac{2}{100}\right) = \frac{km^2 t}{L}\left(1-\frac{3}{100}\right)$$

$= Q\left(1-\frac{3}{100}\right)$, Therefore the **percentage error in Q is - 3%**

9. Let δ' be the new deflection

$$\delta' = \frac{Kw\left(1+\frac{2}{100}\right)\left[L\left(1-\frac{2}{100}\right)\right]^3}{\left[d\left(1-\frac{1.5}{100}\right)\right]^4}$$

$$= \frac{KwL^3}{d^4}\left(1+\frac{2}{100}\right)^1\left(1-\frac{2}{100}\right)^3\left(1-\frac{1.5}{100}\right)^{-4}$$

$$= \frac{KwL^3}{d^4}\left(1+\frac{2}{100}-\frac{2\times3}{100}-\frac{1.5\times(-4)}{100}\right)$$

$$= \delta\left(1+\frac{2}{100}-\frac{6}{100}+\frac{6}{100}\right)=\delta\left(1+\frac{2}{100}\right)$$

From the above answer it can be concluded that the deflection **increases by 2%**.

10. $A = \dfrac{M}{tjd}$ and $A' = \dfrac{M\left(1+\frac{3}{100}\right)}{t\left(1-\frac{2}{100}\right)jd}$

$$A' = \frac{M}{tjd}\left(1+\frac{3}{100}\right)^{1}\left(1-\frac{2}{100}\right)^{-1}$$

$$= \frac{M}{tjd}\left(1+\frac{3}{100}\right)\left(1+\frac{2}{100}\right)=\frac{M}{tjd}\left(1+\frac{3}{100}+\frac{2}{100}\right)$$

$= A\left(1+\dfrac{5}{100}\right)$, hence, area A **increases by 5%**

Solutions – Exercise 4.1

1. a) $x + y = 3$ ......(1) and $x - y = -1$ ....(2)

From equation (1) $x = 3 - y$

Substitute value of x in equation (2), $3 - y - y = -1$

$-2y = --4$ or $\mathbf{y = 2}$

Substitute the value of y in equation 1 to give $\mathbf{x = 1}$

b) $2x + y = -1$ ......(1) and $x + 2y - 1$ ....(2)

From equation (2) $x = 1 - 2y$

Substitute value of x in equation (1), $2(1 - 2y) + y = -1$

$2 - 4y + y = -1$ or $\mathbf{y = 1}$

Substitute the value of y in equation 1 to give $\mathbf{x = -1}$

2. a) $x + y = 5$ ....(1) $3x - 2y = -5$ ....(2)

Multiply equation (1) by 3 and subtract from equation (2)

$3x - 2y = -5$

$3x + 3y = 15$ (Eq.1× 3) (the new signs are shown in bold; they supersede the original signs)

− − −

$-5y = -20$

$\therefore$ $\mathbf{y = -20/-5 = 4}$

Substitute the value of y in any equation, say equation (1):

$x + 4 = 5$, Hence $\mathbf{x = 1}$

b) $x + y = 1$ ....(1) $3x - y = -5$ ....(2)

+y will cancel – y, therefore add equations (1) and (2)

$x + y = 1$

$3x - y = -5$

− − −

$4x \quad = -4$

$\therefore$ $\mathbf{x = -4/4 = -1}$

Substitute the value of x in any equation, say equation (1):

$-1 + y = 1$, Hence $\mathbf{y = 2}$

c) $2x - y = 1$ ....(1) $x - 2y = -2.5$ ....(2)

Multiply equation (2) by 2 and subtract from equation (1)

$2x - y = 1$

$2x - 4y = -5$ (Eq.2 × 2) (the new signs are shown in bold; they supersede the original signs)

$-\quad +\quad +$

$3y = 6$

$\therefore \mathbf{y = 2}$

Substitute the value of y in any equation, say equation (1):

$2x - 2 = 1$, hence $\mathbf{x = 1.5}$

d) Rearrange the equations: $4x + 5y = 0.5$ ......(1), $2x + 4y = 4$ ......(2)

Multiply equation (2) by 2 and subtract from equation (1)

$4x + 5y = 0.5$

$4x + 8y = 8$ (Eq.2 × 2)

$-\quad -\quad -$

$-3y = -7.5$

$\therefore \mathbf{y = -7.5/-3 = 2.5}$

Substitute the value of y in any equation, say equation (1):

$4x + 5 \times 2.5 = 0.5$, Hence $\mathbf{x = -3}$

e) $1.5a + 2.5b = 3$ ....(1) $2a + 3.5b = 5$ ....(2)

Multiply equation (1) by 4 and equation (2) by 3 and subtract

$6a + 10b = 12$

$6a + 10.5b = 15$

$-\quad -\quad -$

$-0.5b = -3$

$\therefore \mathbf{b = -3/-0.5 = 6}$

Substitute the value of b in any equation, say equation (1):

$1.5a + 2.5 \times 6 = 3$, Hence $\mathbf{a = -8}$

3. $2d + 1w = 260$ .... (1)

$3d + 2w = 430$ .... (2) (d= door; w = window)

Multiply equation (1) by 2 and subtract equation (2) from it:

$4d + 2w = 520$

$3d + 2w = 430$

$-\quad -\quad -$

$d = 90$

Substitute the value of d in any equation: $w = 80$

One door costs £90.00, and one window costs £80.00

4. $P + 29\,Q = 800\,000$ (1)

$P - 27\,Q = 100\,000$ (2)

$-\quad +\quad\quad -$

$56\,Q = 700\,000$ or $\mathbf{Q = 12\,500}$

Substitute the value of Q in equation (1)

$P = 800\,000 - 29 \times 12\,500 =$ **£437 500**

5. Test 1: $320 = c + 310\,R$ ...... (1)

Test 2: $200 = c + 110\,R$ ...... (2)

These are simultaneous equations. Subtract equation (2) from equation (1)

$120 = 200\,R$ or $\mathbf{R = 0.6}$

Substitute the value of R in equation (1)

$320 = c + 310 \times 0.6$ or **c = 134 kN/m²**

6. $N = a\,D + e$ (1)

After 2 days: $22\,000 = a \times 2 + e$ or $22\,000 = 2a + e$ (2)

After 5 days: $40\,000 = a \times 5 + e$ or $40\,000 = 5a + e$ (3)

Solution of equations (2) and (3) gives **a = 6000 and e = 10 000**

Substitute the values of a and e in equation (1): $N = 6000\,D + 10\,000$

After 8 days, number of bricks $N = 6000 \times 8 + 10\,000 =$ **58 000**

Solutions – Exercise 4.2

1. a) $x^2 - 4x + 3 = (x - 3)(x - 1) = 0$

   $x - 3 = 0$, or $\mathbf{x = 3}$; $x - 1 = 0$, or $\mathbf{x = 1}$

   b) $x^2 + x - 20 = (x + 5)(x - 4) = 0$

   $x + 5 = 0$, or $\mathbf{x = -5}$; $x - 4 = 0$, or $\mathbf{x = 4}$

   c) $x^2 + 8x + 12 = (x + 6)(x + 2) = 0$

   $x + 6 = 0$, or $\mathbf{x = -6}$; $x + 2 = 0$, or $\mathbf{x = -2}$

   d) $4x^2 - 4x - 3 = (2x - 3)(2x + 1) = 0$

   $2x - 3 = 0$, or $\mathbf{x = 1.5}$; $2x + 1 = 0$, or $\mathbf{x = -0.5}$

2. a) $a = 3, b = 10, c = -8$

   $$x = \frac{-10 \pm \sqrt{10^2 - 4(3)(-8)}}{2 \times 3} = \frac{-10 \pm 14}{6}$$

   $\mathbf{x} = 4/6 = \mathbf{2/3}$ or $\mathbf{x} = -24/6 = \mathbf{-4}$

   b) $a = 6, b = 9, c = -6$

   $$x = \frac{-9 \pm \sqrt{9^2 - 4(6)(-6)}}{2 \times 6} = \frac{-9 \pm 15}{12}$$

   $\mathbf{x} = 6/12 = \mathbf{0.5}$ or $\mathbf{x} = -24/12 = \mathbf{-2}$

c) a = 2, b = -3, c = –5

$$x = \frac{-(-3) \pm \sqrt{(-3)^2 - 4(2)(-5)}}{2 \times 2} = \frac{3 \pm 7}{4}$$

x = 10/4 = **2.5** or **x** = –4/4 = **–1**

d) a = 4, b = 8, c = 3

$$x = \frac{-8 \pm \sqrt{8^2 - 4(4)(3)}}{2 \times 4} = \frac{-8 \pm 4}{8}$$

x = –4/8 = – **0.5** or **x** = -12/8 = **-1.5**

e) a = 1, b = 5, c = 6

$$x = \frac{-5 \pm \sqrt{5^2 - 4(1)(6)}}{2 \times 1} = \frac{-5 \pm 1}{2}$$

x = –4/2 = **-2** or **x** = –6/2 = **-3**

3. a) $(x + 2)^2 - 4 = 5$
   $(x + 2)^2 = 5 + 4 = 9$
   $(x + 2)^2 = 3^2$ or $x + 2 = \pm 3$
   x = +3 – 2 = **1** or **x** = -3 -2 = **–5**

   b) $(x + 1)^2 - 1 = 3$
   $(x + 1)^2 = 3 + 1 = 4$
   $(x + 1)^2 = 2^2$ or $x + 1 = \pm 2$
   x = +2 – 1 = **1** or **x** = -2 -1 = **-3**

   c) $(x - 3)^2 - 9 = 7$
   $(x - 3)^2 = 7 + 9 = 16$
   $(x - 3)^2 = 4^2$ or $x - 3 = \pm 4$
   x = +4 + 3 = **7** or **x** = –4 +3 = **-1**

4. Perimeter = 2l + 2w = 84 (l = length; w = width)
   l + w = 42 or w = 42 – l .... (1)
   Also, $l^2 + w^2 = 30^2$ .... (2) (Pythagoras' theorem)
   From equations (1) and (2) $l^2 + (42 - l)^2 = 900$
   $2l^2 - 84l + 864 = 0$

$$l = \frac{-(-84) \pm \sqrt{(-84)^2 - 4(2)(864)}}{2 \times 2} = \frac{84 \pm 12}{4}$$

   Either **l = 24 m** or **l = 18 m**
   length, l = 24 m; and width, **w** = 42 – l = **18 m.**

5. Perimeter = 2l + 2w = 21 (l = length; w = width)
   l + w = 10.5 .... (1)
   Also, l × w = 24.5 or l = 24.5/w .... (2)
   From equations (1) and (2): $\frac{24.5}{w} + w = 10.5$
   Multiply both sides by w: $24.5 + w^2 = 10.5w$
   After transposition, $w^2 - 10.5w + 24.5 = 0$

$$w = \frac{-(-10.5) \pm \sqrt{(-10.5)^2 - 4(1)(24.5)}}{2 \times 1} = \frac{10.5 \pm 3.5}{2}$$

   w = 7 m or **w** = **3.5 m**
   l = 10.5 – w = 10.5 – 3.5 = **7 m**

6. length = x; width = y (refer to Figure 4.2)
   2x + y = 121 .... (1)

$x \times y = 605;$ or $y = 605/x$ ....(2)

From equations (1) and (2): $2x + \frac{605}{x} = 121$

Multiply both sides by x, $2x^2 + 605 = 121x$

After transposition, $2x^2 - 121x + 605 = 0$

$$x = \frac{-(-121) \pm \sqrt{(-121)^2 - 4(2)(605)}}{2 \times 2} = \frac{121 \pm 99}{4}$$

x = 55 m (length) or x = 5.5 m

y = width = 605/55 = **11 m**

7. $M = 12 = 9x - 1.5x^2$

   After transposition, $1.5x^2 - 9x + 12 = 0$

   $$x = \frac{-(-9) \pm \sqrt{(-9)^2 - 4(1.5)(12)}}{2 \times 1.5} = \frac{9 \pm 3}{3}$$

   x = 4 m or x = 2 m

8. a) Refer to Figure S4.1: $\frac{OC}{OA} = \frac{FE}{FA}$

   $$\frac{3}{5} = \frac{FE}{5 - x}$$

   $3(5 - x) = 5 \times FE$ or $FE = 3 - 0.6x$

   $DE = 2(3 - 0.6x) = 6 - 1.2x$

   Cross-sectional area of flow $= \frac{DE \times FA}{2}$ *(Area of Δ = Base × Height / 2*

   $$= \frac{(6 - 1.2x)(5 - x)}{2}$$

   $$= 0.6x^2 - 6x + 15$$

   b) $0.6x^2 - 6x + 15 = 5.4$

   After transposition, $0.6x^2 - 6x + 9.6 = 0$

   This is a quadratic equation; it can be solved by the quadratic formula. The solution gives two values of x, i.e. x = 8 m and x = 2 m. Rejecting x = 8m, as x cannot be more than 5 m, the answer is **x = 2 m.**

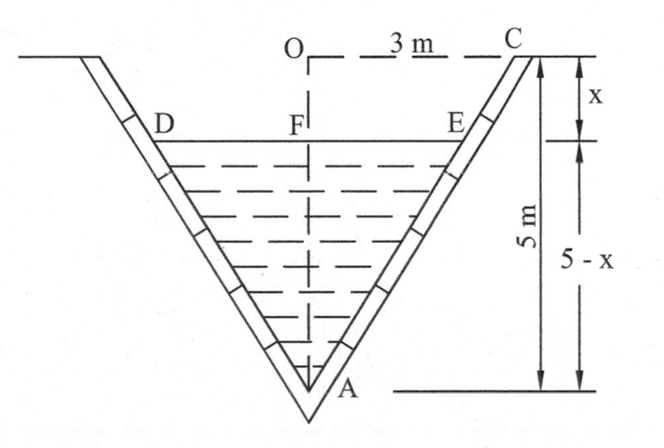

Figure S4.1 Irrigation canal

Solutions – Exercise 5.1

1. See Figure S5.1

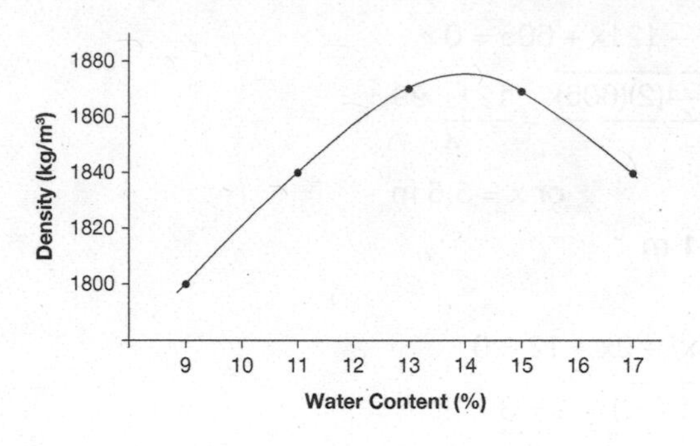

Figure S5.1

2. Co-ordinates of points for equation: $y = 2x - 3$

x	–1	1	4
y	–5	–1	5

Co-ordinates of points for equation: $y = 4 - 0.5x$

x	–2	2	4
y	5	3	2

These points are plotted to get a straight line graph, as shown in Figure S5.2.

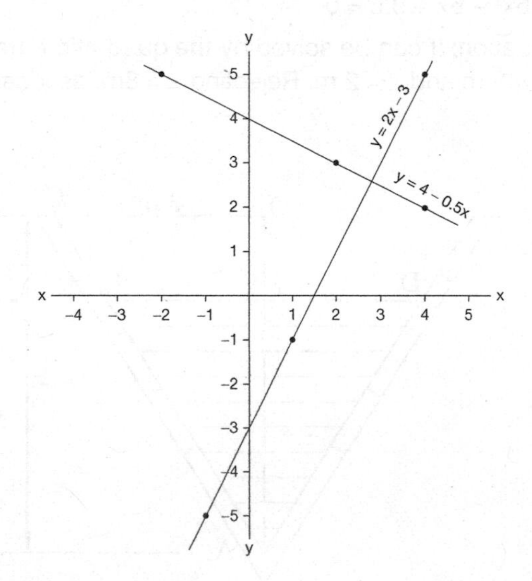

Figure S5.2

3.

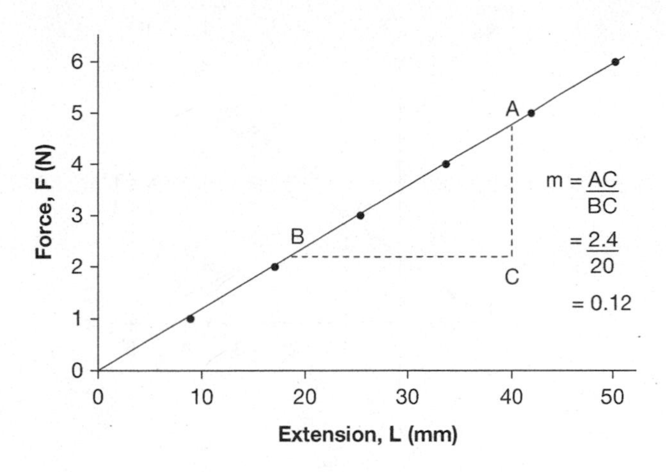

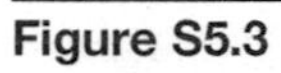
Figure S5.3

a) From Figure S5.3, gradient, **m = 0.12**

b) As the straight line passes through the origin, intercept c = 0

Therefore the law connecting F and L is: F = mL

Replacing m by its value, we have **F = 0.12L**

4. a) Equation 1: $x + y = 5$ or $y = 5 - x$

Equation 1	x	1	3	5
	$y = 5 - x$	4	2	0

Equation 2: $3x - 2y = -5$, $3x + 5 - 2y$, $y = \frac{3x+5}{2}$

Equation 2	x	−1	1	3
	$y = \frac{3x+5}{2}$	1	4	7

The graphs are shown in Figure S5.4a. **Answer: x = 1, y = 4**

b) Equation 1: $x + y = 1$, $y = 1 - x$

Equation 1	x	−3	−1	2
	$y = 1 - x$	4	2	−1

Equation 2: $3x - y = -5$, $y = 3x + 5$

Equation 2	x	−3	−1	1
	$y = 3x + 5$	−4	2	8

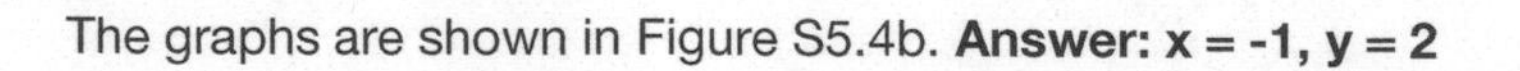

The graphs are shown in Figure S5.4b. **Answer: x = -1, y = 2**

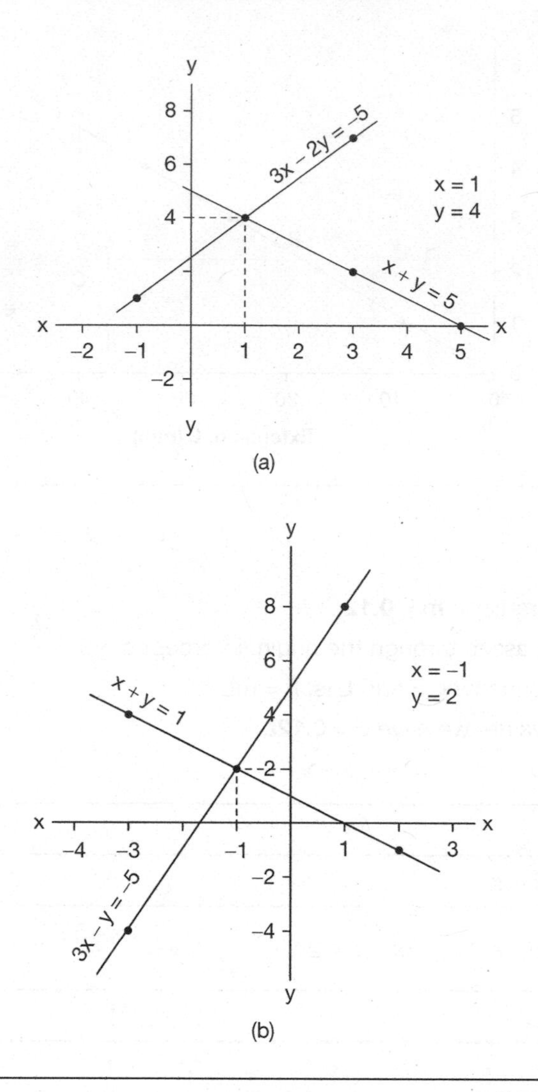

Figure S5.4

5. a) Assume any reasonable values of x, for example we can start off by assuming –6, –5, –4, etc., and then the corresponding values of y are calculated. Depending on the shape of the graph further points may be necessary to find the solution.

x	**–6**	**–5**	**–4**	**–2**	**0**	**2**	**4**	**5**
x2	36	25	16	4	0	4	16	25
x	–6	–5	–4	–2	0	2	4	5
–20	–20	–20	–20	–20	–20	–20	–20	–20
$y = x^2 + x - 20$	**10**	**0**	**–8**	**–18**	**–20**	**–14**	**0**	**10**

We have 8 points: (–6, 10), (–5, 0), (–4, –8), (-2, -18), (0, -20), (2, -14), (4, 0) and (5, 10). The points are joined by a smooth curve (see Figure S5.5a). The curve crosses the x-axis at x = –5 and x = 4. Therefore the solution of $x^2 + x - 20 = 0$ is: either **x = –5 or x = 4**

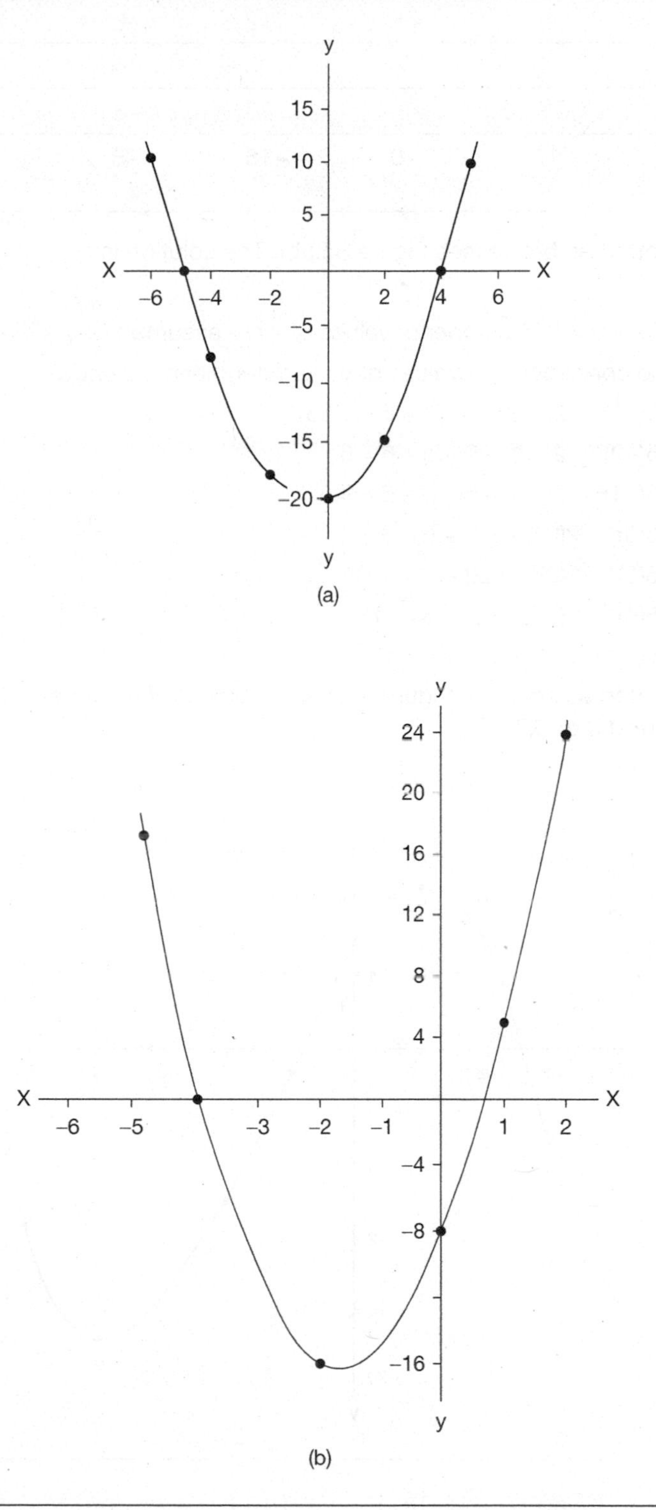

Figure S5.5

b)

x	–6	–5	–4	–2	0	1	2
$3x^2$	108	75	48	12	0	3	12
10x	–60	–50	–40	–20	0	10	20
–8	–8	–8	–8	–8	–8	–8	–8
$y = 3x^2 + 10x - 8$	**40**	**17**	**0**	**–16**	**–8**	**5**	**24**

The points are plotted as before (see Figure S5.5b). The solution is:

$x = -4$, or $x = 0.7$

6. $y = 0.5x^3 - 2x^2 - 6x + 6 = 0$. A number of values of x are assumed (say –3, –2, –1, 0, etc.) and the corresponding values of y are determined, as shown:

$x = -3$, $y = 0.5(-3)^3 - 2(-3)^2 - 6(-3) + 6 = -7.5$

$x = -2$, $y = 0.5(-2)^3 - 2(-2)^2 - 6(-2) + 6 = 6$

$x = -1$, $y = 0.5(-1)^3 - 2(-1)^2 - 6(-1) + 6 = 9.5$

$x = 0$, $y = 0.5(0)^3 - 2(0)^2 - 6(0) + 6 = 6$

$x = 2$, $y = 0.5(2)^3 - 2(2)^2 - 6(2) + 6 = -10$

$x = 4$, $y = 0.5(4)^3 - 2(4)^2 - 6(4) + 6 = -18$

$x = 6$, $y = 0.5(6)^3 - 2(6)^2 - 6(6) + 6 = 6$

The points are plotted as shown in Figure S5.6 and a smooth curve drawn through the points.
Answer: **x = -2.5 or 0.8 or 5.7**

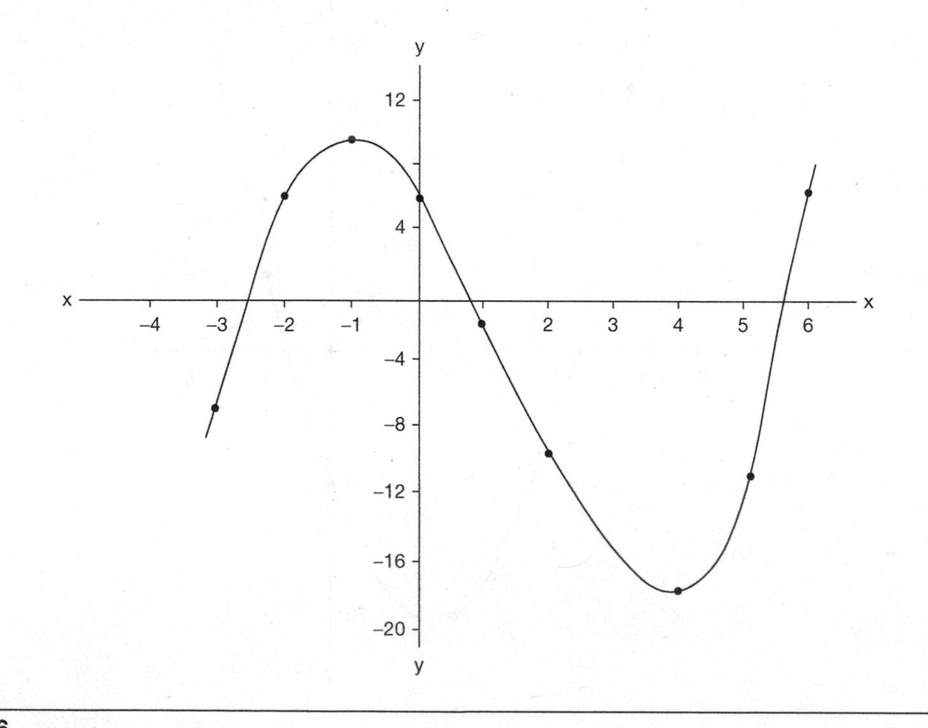

Figure S5.6

7. a) The scatter diagram is shown in **Figure S5.7**

b) $\Sigma Y = a N + b \Sigma X$ .... (1); $\Sigma XY = a \Sigma X + b \Sigma X^2$ .... (2)

The Table below shows the calculation of ΣX, ΣY, ΣXY and ΣX^2:

X	Y	X^2	XY
0	110	0	0
100	190	10000	19000
200	220	40000	44000
300	280	90000	84000
400	380	160000	152000
500	430	250000	215000
$\Sigma X = 1500$	$\Sigma Y = 1610$	$\Sigma X^2 = 550\,000$	$\Sigma XY = 514\,000$

The above values are substituted in equations 1 and 2:

$1610 = a \times 6 + b \times 1500$

$514\,000 = a \times 1500 + b \times 550\,000$

Solving these equations simultaneously, $a = 109.08$ and, $b = 0.637$

The equation of the least squares line is: $Y = 109.08 + 0.637\,X$

c) We need the co-ordinates of two points to draw the best-fit line:

When $X = 100$, $Y = 172.78$

When $X = 450$, $Y = 395.73$

The points are plotted on the graph and a straight line drawn. This line is the least squares line (or Regression line), as shown in Figure S5.7.

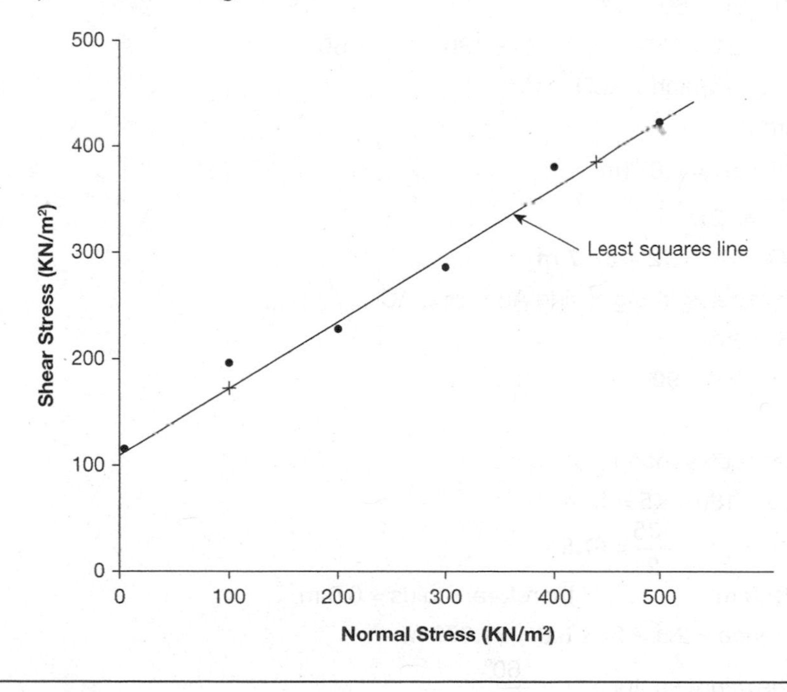

Figure S5.7

Solutions – Exercise 6.1

1. a) $1° = \frac{\pi}{180}$ radians

   $35.5° = 35.5 \times \frac{\pi}{180} = \mathbf{0.6196\ radian}$

   b) $1\ \text{radian} = \frac{180°}{\pi}$ degrees

   $\frac{2\pi}{5}\text{radians} = \frac{2\pi}{5} \times \frac{180°}{\pi} = \mathbf{72}°$

2. Lines AB and CD are parallel

   $\angle c = \angle b = \mathbf{80°}$

   $\angle d = 180° - \angle c = 180 - 80 = \mathbf{100°}$

   $\angle f = \angle a = \mathbf{120°}$

   $\angle e = 180° - 120° = \mathbf{60°}$

   $\angle g = \angle e = \mathbf{60°}$

3. In the isosceles ΔDCF, $\angle 5 = \angle F = \mathbf{30°}$

   $\angle 7 = \angle 8$; and $\angle 7 + \angle 8 = 180 - 30 - 30 = 120°$

   Therefore, $\angle 7 = \angle 8 = \mathbf{60°}$

   $\angle 6 = 180 - (\angle 7 + \angle 8) = 180 - 120 = \mathbf{60°}$

   $\angle 4 = 90 - \angle 5 = 90 - 30 = \mathbf{60°}$

   In ΔACD, $\angle 3 = 180 - (\angle 4 + \angle 6) = 180 - (60 + 60) = \mathbf{60°}$

   $\angle 2 = 90 - \angle 3 = 90 - 60 = \mathbf{30°}$

   In ΔABC, $\angle 1 = 180 - (\angle 2 + 90) = 180 - 120 = \mathbf{60°}$

4. Imagine ΔACD, in which $\angle$ACD = 45°

   AD = CD = 5m

   Side $AC = \sqrt{5^2 + 5^2} = 7.071\text{m}$

   $BD = \sqrt{13^2 - 5^2} = 12\,\text{m}$

   Side BC = BD – CD = 12 – 5 = **7 m**

5. ΔABC is an isosceles triangle; side AB = side AC

   and $\angle 2 = \angle B = \mathbf{45°}$

   $\angle 1 = 180 - 45 - 45 = \mathbf{90°}$

   $\angle 3 = \angle 2 = \mathbf{45°}$

   ΔCED is an isosceles triangle, $\angle 4 = \angle 5$

   Also, $\angle 4 + \angle 5 = 180 - 45 = 135°$

   Therefore, $\angle 4 = \angle 5 = \frac{135}{2} = \mathbf{67.5}°$

6. Diameter = 12.0 m; therefore, radius = 6.0 m

   a) Circumference $= 2\pi r = 2\pi \times 6.0 = \mathbf{37.699\ m}$

   b) Area of the minor sector $= \pi r^2 \times \frac{60°}{360°}$

$$= \pi(6.0)^2 \times \frac{60^\circ}{360^\circ} = \mathbf{18.85\ m^2}$$

$$\text{Area of the major sector} = \pi r^2 \times \frac{300^\circ}{360^\circ}$$

$$= p(6.0)^2 \times \frac{300^\circ}{360^\circ} = \mathbf{94.248\ m^2}$$

c) $$\text{Length of arc ACB} = \text{Circumference} \times \frac{60^\circ}{360^\circ}$$

$$= 2\pi r \times \frac{60^\circ}{360^\circ} = \mathbf{6.283\ m}$$

7. **Method 1:** Area of a triangle $= \dfrac{\text{base} \times \text{height}}{2}$

Base = 12 cm; Vertical height $= \sqrt{15^2 - 6^2} = 13.748$ cm

$$\text{Area} = \frac{12 \times 13.748}{2} = \mathbf{82.49\ cm^2}$$

Method 2 : Area of a triangle $= \sqrt{s(s-a)(s-b)(s-c)}$

$$s = \frac{12+15+15}{2} = 21$$

$$\text{Area of the triangle} = \sqrt{21(21-12)(21-15)(21-15)}$$

$$= \sqrt{21(9)(6)(6)} = \sqrt{6804} = \mathbf{82.49\ cm^2}$$

Method 3 : Area of the triangle $= \dfrac{1}{2} ac \sin\theta = \dfrac{1}{2} \times 12 \times 15 \times \dfrac{13.748}{15}$

$$= \mathbf{82.49\ cm^2} \qquad (a = 12;\ c = 15;\ \sin\theta = \frac{13.748}{15})$$

8. Area of top and bottom plates = 2 × (500 × 20) = 20 000 mm^2

Area of the vertical plate = 800 × 30 = 24 000 mm^2

Area of one angle section = 150 × 18 + (150 − 18) × 18 = 5076 mm^2

Area of 4 angle section (Figure 6.27b) = 4 × 5076 = 20 304 mm^2

Total area = 20 000 + 24 000 + 20 304 = **64304 mm^2 or 643.04 cm^2**

9. Cross-sectional area of the dry part of the drain

= Area of sector OAB − Area of ΔOAB

$$= \frac{1}{2} r^2(\theta - \sin\theta)$$

For calculating angle θ, consider right-angled triangle OAD:

$$\cos\frac{\theta}{2} = \frac{OD}{OA} = \frac{100}{225} = 0.44444 \qquad (OD = 225 - 125 = 100\ mm)$$

$$\frac{\theta}{2} = \cos^{-1} 0.44444 = 1.11024\ \text{radians}$$

Therefore, $\angle\theta = 2.2205$ radians

Cross-sectional area of the dry part of the drain $= \dfrac{1}{2} \times 225^2 \times (2.2205 - \sin 2.2205)$

$$= 25312.5 \times (2.2205 - 0.79626) = 36051.08\ mm^2$$

Cross-sectional area of water flow = Cross-sectional area of the drain – 36 051.08

$= \pi \times 225^2 - 36\,051.08$

$= 159\,043.13 - 36\,051.08 = \mathbf{122\,992.05\ mm^2}$

10. Let R be the radius of the arch

OA = R metres; OD = (R – 1) metres

$(OA)^2 - (OD)^2 = 4^2$; or $R^2 - (R-1)^2 = 16$

$R^2 - (R^2 - 2R + 1) = 16$; Therefore R = 8.5 m

$$\text{Area of sector OABC} = \pi \times (8.5)^2 \times \frac{56}{360} = 35.308\ m^2$$

$$\text{Area of } \Delta OAC = \frac{8 \times 7.5}{2} = 30\ m^2$$

Area of the shaded part = 35.308 – 30.0 = **5.308 m²**

11. $y_1 = 30 + 16 = 46$ m, $y_2 = 29 + 18 = 47$ m, $y_3 = 52$ m, $y_4 = 55$ m, $y_5 = 51$ m

$y_6 = 48$ m, $y_7 = 42$ m

Mid-ordinate Rule: Area = 10[46.5 + 49.5 + 53.5 + 53 + 49.5 + 45]

= 10[297] = **2970 m²**

$$\text{Trapezoidal Rule : Area} = 10[\frac{46+42}{2} + 47 + 52 + 55 + 51 + 48]$$

= 10[297] = **2970 m²**

$$\text{Simpson's rule: Area} = \frac{10}{3}\left[(46+42) + 4(47+55+48) + 2(52+51)\right]$$

$$= \frac{10}{3}\left[88+600+206\right] = \mathbf{2980\ m^2}$$

12. Radius of the top hemispherical portion = 600 ÷ 2 = 300 mm or 0.3 m

$$\text{Volume} = \frac{1}{2}\left(\frac{4}{3}\pi r^3\right) = \frac{1}{2}\left(\frac{4}{3}\pi (0.3)^3\right) = 0.05655\ m^3$$

Volume of the cylindrical portion = $\pi r^2 h = \pi(0.3)^2 \times 0.9 = 0.25447\ m^3$

Volume of the cylinder = 0.05655 + 0.25447 = **0.311 m³** (3 d.p.)

= 0.31102 × 1000 = **311.02 litres**

13. $A_1 = 1.7\ m^2$, $A_2 = 1.68\ m^2$, $A_3 = 1.63\ m^2$, $A_4 = 1.75\ m^2$, $A_5 = 1.8\ m^2$

$$\text{Trapezoidal Rule : Volume} = 10[\frac{1.7+1.8}{2} + 1.68 + 1.63 + 1.75]$$

= 10[6.81] = **68.1 m³**

$$\text{Simpson's rule : Area} = \frac{10}{3}\left[(1.7+1.8) + 4(1.68+1.75) + 2(1.63)\right]$$

$$= \frac{10}{3}\left[3.5+13.72+3.26\right] = \mathbf{68.27\ m^3}$$

14. There are 5 sections in the vertical direction and 5 in the horizontal direction. The solution here is based on the sections in the vertical direction, but sections in the horizontal direction may also be used. Figure S6.1 shows the spot levels and the depth of the excavation for 2 sections, Sections A–A and B–B. Similar figures may be produced for sections C–C, D–D and E–E.

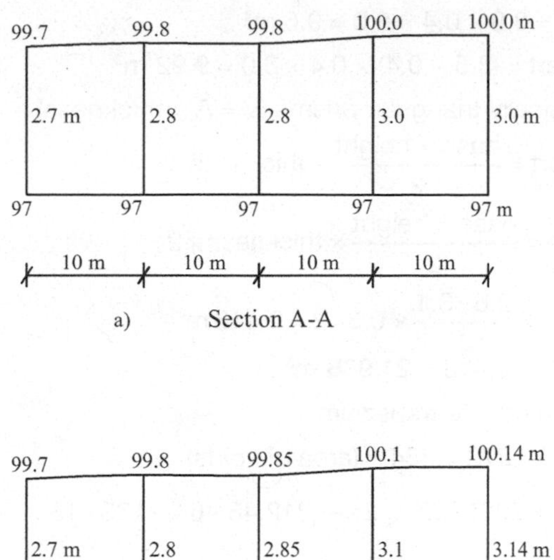

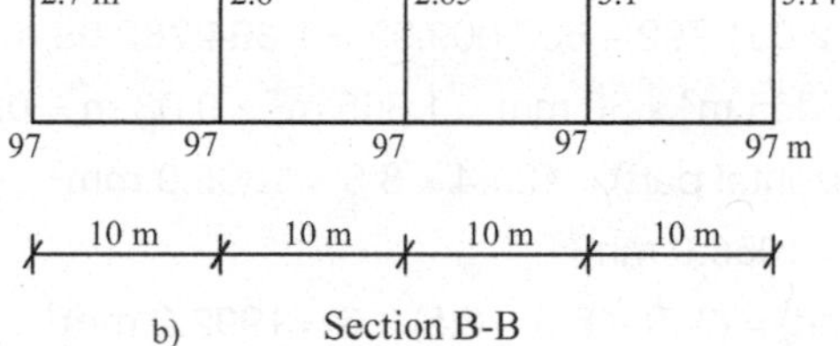

Figure S6.1

The table below shows the depth of the excavation at points where the ground level have been taken (for all sections):

Section A–A	2.7	2.8	2.8	3.0	3.0
Section B–B	2.7	2.8	2.85	3.1	3.14
Section C–C	2.7	2.9	2.9	3.1	3.16
Section D–D	2.6	2.95	3.0	3.15	3.2
Section E–E	2.6	3.0	3.08	3.2	3.3

The area of each section could be calculated by any method (mid-ordinate rule, trapezoidal rule or Simpson's rule). Here trapezoidal rule will be used for area as well as volume calculation.

Section A - A : Area $= 10[\frac{2.7+3.0}{2} + 2.8 + 2.8 + 3.0] = 114.5\,\text{m}^2$

Similarly the areas of Sections B–B, C–C, D–D and E–E can be calculated. These are: 116.7, 118.3, 120.0 and 122.3 m^2, respectively.

The volume (V) of the excavation can also be calculated by any method (mid-ordinate rule, trapezoidal rule or Simpson's rule). Here, the trapezoidal rule is used:

$$V = 10[\frac{\text{Area of first + Area of last section}}{2} + \text{Sum of areas of remaining sections}]$$

$$= 10[\frac{114.5+122.3}{2} + 116.7 + 118.3 + 120.0] = \mathbf{4734.0\,m^3}$$

15. Volume of the wall base = 3.0 × 0.4 × 8.0 = 9.6 m^3

Volume of the vertical part = (3.5 – 0.4) × 0.4 × 8.0 = 9.92 m^3

The counterforts are basically triangular prisms (V = A × thickness)

$$\text{Volume of one counterfort} = \frac{\text{base} \times \text{height}}{2} \times \text{thickness}$$

$$\text{Volume of 2 counterforts} = \frac{\text{base} \times \text{height}}{2} \times \text{thickness} \times 2$$

$$= \frac{2.6 \times 3.1}{2} \times 0.3 \times 2 = 2.418\,\text{m}^3$$

Total volume = 9.6 + 9.92 + 2.418 = **21.938 m^3**

16. a) The void is basically a double trapezium

Net area of the plate = 5213 × 384 – (area of voids)

$$= 2001792 - \left(2 \times \frac{1}{2}(212.48 + 64) \times 128 \times 18\right)$$

= 2 001 792 – 637 009.92 = **1 364 782.08 mm^2** = **1.365 m^2**

b) Volume of the plate = 1.365 m^2 × 30 mm = 1.365 m^2 × 0.03 m = **0.041 m^3**

17. Area of the top flange (horizontal part) = 125.4 × 8.5 = 1065.9 mm^2

Area of the bottom flange = 1065.9 mm^2

Area of the web (vertical part) = (349 – 8.5 – 8.5) × 6 = 1992.0 mm^2

Extra steel where the web meets the flange (as shown in Figure 9.35b)

$$= 10.2 \times 10.2 - \frac{\pi \times 10.2^2}{4} = 22.327\,\text{mm}^2$$

As there are 4 points where the web meets the flange:

Extra steel area = 4 × 22.327 = 89.31 mm^2

Cross-sectional area = 1065.9 + 1065.9 + 1992.0 + 89.31 = **4213.11 mm^2**

Mass/metre = Volume × density

$$= \frac{4213.11}{1000 \times 1000} \times 1.0 \times 7850 = \mathbf{33.073\,kg}$$

18. The solid wall is generated by rotating trapezium ABCD about axis x-x through 90° as shown in Figure 6.36.

The position of the centroid $(\bar{x})$ of the wall, from point D, can be determined by taking the moment of areas about point D:

$$\text{Total area} \times \bar{x} = \text{Area of the rectangular part} \times 0.8 + \text{Area of the triangular part} \times \left(\frac{2}{3} \times 0.6\right)$$

$$\frac{1}{2}(1.0 + 0.4) \times 2 \times \bar{x} = 0.4 \times 2 \times 0.8 + \frac{1}{2} \times 0.6 \times 2 \times \left(\frac{2}{3} \times 0.6\right)$$

$$1.4\bar{x} = 0.88$$

$$\bar{x} = 0.629$$

Distance of the centroid from face BC of the wall is:

1.0 m – 0.629 m = 0.371 m

Distance between axis x-x and the centroid of the wall = 75 – 0.371 = 74.629 m

$$\text{Area of the wall} = \frac{1}{2}(1.0 + 0.4) \times 2 = 1.4\,\text{m}^2$$

$$\text{Volume} = 2\pi\bar{y}A \times \frac{90^\circ}{360^\circ} \qquad (\bar{y} = 74.629\,\text{m})$$

$$= 2\pi \times 74.629 \times 1.4 \times \frac{90^\circ}{360^\circ} = \mathbf{164.118\ m^3}$$

Solutions – Exercise 7.1

1. Refer to Figure S7.1

$$\frac{CB}{AB} = \tan 7^\circ \qquad (CB = 8.0 - 1.35 = 6.65\ \text{m})$$

$$\frac{6.65}{AB} = \tan 7^\circ,\ AB = \frac{6.65}{\tan 7^\circ} = \mathbf{54.16\ m}$$

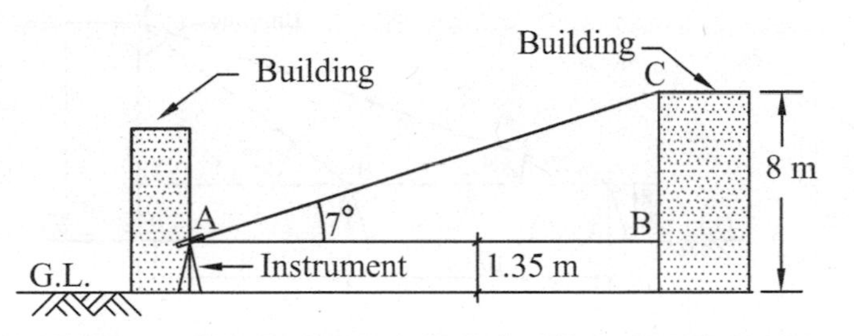

Figure S7.1

2. Refer to Figure S7.2; Let BC be the distance between the foot of the ladder and the wall.

$$\frac{BC}{AB} = \cos 75^\circ$$

$$\frac{BC}{6} = \cos 75^\circ, \qquad \text{Therefore } \mathbf{BC} = 6 \times \cos 75^\circ = \mathbf{1.553\ m}$$

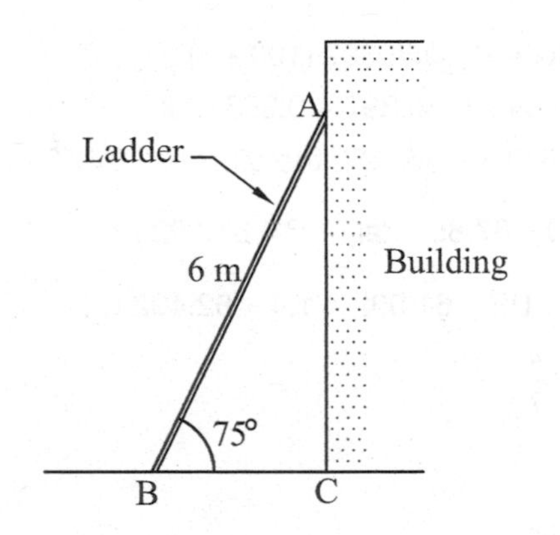

Figure S7.2

3. Width of the river = distance TS, as shown in Figure 7.17.

$$\frac{TS}{60} = \tan 60^\circ$$

TS = 60 × tan 60° = **103.923 m**

4. Refer to Figure 7.18.

Height of the building = AB

$$\frac{BC}{100} = \tan 20^\circ, \qquad BC = 100 \times \tan 20^\circ = 36.397\,m$$

$$\frac{AC}{100} = \tan 60^\circ, \qquad AC = 100 \times \tan 60^\circ = 173.205\,m$$

Height of the building (**AB**) = AC – BC = 173.205 – 36.397 = **136.808 m**

5. Height of the building = CE (as shown in Figure S7.3)

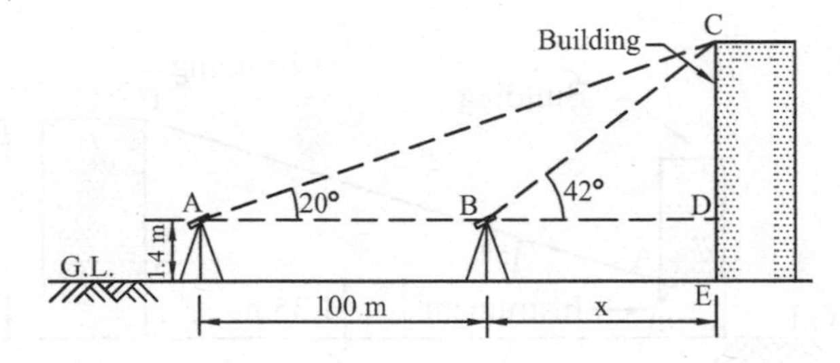

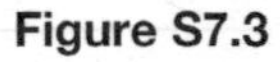

Figure S7.3

Let distance BD = x metres

In ΔCAD, $\frac{CD}{(100+x)} = \tan 20^\circ$

Therefore, CD = 0.36397 (100 + x) (1)

In ΔCBD, $\frac{CD}{x} = \tan 42^\circ$

Therefore, CD = 0.9004 x (2)

From (1) and (2)

$$0.9004\,x = 0.36397\,(100 + x)$$

$$0.9004\,x = 36.397 + 0.36397\,x$$

$$0.53643\,x = 36.397, \text{ therefore } x = 67.85\,m$$

$$\frac{CD}{67.85} = \tan 42^\circ; \qquad CD = 67.85 \times \tan 42^\circ = 61.092\,m$$

Height of the building = CD + DE = 61.092 + 1.4 = **62.492 m**

6. Refer to Figure S7.4; $\tan x = \frac{4}{7}$

$$\angle x = \tan^{-1}\left(\frac{4}{7}\right) = 29.745^\circ$$

a) Pitch of the roof = **29.745°**

b) Let L be the length of common rafters

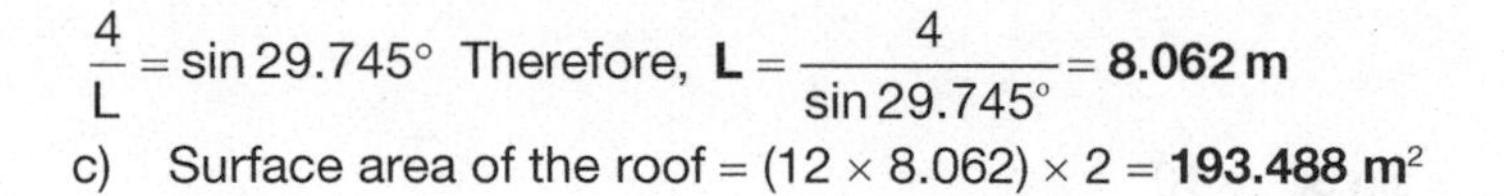

$\frac{4}{L} = \sin 29.745°$ Therefore, $\mathbf{L} = \frac{4}{\sin 29.745°} = \mathbf{8.062\ m}$

c) Surface area of the roof = (12 × 8.062) × 2 = **193.488 m²**

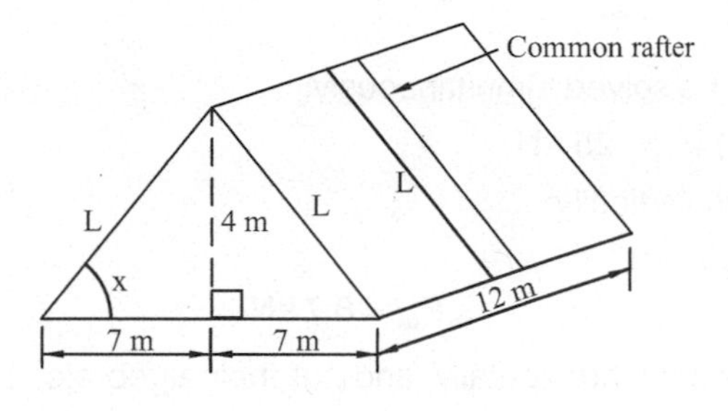

Figure S7.4

7. Refer to Figure 7.19

In ΔABJ, $\frac{AJ}{5} = \tan 30°$, therefore AJ = 5 × tan 30° = 2.887 m

In ΔAHJ, $\frac{AJ}{AH} = \sin 60°$ or $\frac{AJ}{\sin 60°} = AH$

Length of member **AH** $= \frac{2.887}{\sin 60°} = \mathbf{3.334\ m}$

In ΔAHJ, $\frac{2.887}{HJ} = \tan 60°$, or $HJ = \frac{2.887}{\tan 60°} = 1.667\ m$

BH = 5.0 − HJ = 5.0 − 1.667 = 3.333 m

In ΔDBH, $\frac{HD}{3.333} = \sin 30°$, or **HD** = 3.333 × sin 30° = **1.667 m**

Member HE = 3.333 × tan 30° = **1.924 m**

8. Refer to Figure 7.20.

At joint (JT.) 1: Resolve forces in the members vertically, and put their algebraic sum equal to 0:

Vertical component of $F_{BF} - R_1 = 0$ (1)

Vertical component of $F_{BF} = F_{BF} \times \sin\theta$ (explained earlier in section 7.6)

$= F_{BF} \times \sin 30° = 0.5\ F_{BF}$

Putting values in equation (1), $0.5\ F_{BF} - 15 = 0$ or $\mathbf{F_{BF} = 30\ kN}$

Resolve forces in the members horizontally, and put their algebraic sum equal to 0:

Horizontal component of $F_{BF} - F_{FA} = 0$ (2)

Horizontal component of $F_{BF} = F_{BF} \times \cos\theta$

$= 30 \times \cos 30° = 25.98$ or 26.0 kN (1 d.p.)

From equation (2), $26 - F_{FA} = 0$ or $\mathbf{F_{FA} = 26\ kN}$

At joint 2: Resolve forces in the members horizontally, and put their algebraic sum equal to 0:

Horizontal component of F_{BF} = Horizontal component of F_{CG} + Horizontal component of F_{GF}

$F_{BF} \times \cos 30° = F_{CG} \times \cos 30° + F_{GF} \times \cos 60°$

$30 \times 0.866 = F_{CG} \times 0.866 + F_{GF} \times 0.5$

or $0.866\, F_{CG} + 0.5\, F_{GF} = 26$ (3)

Resolve forces vertically:

$F_{BF} \times \sin 30° + F_{GF} \times \sin 60° = F_{CG} \times \sin 30° + 10$

$30 \times 0.5 + F_{GF} \times 0.866 = F_{CG} \times 0.5 + 10$

$0.5\, F_{CG} - 0.866\, F_{GF} = 5$ (4)

Equations (3) and (4) can be solved simultaneously:

$F_{GF} = 8.67$ (8.7 kN: 1 d.p.) $F_{CG} = 25$ kN

At Joint 3: Resolve forces vertically:

$F_{GF} \times \sin 60° - F_{GH} \times \sin 60° = 0$

$8.7 \times 0.866 - F_{GH} \times \sin 60° = 0, \quad \therefore \mathbf{F_{GH} = 8.7\ kN}$

Resolve forces in the members horizontally, and put their algebraic sum equal to 0:

$F_{FA} = F_{GF} \times \cos 60° + F_{GH} \times \cos 60° + F_{HA}$

$26 = 8.7 \times 0.5 + 8.7 \times 0.5 + F_{HA} \quad \therefore \mathbf{F_{HA} = 17.3\ kN}$

As the frame is symmetrical, force in member EK is equal to the force in member BF; force in member DJ is equal to the force in member CG. Similarly, the forces in other members can be deduced as well.

Solutions – Exercise 7.2

1. $\angle P = 180° - 50° - 70° = \mathbf{60°}$

   Using the sine rule, $\dfrac{p}{\sin 60°} = \dfrac{q}{\sin 50°} = \dfrac{75}{\sin 70°}$

   $$p(\text{side QR}) = \frac{75 \times \sin 60°}{\sin 70°} = \mathbf{69.12\ cm}$$

   $$q(\text{side PR}) = \frac{75 \times \sin 50°}{\sin 70°} = \mathbf{61.14\ cm}$$

2. a = 8 cm, b = 7 cm, c = 5 cm

   Using the cosine rule, $b^2 = c^2 + a^2 - 2ac \cos B$

   $7^2 = 5^2 + 8^2 - 2 \times 8 \times 5 \times \cos B$

   $49 = 25 + 64 - 80 \cos B$

   $80 \cos B = 40, \quad \cos B = 0.5$

   Therefore $\angle B = \cos^{-1} 0.5 = \mathbf{60°}$

3. Let AB be the distance between the two buildings. Another point C will form a triangle with A and B. Side c of triangle CAB will represent the distance between the 2 buildings.

   From the cosine rule, $c^2 = a^2 + b^2 - 2ab \cos C$

   $$c^2 = 260^2 + 220^2 - 2 \times 260 \times 220 \times \cos 120°$$

   $$= 67\,600 + 48\,400 - (-57\,200) = 173\,200$$

   $$c = \sqrt{173200} = \mathbf{416.173\ m}$$

4. Refer to Figure 7.21.

   Using the sine rule, $\dfrac{3.5}{\sin 55°} = \dfrac{b}{\sin B} = \dfrac{2.5}{\sin C}$

$$\sin C = \frac{2.5 \times \sin 55^\circ}{3.5} = 0.58511$$

$\angle C = \sin^{-1} 0.58511 = 35.81^\circ$ or 144.19° (144.19° is not possible)

The strut makes an angle of **35.81° with the wall**

$\angle B = 180^\circ - 55^\circ - 35.81^\circ = 89.19^\circ$

$$\frac{3.5}{\sin 55^\circ} = \frac{b}{\sin 89.19^\circ}$$

$$b\,(\text{Length of the strut}) = \frac{3.5 \times \sin 89.19^\circ}{\sin 55^\circ} = \mathbf{4.272\,m}$$

5. Refer to Figure 7.22, which shows the parallelogram of forces.

$\angle C + \angle D = 180^\circ$, therefore $\angle C = 180^\circ - 70^\circ = 110^\circ$

In ΔBDC, $c^2 = d^2 + b^2 - 2db \cos C$

$= 40^2 + 70^2 - 2 \times 40 \times 70 \times \cos 110^\circ$

$= 8415.3128$

c (Resultant) = $\sqrt{8415.3128}$ = **91.735 kN**

In ΔBDC, $\dfrac{b}{\sin B} = \dfrac{c}{\sin C} = \dfrac{d}{\sin X}$

From above $\dfrac{c}{\sin C} = \dfrac{d}{\sin X}$, or $\sin X = \dfrac{d \times \sin C}{c}$

$$= \frac{40 \times \sin 110^\circ}{91.735} = 0.40974$$

$\angle X - \sin^{-1} 0.40974 = \mathbf{24.19^\circ}$

6. Refer to Figure 7.23 and Figure S7.5.

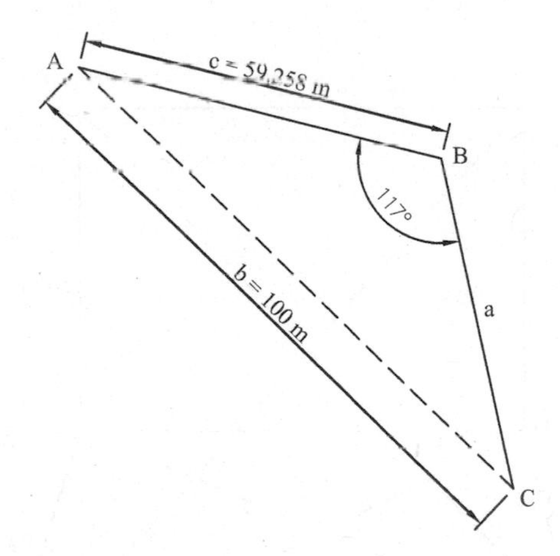

Figure S7.5

ΔADC is a right-angled triangle

$(AC)^2 = 80^2 + 60^2 = 10\,000$

AC = $\sqrt{10\,000}$ = 100 m

In ΔABC, $\dfrac{100}{\sin 117^\circ} = \dfrac{59.258}{\sin C}$

$\sin C = \dfrac{59.258 \times \sin 117^\circ}{100} = 0.528$

$\angle C = 31.87^\circ$ and $\angle A = 180^\circ - 117^\circ - 31.87^\circ = 31.13^\circ$

Area of $\Delta ADC = \dfrac{60 \times 80}{2} = 2400\ \text{m}^2$

Area of $\Delta ABC = \dfrac{1}{2} \times bc \sin A = \dfrac{1}{2} \times 100 \times 59.258 \times \sin 31.13^\circ = 1531.765\ \text{m}^2$

Area of the building site (ABCD) = 2400 + 1531.765 = **3931.765 m²**

7. Refer to Figure S7.6: Distance AB = c.

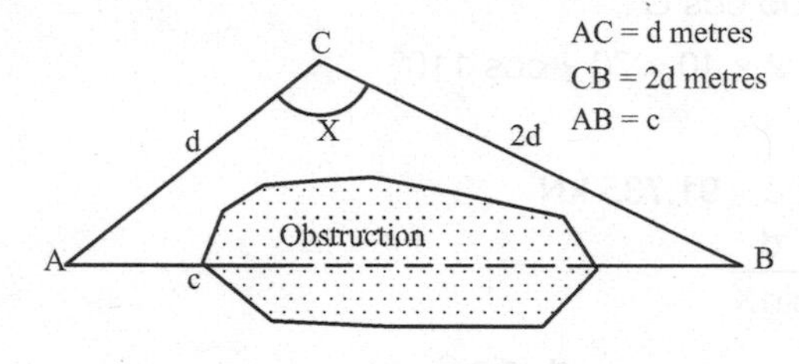

Figure S7.6

From Cosine rule: $c^2 = d^2 + (2d)^2 - 2 \times d \times 2d \times \cos X^\circ$

$c^2 = 5\,d^2 - 4\,d^2 \cos X^\circ$

$= d^2\,(5 - 4\cos X^\circ)$

(AB) or $c = d\sqrt{(5 - 4\cos X^\circ}$

8. Refer to Figure S7.7.

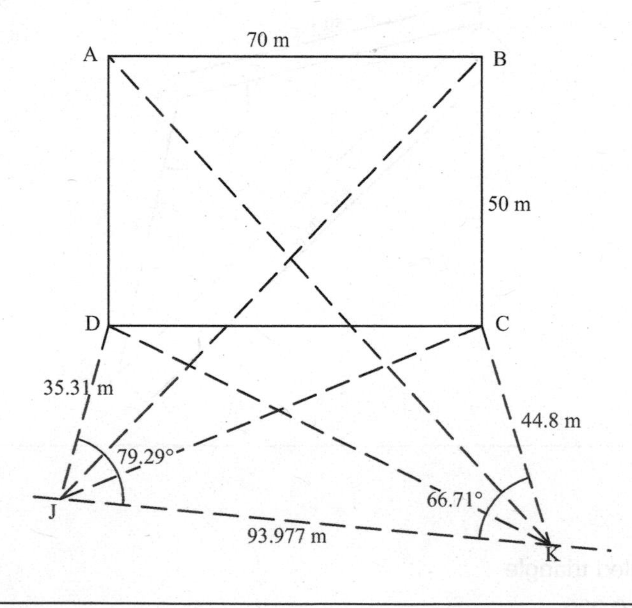

Figure S7.7

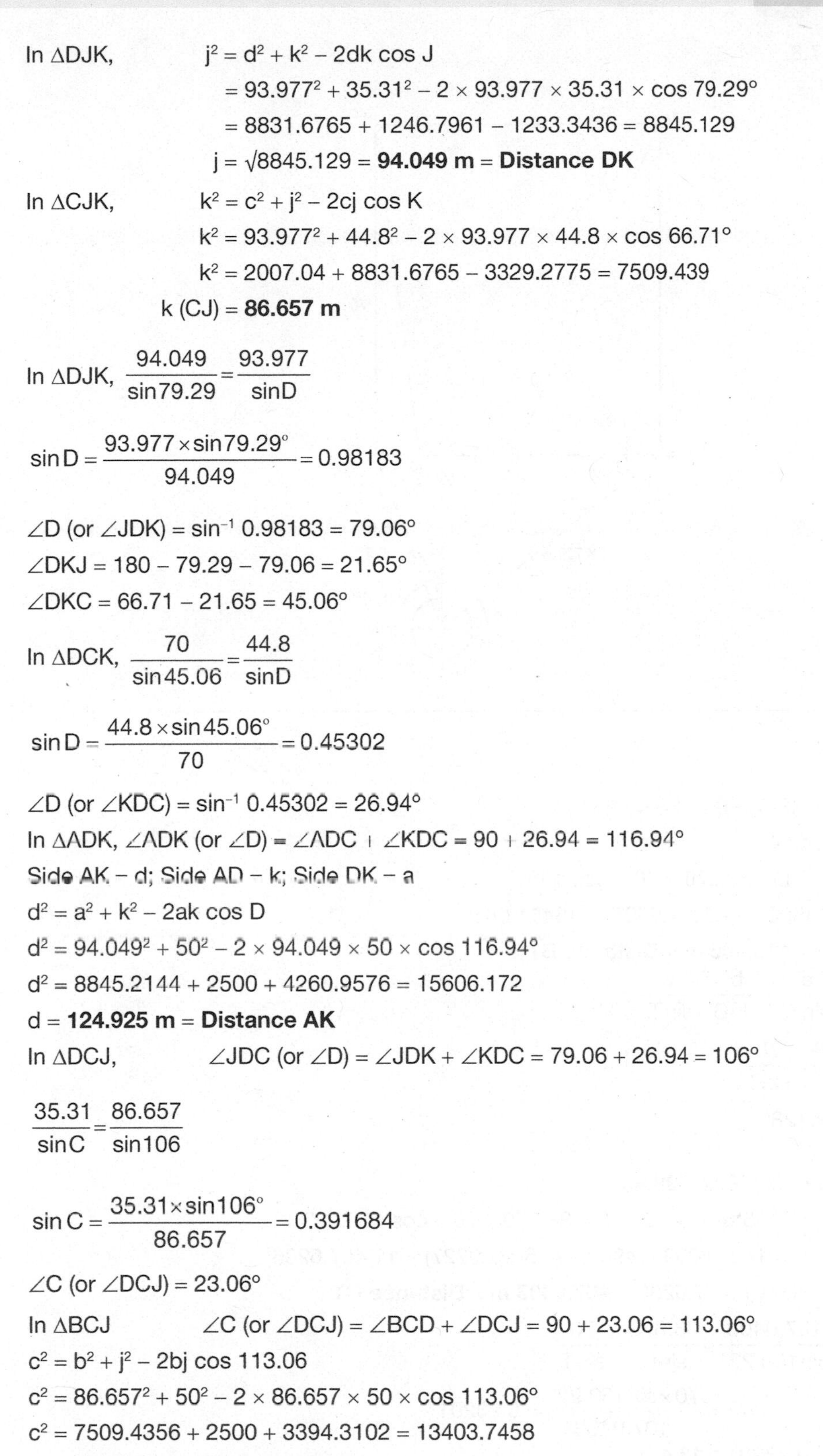

In ΔDJK, $j^2 = d^2 + k^2 - 2dk \cos J$

$= 93.977^2 + 35.31^2 - 2 \times 93.977 \times 35.31 \times \cos 79.29°$

$= 8831.6765 + 1246.7961 - 1233.3436 = 8845.129$

$j = \sqrt{8845.129} =$ **94.049 m = Distance DK**

In ΔCJK, $k^2 = c^2 + j^2 - 2cj \cos K$

$k^2 = 93.977^2 + 44.8^2 - 2 \times 93.977 \times 44.8 \times \cos 66.71°$

$k^2 = 2007.04 + 8831.6765 - 3329.2775 = 7509.439$

k (CJ) = **86.657 m**

In ΔDJK, $\dfrac{94.049}{\sin 79.29} = \dfrac{93.977}{\sin D}$

$$\sin D = \frac{93.977 \times \sin 79.29°}{94.049} = 0.98183$$

∠D (or ∠JDK) = $\sin^{-1} 0.98183 = 79.06°$

∠DKJ = 180 − 79.29 − 79.06 = 21.65°

∠DKC = 66.71 − 21.65 = 45.06°

In ΔDCK, $\dfrac{70}{\sin 45.06} = \dfrac{44.8}{\sin D}$

$$\sin D = \frac{44.8 \times \sin 45.06°}{70} = 0.45302$$

∠D (or ∠KDC) = $\sin^{-1} 0.45302 = 26.94°$

In ΔADK, ∠ADK (or ∠D) = ∠ADC + ∠KDC = 90 + 26.94 = 116.94°

Side AK = d; Side AD = k; Side DK = a

$d^2 = a^2 + k^2 - 2ak \cos D$

$d^2 = 94.049^2 + 50^2 - 2 \times 94.049 \times 50 \times \cos 116.94°$

$d^2 = 8845.2144 + 2500 + 4260.9576 = 15606.172$

d = **124.925 m** = **Distance AK**

In ΔDCJ, ∠JDC (or ∠D) = ∠JDK + ∠KDC = 79.06 + 26.94 = 106°

$$\frac{35.31}{\sin C} = \frac{86.657}{\sin 106}$$

$$\sin C = \frac{35.31 \times \sin 106°}{86.657} = 0.391684$$

∠C (or ∠DCJ) = 23.06°

In ΔBCJ ∠C (or ∠DCJ) = ∠BCD + ∠DCJ = 90 + 23.06 = 113.06°

$c^2 = b^2 + j^2 - 2bj \cos 113.06$

$c^2 = 86.657^2 + 50^2 - 2 \times 86.657 \times 50 \times \cos 113.06°$

$c^2 = 7509.4356 + 2500 + 3394.3102 = 13403.7458$

c = **115.775 m** = **Distance BJ**

9. Refer to Figure S7.8.

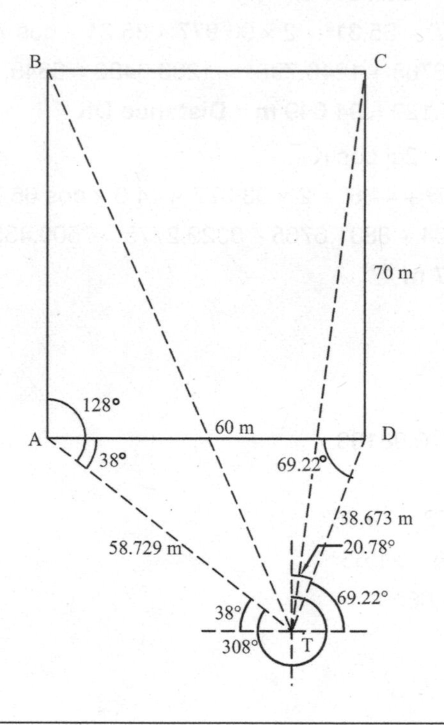

Figure S7.8

In ΔABT, $\angle BAT = 90 + 38 = 128^\circ$

$a^2 = b^2 + t^2 - 2bt \cos A$

$= 58.279^2 + 70^2 - 2 \times 58.279 \times 70 \times \cos 128^\circ$

$= 3449.0954 + 4900 - (-5061.9935) = 13411.0889$

$a = \sqrt{13411.0889}$ = **115.806 m = Distance BT**

Again in ΔABT, $\dfrac{a}{\sin A} = \dfrac{b}{\sin B} = \dfrac{t}{\sin T}$

$$\frac{115.806}{\sin 128^\circ} = \frac{58.729}{\sin B} = \frac{70}{\sin T}$$

$$\sin T = \frac{70 \times \sin 128^\circ}{115.806}$$

$\angle T$ (or $\angle ATB$) = $\sin^{-1} 0.47632$ = **28.45°**

In ΔCTD, $d^2 = 38.673^2 + 70^2 - 2 \times 38.673 \times 70 \times \cos 159.22^\circ$

$= 1495.6009 + 4900 - (-5062.0227) = 11\,457.6236$

$d = \sqrt{11457.6236}$ = **107.0403 m = Distance CT**

Again in ΔCTD, $\dfrac{107.0403}{\sin 159.22^\circ} = \dfrac{38.673}{\sin C} = \dfrac{70}{\sin T}$

$$\sin T = \frac{70 \times \sin 159.22^\circ}{107.0403} = 0.23201$$

$\angle T$ ($\angle CTD$) = $\sin^{-1} 0.23201$ = **13.42°**

Solutions – Exercise 8.1

1. a) $\dfrac{\cos A \tan A}{\sin^2 A} = \dfrac{\cos A \times \dfrac{\sin A}{\cos A}}{\sin A \times \sin A}$

$$= \frac{\sin A}{\sin A \times \sin A} = \frac{1}{\sin A} = \mathbf{cosec\ A}$$

b) $\sin A \cos A \cot A = \sin A \cos A \dfrac{\cos A}{\sin A}$

$$= \cos A \times \cos A = \mathbf{\cos^2 A}$$

2. Refer to Figure 8.2 that shows a right-angled triangle ABC. Using Pythagoras' theorem:

$a^2 + b^2 = c^2$

Divide both sides by b^2: $\dfrac{a^2}{b^2} + \dfrac{b^2}{b^2} = \dfrac{c^2}{b^2}$

$$\left(\frac{a}{b}\right)^2 + 1 = \left(\frac{c}{b}\right)^2$$

$$\left(\frac{1}{\tan\theta}\right)^2 + 1 = \left(\frac{1}{\sin\theta}\right)^2$$

$\cot^2\theta + 1 = \text{cosec}^2\theta$ or $\mathbf{1 + \cot^2\theta = cosec^2\theta}$

3. $\dfrac{\sin\theta + \tan\theta}{\cot\theta + \text{cosec}\,\theta} = \dfrac{\dfrac{b}{c} + \dfrac{b}{a}}{\dfrac{a}{b} + \dfrac{c}{b}} = \dfrac{\dfrac{ab+bc}{ca}}{\dfrac{a+c}{b}}$

$$= \frac{ab+bc}{ca} \times \frac{b}{(a+c)}$$

$$= \frac{b(a+c)}{ca} \times \frac{b}{(a+c)} = \frac{b \times b}{c \times a}$$

$$\frac{b}{c} \times \frac{b}{a} = \mathbf{\sin\theta \tan\theta}$$

4. a) $\tan A + \cot A = \dfrac{\sin A}{\cos A} + \dfrac{\cos A}{\sin A}$

$$= \frac{\sin^2 A + \cos^2 A}{\sin A \cos A}$$

$$= \frac{1}{\sin A \cos A} = \mathbf{cosec\ A\ sec\ A}$$

b) $\dfrac{\cos A}{1+\sin A} = \dfrac{\cos A}{1+\sin A} \times \dfrac{1-\sin A}{1-\sin A}$

$$= \frac{\cos A(1-\sin A)}{(1+\sin A)(1-\sin A)}$$

$$= \frac{\cos A(1-\sin A)}{1-\sin^2 A}$$

$$=\frac{\cos A(1-\sin A)}{\cos^2 A}=\frac{1-\sin A}{\cos A}$$

c) $$\frac{\sec A}{\tan A+\cot A}=\frac{\sec A}{\tan A+\dfrac{1}{\tan A}}$$

$$=\frac{\sec A}{\tan A+\dfrac{1}{\tan A}}=\frac{\sec A\times\tan A}{\tan^2 A+1}$$

$$=\frac{\sec A\times\tan A}{\tan^2 A+1}=\frac{\sec A\times\tan A}{\sec^2 A}\qquad (1+\tan^2 A=\sec^2 A)$$

$$=\frac{\tan A}{\sec A}=\frac{\sin A\times\cos A}{\cos A}=\mathbf{\sin A}$$

5. $$\frac{1+\cos\theta}{\sin\theta}+\frac{\sin\theta}{1+\cos\theta}=\frac{(1+\cos\theta)^2+\sin^2\theta}{(\sin\theta)(1+\cos\theta)}$$

$$=\frac{1+2\cos\theta+\cos^2\theta+\sin^2\theta}{(\sin\theta)(1+\cos\theta)}$$

$$=\frac{1+2\cos\theta+1}{(\sin\theta)(1+\cos\theta)}$$

$$=\frac{2+2\cos\theta}{(\sin\theta)(1+\cos\theta)}=\frac{2(1+\cos\theta)}{(\sin\theta)(1+\cos\theta)}$$

$$=\frac{2}{\sin\theta}$$

6. $$\tan(A+B)=\frac{\sin(A+B)}{\cos(A+B)}$$

$$=\frac{\sin A\cos B+\cos A\sin B}{\cos A\cos B-\sin A\sin B}$$

$$=\frac{\dfrac{\sin A\cos B}{\cos A\cos B}+\dfrac{\cos A\sin B}{\cos A\cos B}}{\dfrac{\cos A\cos B}{\cos A\cos B}-\dfrac{\sin A\sin B}{\cos A\cos B}}\qquad \text{(divide all terms by cos A cos B)}$$

$$=\frac{\dfrac{\sin A}{\cos A}+\dfrac{\sin B}{\cos B}}{1-\dfrac{\sin A\sin B}{\cos A\cos B}}=\frac{\tan A+\tan B}{1-\tan A\tan B}$$

7. a) Let $\theta = 2x$, therefore, cos 2x becomes $\cos\theta$

Now $\cos\theta = 0.5$

Therefore $\theta = \cos^{-1} 0.5$; hence $\theta = 60°$

$$\angle x=\frac{\theta}{2}=30°$$

Since the cosine is positive, the solution must be in quadrants 1 and 4.

$\therefore$ $\theta =$ **30°** and $\theta = 360 - 30° =$ **330°**

b) After transposition, $2.5 \tan x = 2.5$

$\therefore \tan x = 1$, or $x = \tan^{-1} 1$

Angle $x = 45°$

Tangent is positive in the first and third quadrants

Therefore $x = \mathbf{45°}$

and $x = 180 + 45 = \mathbf{225°}$

c) After transposition $4.5 \sec x = 5.2$

$\therefore \sec x = \frac{5.2}{4.5} = 1.1556$

$$\cos x = \frac{1}{\sec x} = \frac{1}{1.1556} = 0.8654$$

Angle $x = \cos^{-1} 0.8654 = 30.07°$

As cosine (and secant) is positive in the first and fourth quadrants, $x = \mathbf{30.07°}$ and $x = 360 - 30.07 = \mathbf{329.93°}$

8. After transposition, $2 \operatorname{cosec}^2 B = 6 - 4 = 2$

or $\operatorname{cosec}^2 B = 1$

From the above $\frac{1}{\sin^2 B} = 1$, $\sin^2 B = 1$

$\sin B = \sqrt{1} = \pm 1$, $\angle B = \pm 90°$

sine (and cosec of $\angle B$) is positive in first and second quadrants, and negative in the third and fourth quadrants

Therefore, $B = \mathbf{90°}$, and $B = 360 - 90 = \mathbf{270°}$

9. $2\cos^2 x + \sin^2 x = 2(1 - \sin^2 x) + \sin^2 x = 1$

$2 - 2\sin^2 x + \sin^2 x = 1$

$2 - 1 = 2\sin^2 x - \sin^2 x$

$\sin^2 x = 1$, therefore $\sin x = \sqrt{1} = \pm 1$

When $\sin x = +1$, $x = \mathbf{90°}$

When $\sin x = -1$, $x = -90°$, or $x = 360 - 90 = \mathbf{270°}$

10. $2(1 + \tan^2 \theta) + \tan \theta - 3 = 0$ $(1 + \tan^2 \theta = \sec^2 \theta)$

$2 + 2\tan^2 \theta + \tan \theta - 3 = 0$

After simplifying the above, $2\tan^2 \theta + \tan \theta - 1 = 0$

Solving the above equation by the quadratic formula:

$$\tan \theta = \frac{-1 \pm \sqrt{1 - 4(2)(-1)}}{2 \times 2}$$

$$= \frac{-1 \pm \sqrt{9}}{4} = \frac{-1 \pm 3}{4}$$

$\tan \theta = \frac{-1 - 3}{4} = -1$, and $\tan \theta = \frac{-1 + 3}{4} = 0.5$

If $\tan \theta = -1$, then $\theta = -45°$

tangent is negative in the second and fourth quadrants

Therefore, $\theta = 180 - 45 =$ **135°**, and $\theta = 360 - 45 =$ **315°**

If $\tan\theta = 0.5$, $\theta = 26.57°$

tangent is positive in the first and third quadrants

Therefore, $\theta =$ **26.57°**, and $\theta = 180 + 26.57 =$ **206.57°**

11. a) Amplitude = **330 V**

  b) Angular velocity, $\omega = 100\,\pi$

  Periodic time, $T = \frac{2\pi}{\omega} = \frac{2\pi}{100\,\pi} = \mathbf{0.02\,s}$

  c) Frequency, $f = \frac{1}{T} = \frac{1}{0.02} = \mathbf{50\,Hz}$

  d) Phase angle, $\boldsymbol{\alpha}$ = **0.5 rad. (or 28.65°)** e) **lagging**

  (angle in degrees = radians $\times \frac{180}{\pi}$)

12. Maximum displacement = Amplitude = 0.5 m

  Angular velocity, $\omega = 2\pi f = 2\pi \times 40 = 80\pi$ rad/s

  Now, displacement = $0.5\sin(80\pi t + \alpha)$

  $0.2 = 0.5\sin(80\pi \times 0 + \alpha)$ (20 cm = 0.2 m)

  $0.2 = 0.5\sin\alpha$, from which $\alpha = 0.412$ rad.

  Thus, **displacement** = $\mathbf{0.5\sin(80\pi t + 0.412)\,m}$

13. a) Angular velocity, $\omega = 2\pi f = 2\pi \times 0.4 = 0.8\pi$ rad/s

  b) At time t = 0.4 s, displacement of the building = 6.75 cm (or 0.0675 m)

  $0.0675 = A\sin(0.8\pi \times 0.4) = A\sin(0.32\pi)$

  $0.0675 = A \times 0.844$

  Therefore, maximum displacement, $A = \frac{0.0675}{0.8443} = \mathbf{0.08\,m}$

  c) Displacement of the building = $A\sin(\omega t)$ m

  $= \mathbf{0.08\sin(0.8\pi t)\,m}$

14. The graphs of y_1 and y_2 are shown in Figure S8.1; the data have been plotted for 1 cycle of 360°. The ordinates of the combined graph can be determined either from the graph or from Table S8.1. Because the 2 sine waves have the same frequency, the combined graph will be a sine wave with the same frequency. The phase angle of the combined graph is 13.3° and its peak value is 14.75 mm. The formula of the combined wave is: $y_3 = 14.75\sin(\theta + 13.3°)$.

Table S8.1

θ°	0	30	60	90	120	150	180	210	240	270	300	330	360
sin θ	0	0.5	0.866	1.0	0.866	0.5	0	–0.5	–0.866	–1.0	–0.866	–0.5	0
5 sin θ	0	2.5	4.33	5	4.33	2.5	0	–2.5	–4.33	–5	–4.33	–2.5	0
(θ + 20°)	20	50	80	110	140	170	200	230	260	290	320	350	380
sin (θ + 20)	0.342	0.766	0.985	0.94	0.643	0.174	–0.342	–0.766	–0.985	–0.94	–0.643	–0.174	0.342
10 sin (θ + 20)	3.42	7.66	9.85	9.4	6.43	1.74	-3.42	-7.66	-7.66	-12.99	-15	-1.74	-7.5
y1 + y2	–7.5	5	16.16	22.99	23.66	17.99	7.5	–5	–16.16	–22.99	–23.66	–17.99	–7.5

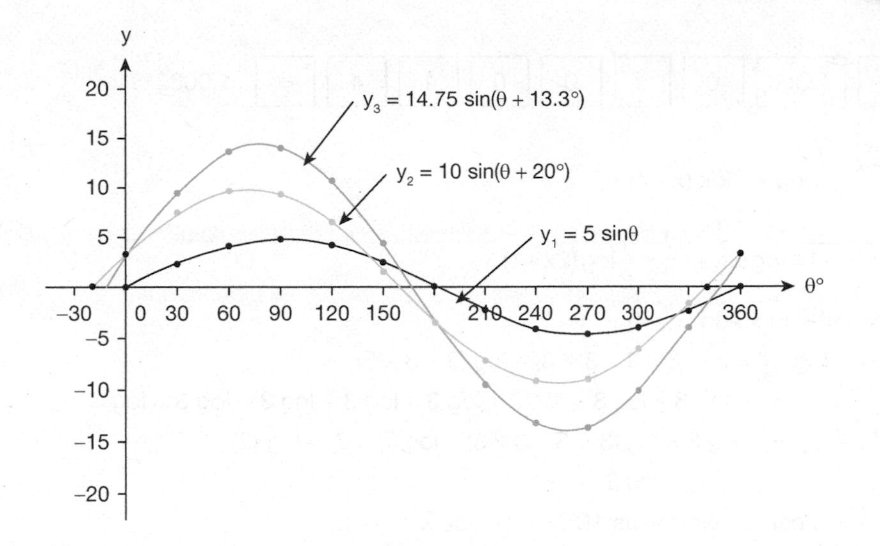

Figure S8.1

Solutions – Exercise 9.1

1. a) From section 9.1, $2^x = 32 = 2^5$

   Therefore $x = 5$

   Hence $\log_2 32 = 5$

   b) $10^x = 10000 = 10^4$

   Therefore $x = 4$

   Hence $\log_{10} 10000 = 4$

2.

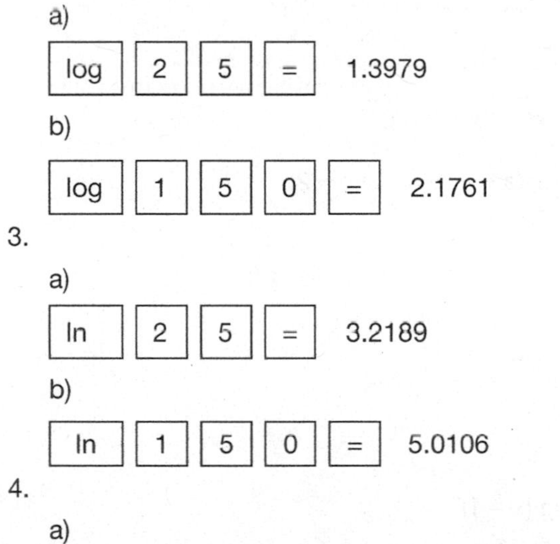

3.

a)

ln 2 5 = 3.2189

b)

ln 1 5 0 = 5.0106

4.

a)

SHIFT LOG 2 . 5 = 316.2278

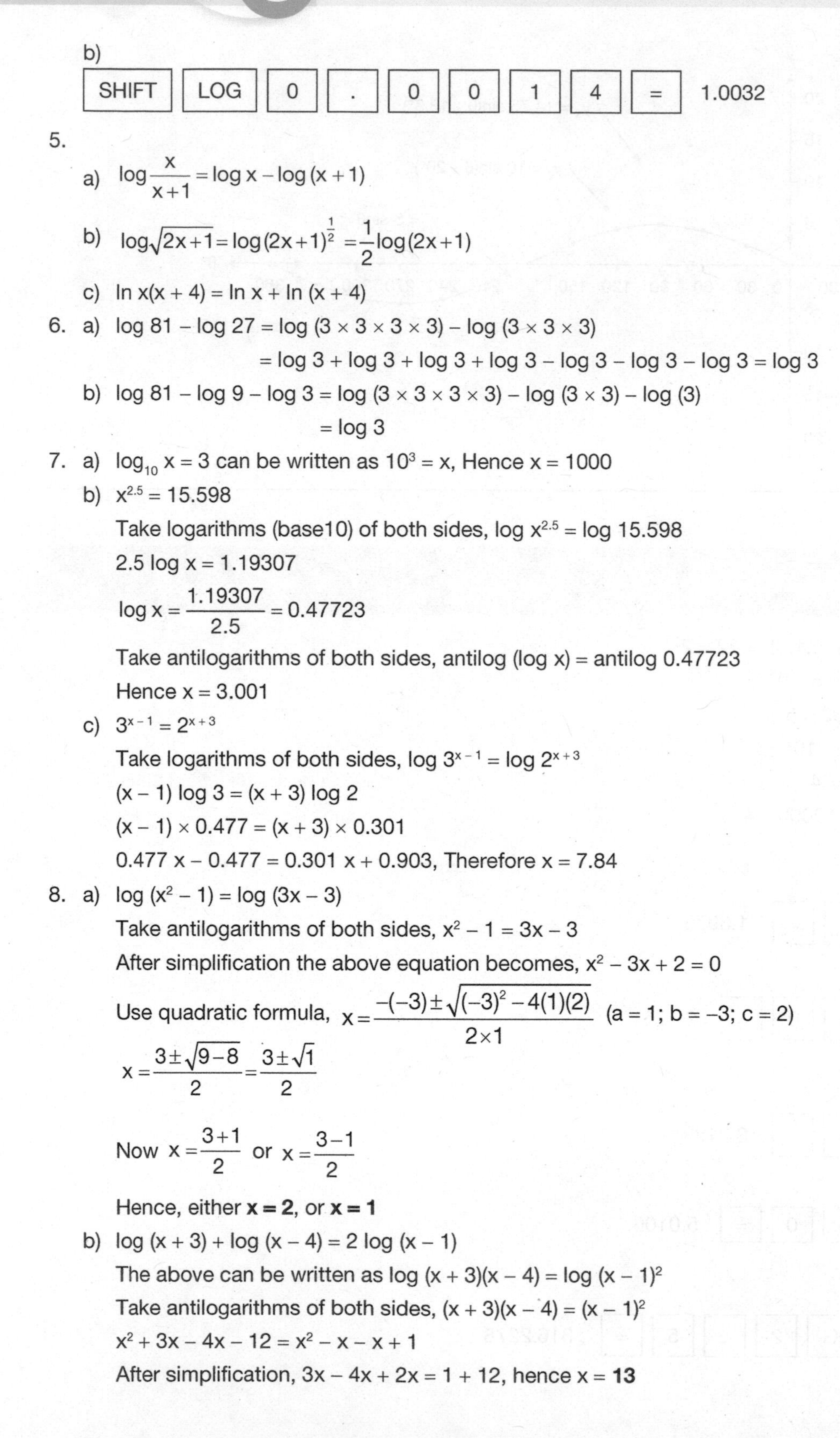

b)

SHIFT	LOG	0	.	0	0	1	4	=	1.0032

5.

a) $\log\frac{x}{x+1} = \log x - \log(x+1)$

b) $\log\sqrt{2x+1} = \log(2x+1)^{\frac{1}{2}} = \frac{1}{2}\log(2x+1)$

c) $\ln x(x+4) = \ln x + \ln(x+4)$

6. a) $\log 81 - \log 27 = \log(3 \times 3 \times 3 \times 3) - \log(3 \times 3 \times 3)$

$= \log 3 + \log 3 + \log 3 + \log 3 - \log 3 - \log 3 - \log 3 = \log 3$

b) $\log 81 - \log 9 - \log 3 = \log(3 \times 3 \times 3 \times 3) - \log(3 \times 3) - \log(3)$

$= \log 3$

7. a) $\log_{10} x = 3$ can be written as $10^3 = x$, Hence $x = 1000$

b) $x^{2.5} = 15.598$

Take logarithms (base10) of both sides, $\log x^{2.5} = \log 15.598$

$2.5 \log x = 1.19307$

$\log x = \frac{1.19307}{2.5} = 0.47723$

Take antilogarithms of both sides, antilog ($\log x$) = antilog 0.47723

Hence $x = 3.001$

c) $3^{x-1} = 2^{x+3}$

Take logarithms of both sides, $\log 3^{x-1} = \log 2^{x+3}$

$(x-1)\log 3 = (x+3)\log 2$

$(x-1) \times 0.477 = (x+3) \times 0.301$

$0.477x - 0.477 = 0.301x + 0.903$, Therefore $x = 7.84$

8. a) $\log(x^2 - 1) = \log(3x - 3)$

Take antilogarithms of both sides, $x^2 - 1 = 3x - 3$

After simplification the above equation becomes, $x^2 - 3x + 2 = 0$

Use quadratic formula, $x = \frac{-(-3) \pm \sqrt{(-3)^2 - 4(1)(2)}}{2 \times 1}$ ($a = 1$; $b = -3$; $c = 2$)

$x = \frac{3 \pm \sqrt{9-8}}{2} = \frac{3 \pm \sqrt{1}}{2}$

Now $x = \frac{3+1}{2}$ or $x = \frac{3-1}{2}$

Hence, either $\mathbf{x = 2}$, or $\mathbf{x = 1}$

b) $\log(x+3) + \log(x-4) = 2\log(x-1)$

The above can be written as $\log(x+3)(x-4) = \log(x-1)^2$

Take antilogarithms of both sides, $(x+3)(x-4) = (x-1)^2$

$x^2 + 3x - 4x - 12 = x^2 - x - x + 1$

After simplification, $3x - 4x + 2x = 1 + 12$, hence $x = \mathbf{13}$

Solutions – Exercise 9.2

1. a) $e^{2t}=\frac{1.2}{3}=0.4$

   Taking logarithms of both sides, $\ln\left(e^{2t}\right)=\ln(0.4)$

   $2t=-0.9163$, therefore $t=-0.458$

   b) $e^{-1.5x}=\frac{5.1}{2.2}$

   $e^{1.5x}=\frac{2.2}{5.1}=0.43137$

   Taking logarithms of both sides, $\ln\left(e^{1.5x}\right)=\ln(0.43137)$

   $1.5x=-0.84079$, therefore, $x=-0.5605$

   c) $1-e^{-\frac{x}{2}}=\frac{10}{14}$

   $e^{-\frac{x}{2}}=1-\frac{10}{14}=\frac{4}{14}$

   $e^{\frac{x}{2}}=\frac{14}{4}=3.5$

   Taking logarithms of both sides, $\ln\left(e^{\frac{x}{2}}\right)=\ln(3.5)$

   $\frac{x}{2}=1.25276$, therefore, $x=2.506$

2. Y = the value of the matured account = £100 000

   P = principal (original) amount = £80 000

   r = 5.5%; n = 4; t = term of the deposit in years

   $100\ 000=80\ 000\left(1+\frac{5.5}{100\times 4}\right)^{4t}$

   $\frac{10\ 0000}{80\ 000}=\left(1.01375\right)^{4t}$

   $1.25=\left(1.01375\right)^{4t}$

   Take logarithms of both sides

   $\ln(1.25)=4t(\ln 1.01375)$

   $0.2231436=4t\times 0.013656$

   $0.2231436=t\times 0.0546253$

   $t=\frac{0.2231436}{0.0546253}=$ **4.08 years**

3. a) $T(t)=T_s+(T_0-T_s)e^{-kt}$

   $T(t)=70$ °C; $T_s=20$ °C; $T_0=86$ °C; $t=6$ minutes

   $70=20+(86-20)e^{-6k}$

   $50=66e^{-6k}$ which gives $\frac{50}{66}=e^{-6k}$

   Take logarithm of both sides, $\ln\left(\frac{50}{66}\right)=\ln e^{-6k}$

   $-0.277632=-6k$, therefore $k=0.04627$

b) $T(t) = 60\ °C$

$60 = 20 + (86 - 20)e^{-0.04627\,t}$

$40 = 66e^{-0.04627\,t}$ which gives $\frac{40}{66} = e^{-0.04627\,t}$

Take logarithm of both sides, $\ln\left(\frac{40}{66}\right) = \ln e^{-0.04627\,t}$

$-0.5007753 = -0.04627\,t$, therefore t = 10.82 minutes

4. $L = L_0 e^{n(t-t_0)}$

$L_0 = 8.0$ m; $\alpha = 7\times10^{-5}/°C$; $t_0 = 15\ °C$; $t = 30\ °C$ $L = 8.0e^{7\times10^{-5}(30-15)} = 8.0e^{1.05\times10^{-3}}$

$L = 8 \times 1.00105 = 8.0084$ m

The linear expansion of the PVC guttering $= L - L_0 = 8.0084 - 8.0$

$= 0.0084$ m or 8.4 mm

5. Population after growth, $P = P_0\, e^{rt}$

$P_0 = 100\ 000$; $r = 1.6\%$ or 0.016; t = 25 years

$P = 100\ 000 \times e^{0.016\times25} = 100\ 000 \times e^{0.4}$

$= 100\ 000 \times 1.4918247 = 149{,}182$

6. Population after reduction, $P = P_0\, e^{-rt}$

$P_0 = 20\ 000\ 000$; $r = -1.3\%$ or -0.013; t = 20 years

$P = 20\ 000\ 000 \times e^{-0.013\times20} = 20\ 000\ 000 \times e^{-0.26}$

$= 20\ 000\ 000 \times 0.771051586 =$ **15 421 032**

7. a) Time, t, between points A and B = 1.0 – 0.2 = 0.8 s

S (at point B) = 46 dB; S_0 (at point A) = 72 dB;

Put these values in the equation, $46 = 72e^{-\frac{0.8}{c}}$

$$\frac{46}{72} = e^{-\frac{0.8}{c}}$$

Take logarithms of both sides, $\ln(0.63888) = \frac{-0.8}{c}$

$$c = \frac{-0.8}{\ln(0.63888)} = \mathbf{1.7856}$$

Therefore, $S = S_0 e^{-\frac{t}{1.7856}}$

b) S = 72 dB; t (between points C and A) = 0.2 – 0 = 0.2

$S = S_0 e^{-\frac{t}{1.7856}}$

$72 = S_0 e^{-\frac{0.2}{1.7856}}$

$72 = S_0 \times 0.894038$; therefore, $S_0 =$ **80.5 dB**

c) t = 0.6 s; $S_0 = 80.5$ dB; S = sound level after 0.6 seconds

$S = S_0 e^{-\frac{t}{1.7856}}$

$S = 80.5\, e^{-\frac{0.6}{1.7856}}$

$S = 80.5 \times 0.7146 =$ **57.5 dB**

Solutions – Exercise 9.3

1. Simplify cosh x – sinh x

$$\cosh x - \sinh x = \left(\frac{e^{x}+e^{-x}}{2}\right) - \left(\frac{e^{x}-e^{-x}}{2}\right)$$

$$= \frac{1}{2}\left(e^{x}+e^{-x}-e^{x}+e^{-x}\right)$$

$$= \frac{1}{2}\left(2e^{-x}\right) = e^{-x}$$

2. Prove that cosh 2x = $(\cosh x)^2 + (\sinh x)^2$

$$\cosh 2x = \frac{e^{2x}+e^{-2x}}{2}$$

$$(\cosh x)^2 + (\sinh x)^2 = \left(\frac{e^{x}+e^{-x}}{2}\right)^2 + \left(\frac{e^{x}-e^{-x}}{2}\right)^2$$

$$= \frac{e^{2x}+2+e^{-2x}}{4} + \frac{e^{2x}-2+e^{-2x}}{4}$$

$$= \frac{1}{4}\left(e^{2x}+2+e^{-2x}+e^{2x}-2+e^{-2x}\right)$$

$$= \frac{1}{4}\left(2e^{2x}+2e^{-2x}\right)$$

$$= \frac{e^{2x}+e^{-2x}}{2}$$

$\frac{e^{2x}+e^{-2x}}{2} = \cosh 2x$, Hence cosh 2x = $(\cosh x)^2 + (\sinh x)^2$

3. Prove that $\coth^2 x - \operatorname{cosech}^2 x = 1$

$$\coth^2 x - \operatorname{cosech}^2 x = \left(\frac{e^{x}+e^{-x}}{e^{x}-e^{-x}}\right)^2 - \left(\frac{2}{e^{x}-e^{-x}}\right)^2$$

$$= \frac{e^{2x}+2+e^{-2x}}{e^{2x}-2+e^{-2x}} - \frac{4}{e^{2x}-2+e^{-2x}}$$

$$= \frac{e^{2x}+2+e^{-2x}-4}{e^{2x}-2+e^{-2x}} = \frac{e^{2x}-2+e^{-2x}}{e^{2x}-2+e^{-2x}} = 1$$

4. $\sinh x = \frac{3}{5}$

$$\sinh x = \left(\frac{e^{x}-e^{-x}}{2}\right) = \frac{3}{5}$$

Multiply the equation by 5, $2.5e^{x} - 2.5\ e^{-x} - 3 = 0$

Multiply the equation by e^x, $2.5\,(e^{x})^2 - 2.5\,(e^{-x})(e^{x}) - 3\,e^{x} = 0$

$$2.5\,(e^{x})^2 - 3\,e^{x} - 2.5 = 0$$

This can be solved as a quadratic equation; a = 2.5, b = –3, c = –2.5

$$e^x = \frac{-(-3)\pm\sqrt{(-3)^2 - 4(2.5)(-2.5)}}{2\times 2.5}$$

$$= \frac{3\pm\sqrt{9+25}}{5} = \frac{3\pm 5.83}{5}$$

$e^x = 1.766$ or -0.566

e^x is always positive, therefore $e^x = 1.766$

Take logarithms of both sides, $\ln(e^x) = \ln(1.766)$

$x = 0.5687$ (logarithm and e^x cancel out to leave behind x)

5. 2 cosh x = 3.4, therefore, cosh x = 1.7

$$\frac{e^x + e^{-x}}{2} = 1.7$$

$$e^x + e^{-x} - 3.4 = 0$$

Multiply the equation by e^x, $(e^x)^2 + (e^{-x})(e^x) - 3.4\,e^x = 0$

$$(e^x)^2 - 3.4\,e^x + 1 = 0$$

This can be solved as a quadratic equation; a = 1, b = –3.4, c = 1

$$e^x = \frac{-(-3.4)\pm\sqrt{(-3.4)^2 - 4(1)(1)}}{1}$$

$$= \frac{3.4\pm\sqrt{7.56}}{2} = \frac{3.4\pm 2.7495}{2}$$

$e^x = 3.0748$ or -0.3253

e^x is always positive, therefore $e^x = 3.0748$

Take logarithms of both sides, $\ln(e^x) = \ln(3.0748)$

$x = 1.123$ (logarithm and e^x cancel out to leave behind x)

6. a) 2 sech x – 1 = 0

2 sech x – 1 = 0

$$2\times\frac{2}{e^x + e^{-x}} - 1 = 0$$

$$\frac{4}{e^x + e^{-x}} = 1, \quad \text{or } 4 = e^x + e^{-x}$$

Multiply by e^x and transpose, $(e^x)^2 + (e^x)(e^{-x}) - 4\,e^x = 0$

$(e^x)^2 - 4\,e^x + 1 = 0$

$$e^x = \frac{-(-4)\pm\sqrt{(-4)^2 - 4(1)(1)}}{2}$$

$$= \frac{4\pm\sqrt{12}}{2} = \frac{4\pm 3.464}{2}$$

$e^x = 3.732$ or $e^x = 0.268$

Take logarithms of both sides of $e^x = 3.732$

$\ln(e^x) = \ln(3.732)$, therefore, $\mathbf{x = 1.317}$

similarly, $\ln(e^x) = \ln(0.268)$, therefore, $\mathbf{x = -1.317}$

b) $2.4 \cosh x + 5 \sinh x = 7.5$

$$2.4 \cosh x + 5 \sinh x = 2.4\left(\frac{e^x + e^{-x}}{2}\right) + 5\left(\frac{e^x - e^{-x}}{2}\right)$$

Therefore, $2.4\left(\frac{e^x + e^{-x}}{2}\right) + 5\left(\frac{e^x - e^{-x}}{2}\right) = 7.5$

$1.2\ e^x + 1.2\ e^{-x} + 2.5e^x - 2.5e^{-x} - 7.5 = 0$

$3.7\ e^x - 7.5 - 1.3\ e^{-x} = 0$

Multiply by e^x: $3.7\ (e^x)^2 - 7.5\ (e^x) - 1.3\ (e^{-x})(e^x) = 0$

$3.7\ (e^x)^2 - 7.5\ (e^x) - 1.3 = 0$

Solve for e^x using the quadratic formula:

$$e^x = \frac{-(-7.5) \pm \sqrt{(-7.5)^2 - 4(3.7)(-1.3)}}{2 \times 3.7}$$

$$= \frac{7.5 \pm \sqrt{75.49}}{7.4} = \frac{7.5 \pm 8.6885}{7.4}$$

$e^x = 2.1876$ or -0.1606

e^x is always positive, therefore $e^x = 2.1876$

Take logarithms of both sides, $\ln(e^x) = \ln(2.1876)$

x = **0.7828** (logarithm and e^x cancel out to leave behind x)

Solutions – Exercise 10.1

1. $y = 2x$

   Let x increase by a small amount δx and the corresponding increase in y be δy

   $y + \delta y = 2(x + \delta x)$

   or $y + \delta y = 2x + 2\delta x$

   As $y = 2x$, the above equation becomes:

   $2x + \delta y = 2x + 2\delta x$

   or $\delta y = 2\delta x$

   Divide both sides by δx

   $$\frac{\delta y}{\delta x} = 2$$

   $$\text{Limit } \delta x \to 0\left(\frac{\delta y}{\delta x}\right) = \frac{dy}{dx} = \mathbf{2}$$

2. $y = 3x + 5$

   Let x increase by a small amount δx and the corresponding increase in y be δy

   $y + \delta y = 3(x + \delta x) + 5$

   or $y + \delta y = 3x + 3\delta x + 5$

   As $y = 3x + 5$, the above equation becomes:

   $3x + 5 + \delta y = 3x + 3\delta x + 5$

   or $\delta y = 3\delta x$

   Divide both sides by δx

$$\frac{\delta y}{\delta x} = 3$$

Limit $\delta x \rightarrow 0(\frac{\delta y}{\delta x}) = \frac{dy}{dx} = \mathbf{3}$

3. $y = 3x^2 + 2$

$$\frac{dy}{dx} = 6x^{2-1} + 0 = \mathbf{6x^1} = \mathbf{6x}$$

4. $y = \frac{-6}{\sqrt{x}} = \frac{-6}{x^{1/2}} = -6x^{-1/2}$

$$\frac{dy}{dx} = -6 \times -\frac{1}{2}x^{-1/2-1}$$

$$= 3x^{-3/2} = \frac{3}{x^{3/2}}$$

5. $y = 5x^2 + 2x + \frac{2}{x} + 5 = 5x^2 + 2x + 2x^{-1} + 5$

$$\frac{dy}{dx} = 10x + 2 + (2 \times -1x^{-1-1}) + 0$$

$$= 10x + 2 - 2x^{-2} = \mathbf{10x + 2 - \frac{2}{x^2}}$$

6. Let $y = 2 \sin 3x$

$a = 3$ in this question

$$\frac{d}{dx}(\sin ax) = a \cos ax$$

Therefore, $\frac{d}{dx}(2 \sin 3x) = 2(3 \cos 3x) = \mathbf{6 \cos 3x}$

7. Let $y = \cos 2\theta - 3 \sin 4\theta$

$\frac{d}{d\theta}(\sin a\theta) = a \cos a\theta$, and $\frac{d}{d\theta}(\cos a\theta) = -a \sin a\theta$

$$\frac{dy}{d\theta} = -2 \sin 2\theta - 3(\cos 4\theta \times 4) = \mathbf{-2 \sin 2\theta - 12 \cos 4\theta}$$

8. $4z^2 - 2 \cos 2z$

Let $y = 4z^2 - 2 \cos 2z$

$$\frac{dy}{dz} = 8z - 2(-\sin 2z \times 2) = \mathbf{8z + 4 \sin 2z}$$

9. Let $y = \frac{1}{2}\log_e 2x$

$$\frac{dy}{dx} = \frac{1}{2} \times \frac{1}{2x} \times 2 = \frac{1}{2x}$$

10. Let $y = e^{7x}$

$$\frac{dy}{dx} = e^{7x} \times 7 = \mathbf{7e^{7x}}$$

11. Let $y = \frac{1}{e^{2x}} = e^{-2x}$

$$\frac{dy}{dx} = e^{-2x} \times (-2) = \mathbf{-2e^{-2x}}$$

Solutions – Exercise 10.2

1. Let $y = (x^2 + x)^5$

   Let $u = x^2 + x$, so that $y = (u)^5$

   $\frac{du}{dx} = 2x + 1, \qquad \text{and } \frac{dy}{du} = 5u^{5-1} = 5u^4$

   $\frac{dy}{dx} = \frac{dy}{du} \times \frac{du}{dx}$

   $= 5u^4 \times (2x + 1)$

   Now replace u with $x^2 + x$

   $\frac{dy}{dx} = \mathbf{5(x^2 + x)^4(2x + 1)}$

2. Let $y = \sqrt{(2x^2 - 3x + 1)} = (2x^2 - 3x + 1)^{½}$

   Using the chain rule, $\frac{dy}{dx} = \frac{1}{2}(2x^2 - 3x + 1)^{½-1}\frac{d}{dx}(2x^2 - 3x + 1)$

   $= \frac{1}{2}\mathbf{(2x^2 - 3x + 1)^{-½}(4x - 3)}$

3. Let $y = \frac{1}{(2x+5)^3} = (2x + 5)^{-3}$

   Let $u = 2x + 5$, so that $y = (u)–^3$

   $\frac{du}{dx} = 2 \quad \text{and } \frac{dy}{du} = -3u^{-3-1} = -3u^{-4}$

   $\frac{dy}{dx} = \frac{dy}{du} \times \frac{du}{dx}$

   $= -3u^{-4} \times 2 = -6u^{-4}$

   Now replace u with $2x + 5$

   $\frac{dy}{dx} = \mathbf{-6(2x + 5)^{-4}}$

4. Let $y = \tan(2x + 4)$

   Using the chain rule, $\frac{dy}{dx} = \sec^2(2x + 4) \times \frac{d}{dx}(2x + 4)$

   $= \sec^2(2x + 4) \times 2 = \mathbf{2\sec^2(2x + 4)}$

5. Let $y = \cos 4x$

   Let $u = 4x$, so that $y = \cos u$

   $\frac{du}{dx} = 4, \quad \text{and } \frac{dy}{du} = -\sin u$

   $\frac{dy}{dx} = \frac{dy}{du} \times \frac{du}{dx}$

   $= -\sin u \times 4 = -4\sin u$

   Now replace u with 4x

$\frac{dy}{dx} = -\mathbf{4\ sin\ 4x}$

6. $y = \sin^5 x = (\sin x)^5$

Using the chain rule, $\frac{dy}{dx} = 5(\sin x)^4 \times \frac{d}{dx}(\sin x)$

$= 5(\sin x)^4 \times \cos x$

$= 5 \cos x \times (\sin x)^4$

$= \mathbf{5 \cos x \sin^4 x}$

7. $y = \cos^3 3x = (\cos 3x)^3$

Using the chain rule, $\frac{dy}{dx} = 3(\cos 3x)^2 \times \frac{d}{dx}(\cos 3x)$

$= 3(\cos 3x)^2 \times (-\sin 3x \times 3)$

$= \mathbf{-9 \cos^2 3x \sin 3x}$

8. Let $y = \sin(3x + 8)$

Let $u = 3x + 8$, so that $y = \sin u$

$\frac{du}{dx} = 3,$ and $\frac{dy}{du} = \cos u$

$\frac{dy}{dx} = \frac{dy}{du} \times \frac{du}{dx}$

$= \cos u \times 3 = 3 \cos u$

Now replace u with 3x + 8

$\frac{dy}{dx} = \mathbf{3 \cos (3x + 8)}$

9. Let $y = \frac{1}{\sin^2 x}$

$= \frac{1}{(\sin x)^2} = (\sin x)^{-2}$

Using the chain rule: $\frac{dy}{dx} = -2(\sin x)^{-3} \times \frac{d}{dx}(\sin x)$

$= -2(\sin x)^{-3} \times \cos x$

$= \frac{-2 \cos x}{\sin^3 x}$

10. Let $y = \log_e (7 - 3x)$

Let $u = 7 - 3x$, so that $y = \log_e u$

$\frac{du}{dx} = -3$, and $\frac{dy}{du} = \frac{1}{u}$

$\frac{dy}{dx} = \frac{dy}{du} \times \frac{du}{dx}$

$= \frac{1}{u} \times (-3) = \frac{-3}{u}$

Now replace u with 7 – 3x

$\frac{dy}{dx} = \frac{-3}{7 - 3x}$

11. Let $y = 4e^{2x-1}$

Let $u = 2x - 1$, so that $y = 4e^u$

$\frac{du}{dx} = 2$, and $\frac{dy}{du} = 4e^u$

$\frac{dy}{dx} = \frac{dy}{du} \times \frac{du}{dx}$

$= 4e^u \times 2 = 8e^u$

Now replace u with $2x - 1$

$\frac{dy}{dx} = \mathbf{8e^{2x-1}}$

12. Let $y = \frac{1}{e^{2x+4}} = e^{-2x-4}$

Using the chain rule: $\frac{dy}{dx} = e^{-2x-4} \times \frac{d}{dx}(-2x-4)$

$= e^{-2x-4} \times (-2)$

$\frac{dy}{dx} = -2e^{-2x-4}$

$= \frac{-2}{e^{2x+4}}$

Solutions – Exercise 10.3

1. $y = x \cos x$

Product rule : $\frac{dy}{dx} = \mathbf{v}\frac{du}{dx} + \mathbf{u}\frac{dv}{dx}$

Let $u = x$ and $v = \cos x$

$\frac{du}{dx} = 1$ and $\frac{dv}{dx} = -\sin x$

$\frac{dy}{dx} = \cos x \times 1 + x \times (-\sin x)$

$= \mathbf{\cos x - x \sin x}$

2. $y = x^2 \log_e x$

Let $u = x^2$ and $v = \log_e x$

$\frac{du}{dx} = 2x$ and $\frac{dv}{dx} = \frac{1}{x}$

$\frac{dy}{dx} = \log_e x \times 2x + x^2 \times \frac{1}{x}$

$= \mathbf{2x \log_e x + x}$

3. $y = e^{2x} \sin 3x$

Let $u = e^{2x}$ and $v = \sin 3x$

$\frac{du}{dx} = 2e^{2x}$ and $\frac{dv}{dx} = 3\cos 3x$

$\frac{dy}{dx} = \sin 3x \times 2e^{2x} + e^{2x} \times 3\cos 3x$

$= \mathbf{e^{2x}(2\sin 3x + 3\cos 3x)}$

4. $y = (2x + 1)\tan x$

   Let $u = 2x + 1$ and $v = \tan x$

   $\frac{du}{dx} = 2$ and $\frac{dv}{dx} = \sec^2 x$

   $\frac{dy}{dx} = \tan x \times 2 + (2x+1) \times \sec^2 x$

   $= \mathbf{2\tan x + (2x + 1)\sec^2 x}$

5. $y = 3(x^2 + 3)\sin x$

   Let $u = 3(x^2 + 3)$ and $v = \sin x$

   $\frac{du}{dx} = 3 \times 2x$ and $\frac{dv}{dx} = \cos x$

   $\frac{dy}{dx} = \sin x \times (3 \times 2x) + 3(x^2 + 3) \times \cos x$

   $= \mathbf{6x\sin x + 3(x^2 + 3)\cos x}$

6. $y = 4\sin\theta\cos\theta$

   Let $u = 4\sin\theta$ and $v = \cos\theta$

   $\frac{du}{d\theta} = 4\cos\theta$ and $\frac{dv}{d\theta} = -\sin\theta$

   $\frac{dy}{d\theta} = \cos\theta \times 4\cos\theta + 4\sin\theta \times (-\sin\theta)$

   $= \mathbf{4\cos^2\theta - 4\sin^2\theta = 4(\cos^2\theta - \sin^2\theta)}$

7. $y = (z^2 + 2z - 2)\sin z$

   Let $u = (z^2 + 2z - 2)$ and $v = \sin z$

   $\frac{du}{dz} = 2z + 2$ and $\frac{dv}{dz} = \cos z$

   $\frac{dy}{dz} = \sin z \times (2z + 2) + (z^2 + 2z - 2)\cos z$

   $= \mathbf{(2z + 2)\sin z + (z^2 + 2z - 2)\cos z}$

8. $y = \frac{\cos x}{x}$

   Let $u = \cos x$ and $v = x$

   $\frac{du}{dx} = -\sin x$ and $\frac{dv}{dx} = 1$

   $\frac{dy}{dx} = \frac{v\frac{du}{dx} - u\frac{dv}{dx}}{v^2}$

$$= \frac{x \times (-\sin x) - \cos x \times 1}{x^2}$$

$$= \frac{-x \sin x - \cos x}{x^2}$$

9. $y = \dfrac{2x}{x+4}$

Let u = 2x and v = x + 4

$$\frac{du}{dx} = 2 \quad \text{and} \quad \frac{dv}{dx} = 1$$

$$\frac{dy}{dx} = \frac{v\frac{du}{dx} - u\frac{dv}{dx}}{v^2}$$

$$= \frac{(x+4) \times 2 - 2x \times 1}{(x+4)^2}$$

$$= \frac{2x + 8 - 2x}{(x+4)^2} = \frac{8}{(x+4)^2}$$

10. $y = \dfrac{x+3}{\sin x}$

Let u = x + 3 and v = sin x

$$\frac{du}{dx} = 1 \quad \text{and} \quad \frac{dv}{dx} = \cos x$$

$$\frac{dy}{dx} = \frac{v\frac{du}{dx} - u\frac{dv}{dx}}{v^2}$$

$$= \frac{\sin x \times 1 - (x+3) \times \cos x}{(\sin x)^2}$$

$$= \frac{\sin x - (x+3)\cos x}{\sin^2 x}$$

11. $y = \sec x = \dfrac{1}{\cos x}$

Let u = 1 and v = cos x

$$\frac{du}{dx} = 0 \text{ and } \frac{dv}{dx} = -\sin x$$

$$\frac{dy}{dx} = \frac{\cos x \times 0 - 1 \times (-\sin x)}{(\cos x)^2}$$

$$= \frac{\sin x}{\cos^2 x} = \frac{\sin x}{\cos x} \times \frac{1}{\cos x}$$

$= \mathbf{\tan x \sec x}$

12. $y = \text{cosec } x = \dfrac{1}{\sin x}$

Let u = 1 and v = sin x

$$\frac{du}{dx} = 0 \text{ and } \frac{dv}{dx} = \cos x$$

$$\frac{dy}{dx} = \frac{\sin x \times 0 - 1 \times (\cos x)}{(\sin x)^2}$$

$$= \frac{-\cos x}{\sin^2 x} = -\frac{\cos x}{\sin x} \times \frac{1}{\sin x}$$

$= \mathbf{-\cot x \operatorname{cosec} x}$

13. $y = \frac{e^{2x}}{x^2}$

Let u = e^{2x} and v = x^2

$$\frac{du}{dx} = 2e^{2x} \text{ and } \frac{dv}{dx} = 2x$$

$$\frac{dy}{dx} = \frac{x^2 \times 2e^{2x} - e^{2x} \times 2x}{(x^2)^2}$$

$$= \frac{2x^2 e^{2x} - 2x e^{2x}}{x^4} = \frac{2x e^{2x}(x-1)}{x^4}$$

14. $y = \frac{2 \log_e x}{\sin 2x}$

Let u = $2 \log_e x$ and v = sin 2x

$$\frac{du}{dx} = \frac{2}{x} \text{ and } \frac{dv}{dx} = 2 \cos 2x$$

$$\frac{dy}{dx} = \frac{\sin 2x \times \frac{2}{x} - 2 \log_e x \times 2 \cos 2x}{(\sin 2x)^2}$$

$$= \frac{\frac{2}{x} \sin 2x - 4 \log_e x \cos 2x}{\sin^2 2x}$$

$$= \frac{2 \sin 2x - 4x \log_e x \cos 2x}{x \sin^2 2x}$$

Solutions – Exercise 10.4

The first step for solving the questions in this exercise is to differentiate them. The questions in this exercise have been taken from Exercises 10.1 and 10.2. So for the first part of the solution, refer to the solutions of questions: 4, 5, 7 and 8 in Exercise 10.1 and questions 10, 11 and 12 in Exercise 10.2.

1. a) $y = \frac{-6}{\sqrt{x}}$

$$\frac{dy}{dx} = \frac{3}{x^{3/2}}$$

At $x = 1.5, \dfrac{dy}{dx} = \dfrac{3}{x^{3/2}} = \dfrac{3}{1.5^{3/2}} = \mathbf{1.633}$

b) $y = 5x^2 + 2x + \dfrac{2}{x} + 5$

$\dfrac{dy}{dx} = \mathbf{10x + 2} - \dfrac{2}{x^2}$

At $x = 1.5, \dfrac{dy}{dx} = 10 \times 1.5 + 2 - \dfrac{2}{1.5^2}$

$= 15 + 2 - 0.889 = \mathbf{16.111}$

2. If $y = 4z^2 - 2\cos 2z$

$\dfrac{dy}{dz} = \mathbf{8z + 4\sin 2z}$

At $z = 0.3, \dfrac{dy}{dz} = 8 \times 0.3 + (4 \times \sin 2 \times 0.3)$

$= 2.4 + (4 \times \sin 0.6)$ $\quad (\sin 0.6 = 0.5646)$

$= 2.4 + 2.259 = \mathbf{4.659}$

3. Let $y = \cos 2\theta - 3\sin 4\theta$

$\dfrac{dy}{d\theta} = -2\sin 2\theta - 3(\cos 4\theta \times 4) = -2\sin 2\theta - 12\cos 4\theta$

At $\theta = \dfrac{\pi}{5}, \dfrac{dy}{d\theta} = -2\sin\dfrac{2 \times \pi}{5} - 12\cos\dfrac{4 \times \pi}{5}$

$= -2\sin 1.2566 - 12\cos 2.5133$

$= -2 \times 0.951 - 12 \times (-0.809) = \mathbf{7.806}$

4. $y = \log_e(7 - 3x)$

$\dfrac{dy}{dx} = \dfrac{-3}{7 - 3x}$

At $x = 0.4, \dfrac{dy}{dx} = \dfrac{-3}{7 - 3 \times 0.4} = -\mathbf{0.5172}$

5. Let $y = 4e^{2x-1}$

$\dfrac{dy}{dx} = 8e^{2x-1}$

At $x = 0.4, \dfrac{dy}{dx} = 8e^{2 \times 0.4 - 1}$

$= 8e^{-0.2} = 8 \times 0.8187 = \mathbf{6.5496}$

6. $y = \dfrac{1}{e^{2x+4}} = e^{-2x-4}$

$\dfrac{dy}{dx} = -2e^{-2x-4}$

$= -2e^{-2 \times 0.4 - 4} = -2e^{-2 \times 0.4 - 4}$

$= -2e^{-4.8} = -2 \times 0.00823 = -\mathbf{0.01646}$

Solutions – Exercise 11.1

1. i) $y = 2x^3 - 3x^2 + 2x$

$$\frac{dy}{dx} = \mathbf{6x^2 - 6x + 2}$$

$$\frac{d^2y}{dx^2} = \mathbf{12x - 6}$$

ii) $y = 3 \sin 2x + \cos 5x$

$$\frac{dy}{dx} = 3 \times \cos 2x \times 2 + (-\sin 5x) \times 5 \text{ (chain rule)}$$

$$= \mathbf{6 \cos 2x - 5 \sin 5x}$$

$$\frac{d^2y}{dx^2} = 6 \times (-\sin 2x) \times 2 - 5 \times \cos 5x \times 5$$

$$= \mathbf{-12 \sin 2x - 25 \cos 5x}$$

iii) $y = e^{2x} - \log_e 7x$

$$\frac{dy}{dx} = 2e^{2x} - \frac{1}{7x} \times 7 = \mathbf{2e^{2x}} - \frac{1}{x} = \mathbf{2e^{2x} - x^{-1}}$$

$$\frac{d^2y}{dx^2} = 2 \times 2e^{2x} - (-1) \times x^{-2}$$

$$= \mathbf{4e^{2x} + x^{-2}}$$

2. $s = 0.5t^3 - 2t^2 + 2t + 5$

Differentiate with respect to s, $\frac{ds}{dt} = 1.5t^2 - 4t + 2$

a) Velocity $= \frac{ds}{dt} = 1.5t^2 - 4t + 2$

When t = 4 s, $\frac{ds}{dt} = 1.5(4)^2 - 4(4) + 2$

$= 24 - 16 + 2 =$ **10 m/s**

b) $\frac{ds}{dt}$ or the velocity is zero when the vehicle comes to rest

Therefore, $\frac{ds}{dt} = 1.5t^2 - 4t + 2 = 0$

This is a quadratic equation, its solution will give the value of time t.

$$t = \frac{-(-4) \pm \sqrt{(-4)^2 - 4 \times 1.5 \times 2}}{2 \times 1.5}$$

$$= \frac{4 \pm \sqrt{4}}{3} = \frac{4 \pm 2}{3}$$

$$t = \frac{4+2}{3} = \mathbf{2.0\ seconds}$$

or $t = \frac{4-2}{3} = \mathbf{0.67\ second}$

c) The second derivative, i.e. $\frac{d^2s}{dt^2}$, gives the acceleration.

From part (a), $\frac{ds}{dt} = 1.5t^2 - 4t + 2$

$\frac{d^2s}{dt^2} = 3t - 4$

Acceleration at the end of t seconds = 8 m/s^2

Therefore, $3t - 4 = 8$

or $t = \frac{12}{3} = \mathbf{4\ seconds}$

After **4 seconds,** the acceleration of the vehicle is 8 m/s^2.

3. Let us assume that the maximum BM occurs at point C, x metres from A, as shown in Figure 11.5. Bending moment at this point is given by:

$M_x = R_1 \times x - 2 \times x \times \frac{x}{2}$

$= 7.5x - x^2$

At maximum BM the slope of the curve is zero, therefore $\frac{dM_x}{dx} = 0$

$\frac{dM_x}{dx} = 7.5 - 2x$

$7.5 - 2x = 0$ or $x = \frac{7.5}{2} = \mathbf{3.75\ m}$

$Mx = 7.5 \times 3.75 - (3.75)^2 = \mathbf{14.06\ kNm}$

4. $y = x^3 - 6x^2 + 9x + 6$

Therefore $\frac{dy}{dx} = 3x^2 - 12x + 9$

$\frac{d^2y}{dx^2} = 6x - 12$

At a turning point $\frac{dy}{dx} = 0$

$\frac{dy}{dx} = 3x^2 - 12x + 9 = 0$

The above equation is a quadratic equation and may solved using the quadratic formula:

$x = \frac{-(-12) \pm \sqrt{(-12)^2 - 4(3)(9)}}{2 \times 3}$

$= \frac{12 \pm 6}{6}$

Either $x = \frac{12+6}{6} = 3$, or $x = \frac{12-6}{6} = 1$

When $x = 3$, $\frac{d^2y}{dx^2} = 6x - 12 = 6 \times 3 - 12 = 6$

Since 6 is positive, the turning point at **x = 3 is a minimum.**

y (minimum) = $(3)^3 - 6(3)^2 + 9 \times 3 + 6 =$ **6**

When $x = 1, \frac{d^2y}{dx^2} = 6x - 12 = 6 \times (1) - 12 = -6$

Since 6 is negative, the turning point at **x = 1 is a maximum.**

y (maximum) = $(1)^3 - 6(1)^2 + 9 \times 1 + 6 =$ **10**

5. Profit, $P = 20x - 0.04x^2$

For finding the best output, $\frac{dP}{dx}$ is determined and equated to 0.

$$\frac{dP}{dx} = 20 - 0.04 \times 2x = 20 - 0.08x$$

$20 - 0.08x = 0$, Therefore, $x = \frac{20}{0.08} = \mathbf{250}$

Test for maximum / minimum: $\frac{d^2P}{dx^2} = -0.08$

As ths value is negative, therefore, x = 250 makes P (profit) a maximum.

6. i) Total cost of producing x number of a roof component = Fixed cost +Variable cost

$= 25000 + 0.05x^2 + 20x$

Average cost (C_{AV}) = Total cost ÷ x

$$= \frac{25000}{x} + \frac{0.05x^2}{x} + 20x$$

$$= \frac{25000}{x} + 0.05x + 20 = 25000x^{-1} + 0.05x + 20$$

Differentiating the above, $\frac{dC_{Av}}{dx} = 25000(-1)x^{-2} + 0.05$

For maximum / minimum, $\frac{dC_{Av}}{dx}$ must be equal to 0

$25000(-1)x^{-2} + 0.05 = 0$

$0.05 = 25000x^{-2}$ or $0.05 = \frac{25000}{x^2}$

Therefore, $x^2 = \frac{25000}{0.05}$ or $x = \sqrt{500000} = \mathbf{707}$

$$\frac{d^2C_{Av}}{dx^2} = -25000(-2)x^{-3} = \frac{50000}{x^3}$$

This will be positive for all values of x, therefore, **x =707 is a minimum.**

ii) Average cost per component $= \frac{25000}{x} + 0.05x + 20$

$$= \frac{25000}{707} + 0.05 \times 707 + 20 = \mathbf{£90.71}$$

7. Perimeter of the beam = 2b + 2h = 700

or b + h = 350, Therefore b = 350 – h

Strength, $S = bh^3$

$= (350 - h)h^3 = 350h^3 - h^4$

$\frac{dS}{dh} = 350 \times 3h^2 - 4h^3 = 1050\ h^2 - 4h^3$

$\frac{d^2S}{dh^2} = 2100\ h - 12\ h^2$

At a turning point, $\frac{dS}{dh} = 0$

$1050\ h^2 - 4h^3 = 0$

After transposition, $4h^3 = 1050\ h^2$

$4h = 1050$, Therefore **h = 262.5 mm**

Test for maximum / minimum: $\frac{d^2S}{dh^2} = 2100\ h - 12\ h^2$

$= 2100 \times 262.5 - 12(262.5)^2 = -275625$

As ths value is negative, therefore h = 262.5 makes S a maximum.

b = 350 – h = 350 – 262.5 = **87.5 mm**

Dimensions of the strongest beam are: **87.5 × 262.5 mm deep**

8. Let **w** be the width and **h** be the depth of the channel (see Figure S11.1).

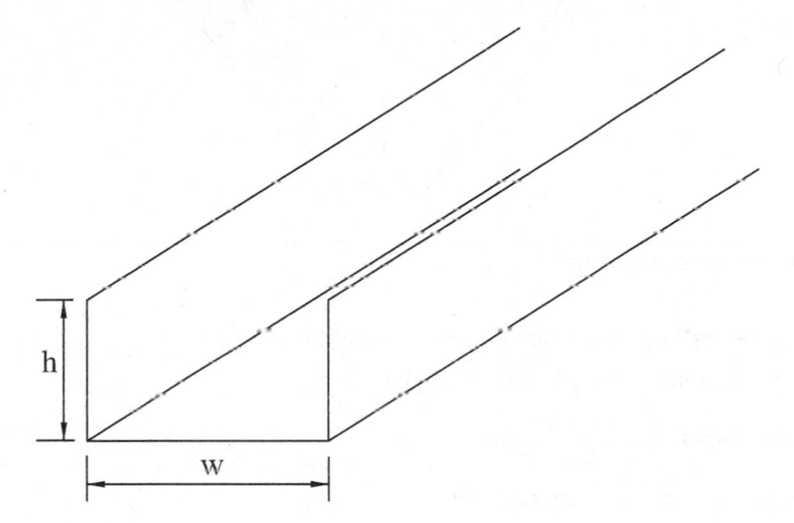

Figure S11.1 Open rectangular channel

Cross-sectional area, A = wh (1)

Perimeter of the channel = 200 = h + w + h

$2h + w = 200$, or $h = \frac{200 - w}{2} = 100 - 0.5w$ (2)

From equations 1 and 2, A = w(100 – 0.5w)

$= 100w - 0.5w^2$

$\frac{dA}{dw} = 100 - 0.5 \times 2w = 100 - w$

At a turning point, $\frac{dA}{dw} = 0$

Therefore 100 – w = 0 or **w = 100 mm**

Test for maximum / minimum: $\frac{d^2A}{dw^2} = -1$, so the turning point is a maximum

Therefore A is maximum when w = 100 mm

$2h + 100 = 200, \qquad h = \frac{100}{2} = \mathbf{50\ mm}$

Width = 100 mm, Depth = 50 mm

9. Let **2x** be the width at the top and **h** be the depth of the channel

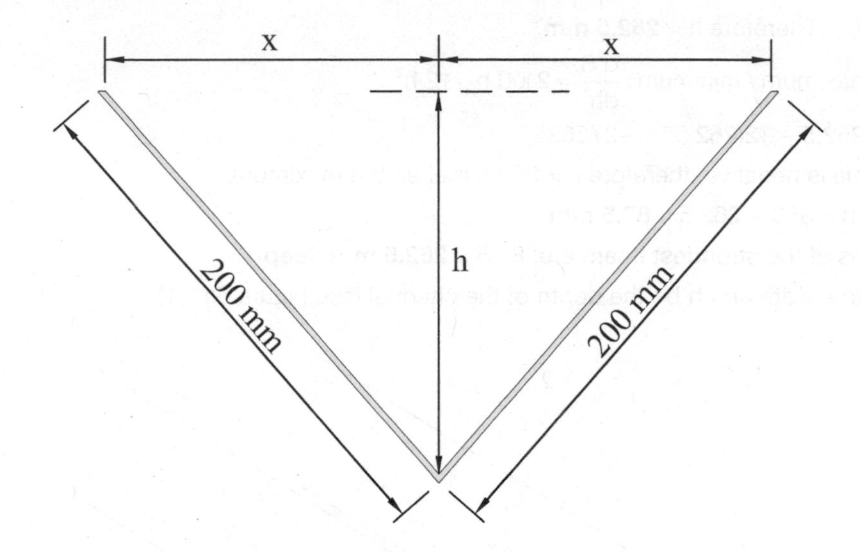

Figure S11.2 Open triangular channel

Using Theorem of Pythagoras, $(200)^2 = x^2 + h^2$ (see Figure S11.2)

$h^2 = 40000 - x^2$, therefore, $h = \sqrt{40000 - x^2} = (40000 - x^2)^{\frac{1}{2}}$ (1)

Cross – sectional area, $A = \frac{2x \times h}{2} = xh$ (2)

From (1) and (2), $\qquad A = x \times (40000 - x^2)^{\frac{1}{2}}$

$\frac{dA}{dx} = x \times \frac{1}{2}(40000 - x^2)^{\frac{1}{2}-1} \times (-2x) + 1 \times (40000 - x^2)^{\frac{1}{2}}$

$= -x^2(40000 - x^2)^{-\frac{1}{2}} + (40000 - x^2)^{\frac{1}{2}}$

For maximum or minimum value of A, $\frac{dA}{dx} = 0$

Therefore, $-x^2(40000 - x^2)^{-\frac{1}{2}} + (40000 - x^2)^{\frac{1}{2}} = 0$

$(40000 - x^2)^{\frac{1}{2}} = x^2(40000 - x^2)^{-\frac{1}{2}}$

$x^2 = \frac{(40000 - x^2)^{\frac{1}{2}}}{(40000 - x^2)^{-\frac{1}{2}}} = (40000 - x^2)^{\frac{1}{2}+\frac{1}{2}}$

$= 40000 - x^2$, or $x^2 = 20000$

Hence, x = 141.42 mm

$$\frac{d^2A}{dx^2} = -x^2 \times \left(-\frac{1}{2}\right)(40000 - x^2)^{-\frac{1}{2}-1} \times (-2x) + (40000 - x^2)^{-\frac{1}{2}} \times (-2x) +$$

$$\frac{1}{2}(40000 - x^2)^{\frac{1}{2}-1}$$

$$= -x^3 \times (40000 - x^2)^{-\frac{3}{2}} - 2x(40000 - x^2)^{-\frac{1}{2}} + \frac{1}{2}(40000 - x^2)^{-\frac{1}{2}}$$

Then above expression gives a negative answer when solved for x = 141.42.

Hence, the turning point at **x = 141.42 is a maximum.**

Width of the channel at the top is **282.84 mm**

$$h = (40000 - x^2)^{\frac{1}{2}} = (40000 - x^2)^{\frac{1}{2}}$$

$$= (40000 - 141.42^2)^{\frac{1}{2}} = \mathbf{141.42\,mm}$$

$$\text{Area of the channel} = \frac{282.84 \times 141.42}{2} = 20000 \text{ mm}^2$$

10. 210 litres = 0.210 m^3

Let r be the radius and h be the height of the cylinder.

Surface area, $S = \pi r^2 + \pi r^2 + 2\pi rh = 2\pi r^2 + 2\pi rh$ .... (1)

Volume, $V = \pi r^2 h = 0.210\ m^3$

$$\text{or } h = \frac{0.210}{\pi r^2} \quad \text{........................ (2)}$$

$$\text{From equations 1 and 2: } S = 2\pi r^2 + 2\pi r\frac{0.210}{\pi r^2} = 2\pi r^2 + 0.42r^{-1}$$

$$\frac{dS}{dr} = 2(2)\pi r + (0.42)(-1)r^{-2} = 4\pi r - \frac{0.42}{r^2}$$

For the surface area to be maximum / minimum, $\frac{dS}{dr} = 0$

$$4\pi r - \frac{0.42}{r^2} = 0, \text{ or } 4\pi r = \frac{0.42}{r^2}$$

$$r^3 = \frac{0.42}{4\pi} \quad \text{or} \quad r = 0.32212 \text{ m} = \mathbf{322.12\,mm}$$

$$h = \frac{0.210}{\pi r^2} = \frac{0.210}{\pi(0.32212)^2} = 0.64422 \text{ m} = \mathbf{644.22\,mm}$$

Diameter of the cylinder = 2r = 2 × 322.12 = **644.24 mm**

$$\text{Test for a minimum: } \frac{d^2S}{dr^2} = 4\pi - 0.42(-2)r^{-3}$$

$= 4\pi + \frac{0.84}{r^3}$; this will be positive for all values of r. Therefore, surface area S of the hot water cylinder is minimum when: **Diameter = 644.24 mm and height = 644.22 mm**

Solutions – Exercise 12.1

1. $(2x+4-\frac{1}{x^2})\,dx = \int(2x+4-\frac{1}{x^2})dx$

$$= \frac{2x^2}{2}+4x-\frac{x^{-1}}{-1}+c$$

$$= \mathbf{x^2+4x+x^{-1}+c \text{ or } x^2+4x+\frac{1}{x}+c}$$

2. $(2x+1)^2\,dx = \int(2x+1)^2dx$

$$= \int(4x^2+4x+1)\,dx$$

$$= \frac{4x^3}{3}+\frac{4x^2}{2}+x+c$$

$$= \frac{4x^3}{3}+2x^2+x+c$$

3. $(\sin 3\theta - \cos 4\theta)\,d\theta = \int(\sin 3\theta-\cos 4\theta)d\theta$

$$= \frac{-\cos 3\theta}{3}-\frac{\sin 4\theta}{4}+c$$

4. $(e^x+\frac{1}{e^{3x}}-\frac{2}{x})\,dx = \int(e^x+\frac{1}{e^{3x}}-\frac{2}{x})\,dx$

$$= \int(e^x+e^{-3x}-\frac{2}{x})\,dx$$

$$= e^x+\frac{e^{-3x}}{-3}-2\log_e x+c$$

$$= e^x-\frac{e^{-3x}}{3}-2\log_e x+c = e^x-\frac{1}{3e^{3x}}-2\log_e x+\mathbf{c}$$

5. $(\sqrt{x}+3)^2\,dx = \int(x+6\sqrt{x}+9)dx$

$$= \frac{x^2}{2}+\frac{6x^{3/2}}{3/2}+9x+c$$

$$= \frac{x^2}{2}+4x^{3/2}+9x+c$$

6. $\int_1^2(2x^2-4x)dx = \left[\frac{2x^3}{3}-\frac{4x^2}{2}\right]_1^2$

$$= \left[\frac{2\times 2^3}{3}-\frac{4\times 2^2}{2}\right]-\left[\frac{2\times 1^3}{3}-\frac{4\times 1^2}{2}\right]$$

$$= \left[-\frac{8}{3}\right]-\left[-\frac{4}{3}\right] = -\frac{4}{3} \text{ or } \mathbf{-1.33}$$

7. $\int_1^2(\sqrt{x}+\frac{1}{\sqrt{x}})dx = \int_1^2(x^{1/2}+x^{-1/2})dx$

$$= \left[\frac{x^{3/2}}{3/2}+\frac{x^{1/2}}{1/2}\right]_1^2 = \left[\frac{2x^{3/2}}{3}+2x^{1/2}\right]_1^2$$

$$= \left[\frac{2\times 2^{3/2}}{3} + 2\times 2^{1/2}\right] - \left[\frac{2\times 1^{3/2}}{3} + 2\times 1^{1/2}\right]$$

$$= [1.886 + 2.828] - [0.667 + 2] = \mathbf{2.047}$$

8. $$\int_0^{\pi/2} 2\sin 3\theta \, d\theta = \left[\frac{2(-\cos 3\theta)}{3}\right]_0^{\pi/2}$$

$$= \left[\frac{2(-\cos 3\times \pi/2)}{3}\right] - \left[\frac{2(-\cos 3\times 0)}{3}\right]$$

$$= \left[\frac{2(-\cos 1.5\pi)}{3}\right] - \left[\frac{2(-\cos 0)}{3}\right]$$

$$= \left[\frac{2(-0)}{3}\right] - \left[\frac{2(-1)}{3}\right] = \frac{2}{3} \text{ or } \mathbf{0.67}$$

9. $$\int_{0.5}^{1} 2\cos\frac{x}{2}\, dx = \left[2\sin\frac{x}{2} \div \frac{1}{2}\right]_{0.5}^{1}$$

$$= \left[4\sin\frac{x}{2}\right]_{0.5}^{1} = \left[4\sin\frac{1}{2} - 4\sin\frac{0.5}{2}\right]$$

$$= 1.918 - 0.990 = \mathbf{0.928}$$

10. $$\int_{-1}^{1} \frac{3}{2e^{3x}}\, dx = \frac{3}{2}\int_{-1}^{1} e^{-3x}\, dx$$

$$= \frac{3}{2}\left[\frac{e^{-3x}}{-3}\right]_{-1}^{1} = -\frac{1}{2}\left[e^{-3\times 1} - e^{-3\times -1}\right]$$

$$= -\frac{1}{2}\left[0.04979 - 20.08554\right] = \mathbf{10.018}$$

11. $$\int_2^3 (e^{x/3} + e^{-x/4})\, dx - \left[\frac{e^{x/3}}{1/3} + \frac{e^{-x/4}}{-1/4}\right]_2^3$$

$$- \left[3e^{x/3} - 4e^{-x/4}\right]_2^3$$

$$= \left[3e^{3/3} - 4e^{-3/4}\right] - \left[3e^{2/3} - 4e^{-2/4}\right]$$

$$= [8.1548 - 1.8895] - [5.8432 - 2.4261] = \mathbf{2.848}$$

12. Find $\int 4x(\sqrt{(1-x^2)})\, dx$

Let $u = 1 - x^2$

Differentiating u, $\frac{du}{dx} = -2x$ or $dx = \frac{du}{-2x}$

$$\int 4x(\sqrt{(1-x^2)})\, dx = \int 4x(u)^{1/2} \frac{du}{-2x}$$

$$= -2\int (u)^{1/2}\, du = -2\times\left[\frac{u^{3/2}}{\frac{3}{2}}\right] + c$$

$$= -\frac{4}{3}(1-x^2)^{3/2} + c$$

13. Evaluate $\int_0^1 \frac{x\,dx}{\sqrt{x^2+1}}$

Let $u = x^2 + 1$

Differentiating u, $\frac{du}{dx} = 2x$ or $dx = \frac{du}{2x}$

After substitution, $\int_0^1 \frac{x\,dx}{\sqrt{x^2+1}} = \int_0^1 \frac{x}{\sqrt{u}} \times \frac{du}{2x}$

$= \frac{1}{2}\int_0^1 u^{-1/2}\,du$

$= \frac{1}{2}\left[\frac{u^{1/2}}{1/2}\right]_0^1 = \left[(x^2+1)^{1/2}\right]_0^1$

$= (1^2 + 1)^{1/2} - (0^2 + 1)^{1/2}$

$= 1.414 - 1 = \mathbf{0.414}$

14. Evaluate $\int_0^1 \tan x \sec^2 x\,dx$

Let $u = \tan x$

Differentiating u, $\frac{du}{dx} = \sec^2 x$ or $dx = \frac{du}{\sec^2 x}$

$\int_0^1 \tan x \sec^2 x\,dx = \int_0^1 u \sec^2 x \frac{du}{\sec^2 x}$

$= \int_0^1 u\,du = \left[\frac{u^2}{2}\right]_0^1 = \frac{1}{2}\left[u^2\right]_0^1$

$= \frac{1}{2}\left[(\tan x)^2\right]_0^1 = \frac{1}{2}\left[(\tan 1)^2 - (\tan 0)^2\right]$

$= \frac{1}{2}\left[(1.5574)^2 - (0)^2\right] = \mathbf{1.213}$

15. Evaluate $\int_1^2 \frac{2\log_e x}{x}\,dx$

Let $u = \log_e x$

Differentiating u, $\frac{du}{dx} = \frac{1}{x}$ or $dx = x\,du$

$\int_1^2 \frac{2\log_e x}{x}\,dx = \int_1^2 \frac{2u}{x}\,x\,du = \int_1^2 2u\,du$

$= \left[2 \times \frac{u^2}{2}\right]_1^2 = \left[u^2\right]_1^2 = \left[(\log_e x)^2\right]_1^2$

$= \left[(\log_e 2)^2 - (\log_e 1)^2\right] = 0.48045 - 0 = \mathbf{0.48}$

16. $\int_1^2 2x e^{x^2-1}\,dx$

Let $u = x^2 - 1$; $\frac{du}{dx} = 2x$ or $dx = \frac{du}{2x}$

Change limits: when $x = 2$, $u = 2^2 - 1 = 3$

When $x = 1$, $u = 1^2 - 1 = 0$

$$\int_1^2 2xe^{x^2-1}dx = \int_0^3 2xe^u \frac{du}{2x} = \int_0^3 e^u du$$

$$= \left[e^u\right]_0^3 = e^3 - e^0 = 20.086 - 1 = \mathbf{19.086}$$

17. Evaluate $\int_0^{0.5} \sin(2x+1)dx$

Let $u = 2x+1; \frac{du}{dx} = 2$ or $dx = \frac{du}{2}$

Change limits: when $x = 0.5$, $u = 2 \times 0.5 + 1 = 2$

When $x = 0$, $u = 2 \times 0 + 1 = 1$

$$\int_0^{0.5} \sin(2x+1)dx = \int_1^2 \sin u \frac{du}{2} = \frac{1}{2}\int_1^2 \sin u\, du$$

$$= \frac{1}{2}\left[-\cos u\right]_1^2 = \frac{1}{2}\left[(-\cos 2) - (-\cos 1)\right]$$

$$= \frac{1}{2}\left[(0.416) - (-0.540)\right] = \mathbf{0.478}$$

18. Evaluate $\int_0^1 x(\sqrt{1+x^2})\,dx$

Let $u = 1+x^2; \frac{du}{dx} = 2x$ or $dx = \frac{du}{2x}$

Change limits: when $x = 1$, $u = 1 + 1^2 = 2$

When $x = 0$, $u = 1 + 0^2 = 1$

$$\int_0^1 x(\sqrt{1+x^2})\,dx = \int_1^2 x(u)^{1/2}\frac{du}{2x} = \frac{1}{2}\int_1^2 (u)^{1/2} du$$

$$= \frac{1}{2}\left[\frac{u^{3/2}}{\frac{3}{2}}\right]_1^2 = \frac{1}{2} \times \frac{2}{3}\left[2^{3/2} - 1^{3/2}\right]$$

$$= \frac{1}{3}\left[2.828 - 1\right] = \mathbf{0.609}$$

19. $\int 2x\log_e x\,dx = \int \log_e x\; 2x\,dx$

On comparing $\int 2x\log_e x dx$ with the left side of the 'integration by parts rule'

$u = \log_e x$ and $dv = 2x\,dx$

Differentiating $u = \log_e x$, we have $\frac{du}{dx} = \frac{1}{x}$

Integrating $\int dv = \int 2x\,dx$, we have $v = x^2$

Substituting the above in the 'integration by parts rule': $\int u\,dv = uv - \int v\,du$

$$\int \log_e x\, 2x\,dx = \log_e x \times x^2 - \int (x^2)(\frac{1}{x})dx$$

$$= x^2 \log_e x - \int x\,dx$$

$$= x^2 \log_e x - \frac{x^2}{2} + c$$

20. $\int x^3 \log_e x\,dx = \int \log_e x\ x^3\,dx$

On comparing $\int \log_e x\ x^3\,dx$ with the left side of the 'integration by parts rule'

$u = \log_e x$ and $dv = x^3\,dx$

Differentiating $u = \log_e x$, we have $\frac{du}{dx} = \frac{1}{x}$

Integrating $\int dv = \int x^3\,dx$, we have $v = \frac{x^4}{4}$

Substituting the above in the 'integration by parts rule': $\int u\,dv = uv - \int v\,du$

$$\int \log_e x\ x^3\,dx = \log_e x \times \frac{x^4}{4} - \int \left(\frac{x^4}{4}\right)\left(\frac{1}{x}\right)dx$$

$$= \frac{x^4}{4}\log_e x - \frac{1}{4}\int x^3\,dx$$

$$= \frac{x^4}{4}\log_e x - \frac{x^4}{16} + c$$

21. $\int x^2 \cos x\,dx$

On comparing $\int x^2 \cos x\,dx$ with the left side of the 'integration by parts rule'

$u = x^2$ and $dv = \cos x\,dx$

Differentiating $u = x^2$, we have $\frac{du}{dx} = 2x$

Integrating $\int dv = \int \cos x\,dx$, we have $v = \sin x$

Substituting the above in the 'integration of parts rule': $\int u\,dv = uv - \int v\,du$

$$\int x^2 \cos x\,dx = x^2(\sin x) - \int (\sin x)(2x)\,dx$$

$$= x^2 \sin x - \int 2x \sin x\,dx \qquad \text{(i)}$$

2x sin x is a product of 2 quantities, so it needs to be integrated again

Let $u = 2x$ and $dv = \sin x\,dx$

Differentiating $u = 2x$, we have $\frac{du}{dx} = 2$

Integrating $\int dv = \int \sin x\,dx$, we have $v = -\cos x$

Substituting the above in the 'integration of parts rule': $\int u\,dv = uv - \int v\,du$

$$\int 2x \sin x\,dx = 2x(-\cos x) - \int(-\cos x)(2)\,dx$$

$$= -2x \cos x - (-2\sin x) = 2\sin x - 2x\cos x \qquad \text{(ii)}$$

From (i) and (ii): $\int x^2 \cos x\,dx = x^2(\sin x) - (2\sin x - 2x\cos x)$

$= \mathbf{(x^2 - 2)\sin x + 2x\cos x + c}$

22. $\int_1^2 \frac{\log_e x}{x^2}dx = \int_1^2 \log_e x \times x^{-2}\,dx$

On comparing $\int \log_e x \times x^{-2}\,dx$ with the left side of the 'integration by parts rule'

$u = \log_e x$ and $dv = x^{-2}\,dx$

Differentiating $u = \log_e x$, we have $\frac{du}{dx} = \frac{1}{x}$

Integrating $\int dv = \int x^{-2}\,dx$, we have $v = \frac{x^{-1}}{-1} = -\frac{1}{x}$

Substituting the above in the 'integration by parts rule': $\int u\,dv = uv - \int v\,du$

$$\int_1^2 \log_e x \times x^{-2}\,dx = \log_e x\left(-\frac{1}{x}\right) - \int -\frac{1}{x}\left(\frac{1}{x}\right)dx$$

$$= -\frac{\log_e x}{x} + \int x^{-2}\,dx$$

$$= \left[-\frac{\log_e x}{x} + \frac{x^{-1}}{-1}\right]_1^2$$

$$= \left[-\frac{\log_e x}{x} - \frac{1}{x}\right]_1^2$$

$$= \left[-\frac{\log_e 2}{2} - \frac{1}{2}\right] - \left[-\frac{\log_e 1}{1} - \frac{1}{1}\right]$$

$= [-0.3466 - 0.5] - [-0 - 1] =$ **0.1534**

23. $\int_0^1 x(x+1)^2\,dx$

Let $u = x$ and $dv = (x + 1)^2\,dx$

Differentiating $u = x$, we have $\frac{du}{dx} = 1$

For integrating $\int dv = \int (x+1)^2\,dx$, substitute: $t = x + 1$

Differentiate $t = x + 1$, $\frac{dt}{dx} = 1$, or $dx = dt$

$\int dv = \int (x+1)^2\,dx$ now becomes $\int dv = \int (t)^2\,dt$

Integration of $\int dv = \int (t)^2\,dt$ gives $v = \frac{t^3}{3} = \frac{(x+1)^3}{3}$

Substituting the above in the 'integration by parts rule': $\int u\,dv = uv - \int v\,du$

$$\int_0^1 x(x+1)^2\,dx = \frac{x(x+1)^3}{3} - \int \frac{(x+1)^3}{3}(1)\,dx$$

$$= \frac{x(x+1)^3}{3} - \frac{1}{3}\int (x+1)^3\,dx$$

After integrating $\frac{1}{3}\int (x+1)^3\,dx$ by substitution, we have:

$$\int_0^1 x(x+1)^2\,dx = \left[\frac{x(x+1)^3}{3} - \frac{1}{12}(x+1)^4\right]_0^1$$

$$= \left[\frac{1(1+1)^3}{3} - \frac{1}{12}(1+1)^4\right] - \left[\frac{0(0+1)^3}{3} - \frac{1}{12}(0+1)^4\right]$$

$= \left[\frac{8}{3} - \frac{16}{12}\right] - \left[\frac{0}{3} - \frac{1}{12}\right] =$ **1.416**

Solutions – Exercise 13.1

1. Area, $A = \int_1^3 y\,dx$

$$= \int_1^3 (3x+1)\,dx \qquad (y = 3x + 1, \text{ is the equation of the curve})$$

$$= \left[\frac{3x^2}{2}+x\right]_1^3 = \left[\frac{3\times3^2}{2}+3\right]-\left[\frac{3\times1^2}{2}+1\right]$$

Therefore, Area = [13.5 + 3] – [1.5 + 1] = **14.0 square units**

2. Area, $A = \int_1^4 y\,dx$

$$= \int_1^4 (2x^2+5)\,dx \qquad (y = 2x^2 + 5, \text{ is the equation of the curve})$$

$$= \left[\frac{2x^3}{3}+5x\right]_1^4 = \left[\frac{2\times4^3}{3}+5\times4\right]-\left[\frac{2\times1^3}{3}+5\times1\right]$$

Therefore, Area = [42.67 + 20] – [0.67 + 5] = **57.0 square units**

3. Area, $A = \int_0^{\pi/2} y\,dx$

$$= \int_0^{\pi/2} \sin\theta\,d\theta \qquad (y = \sin\theta, \text{ is the equation of the curve})$$

$$= \left[-\cos\theta\right]_0^{\pi/2} = -\cos\frac{\pi}{2}-(-\cos 0)$$

Therefore, Area = – 0 + 1 = **1.0 square unit**

4. The points of intersection of the 2 graphs are found by solving the equation:

$7x - x^2 = x + 5$ or $x^2 - 6x + 5 = 0$

$(x - 1)(x - 5) = 0$ Therefore $x = 1$ or $x = 5$

The above values of x are the limits of integration.

Area ABCDEF $= \int_1^5 y\,dx = \int_1^5 (7x - x^2)\,dx$

$$= \left[\frac{7x^2}{2}-\frac{x^3}{3}\right]_1^5$$

$$= \left[\frac{7\times5^2}{2}-\frac{5^3}{3}\right]-\left[\frac{7\times1^2}{2}-\frac{1^3}{3}\right]$$

= [87.5 – 41.67] – [3.5 – 0.33] = 42.66

Area ABCGF $= \int_1^5 y\,dx = \int_1^5 (x+5)\,dx$

$$= \left[\frac{x^2}{2}+5x\right]_1^5$$

$$= \left[\frac{5^2}{2}+5\times5\right]-\left[\frac{1^2}{2}+5\times1\right]$$

= [12.5 + 25] – [0.5 + 5] = 32.0

Enclosed area (shaded) = 42.66 – 32.0 = **10.66 square units** (2 d.p.)

5. y = 3; limits: x = 0 and x = 4

$$\text{Volume of revolution} = \int_0^4 \pi y^2\,dx = \int_0^4 \pi(3)^2\,dx$$
$$= \int_0^4 9\pi\,dx$$
$$= 9\pi\left[x\right]_0^4$$
$$= 9\pi\times 4 - 9\pi\times 0 = \mathbf{113.1\ cubic\ units}$$

6. y = 0.6x; limits: x = 0 and x = 10

$$\text{Volume of revolution} = \int_0^{10} \pi y^2\,dx = \int_0^{10} \pi(0.6x)^2\,dx$$
$$= \int_0^{10} 0.36\pi\,x^2\,dx$$
$$= 0.36\pi\left[\frac{x^3}{3}\right]_0^{10}$$
$$= 0.36\pi\times\frac{10^3}{3} - 0 = \mathbf{376.99\ cubic\ units}$$

7. $y = 2x^2 + 3x$; limits: x = 2 and x = 5

$$\text{Volume of revolution} = \int_2^5 \pi y^2\,dx = \int_2^5 \pi(2x^2+3x)^2\,dx$$
$$= \int_2^5 \pi(4x^4+12x^3+9x^2)\,dx$$
$$= \pi\left[4\frac{x^5}{5}+12\frac{x^4}{4}+9\frac{x^3}{3}\right]_2^5$$
$$= \pi\left[4\frac{5^5}{5}+12\frac{5^4}{4}+9\frac{5^3}{3}\right]-\pi\left[4\frac{2^5}{5}+12\frac{2^4}{4}+9\frac{2^3}{3}\right]$$
$$= \pi[4750 - 97.6]$$
$$= \mathbf{14615.95\ cubic\ units}$$

8. Figure S13.1a shows area OAB; if this area is rotated about the x-axis, it will produce a right circular cone as shown in Figure S13.1b.

The first step is to write the equation of the top edge of area OAB, which is a straight line (OA).

$$\text{Slope of line OA } (m) = \frac{AB}{OB} = \frac{10}{30} = \frac{1}{3}$$

Equation of line OA becomes: y = mx + c

$$\text{or } y = \frac{1}{3}x + 0 = \frac{x}{3}$$

Area OAB generates the cone when it is rotated about the x-axis.

$$\text{Volume of revolution} = \int_0^{30} \pi y^2\,dx = \int_0^{30} \pi\left(\frac{x}{3}\right)^2 dx$$
$$= \frac{\pi}{9}\int_0^{30} x^2\,dx$$
$$= \frac{\pi}{9}\left[\frac{x^3}{3}\right]_0^{30}$$
$$= \frac{\pi}{9}\left[\frac{30^3}{3}-\frac{0^3}{3}\right] = \mathbf{3141.59\ cm^3}$$

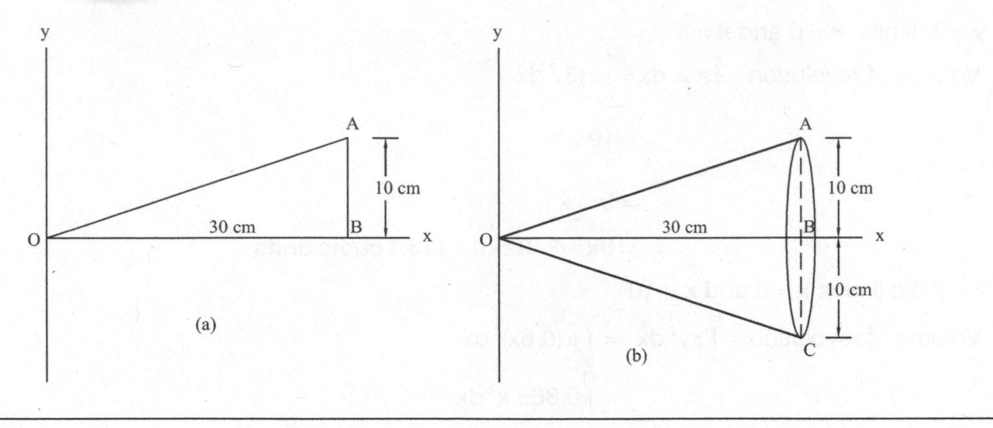

Figure S13.1

9. Figure S13.2 shows area OABD which will produce the slump cone when rotated about the x-axis. The equation of the top edge AB, a straight line, is given by the equation y = mx + c.

Slope of line AB $(m) = \frac{BC}{AC} = \frac{50}{300} = 0.16666$

Equation of line AB becomes: y = 0.16666x + 50   (c = OA = 50)

$$\text{Volume of revolution} = \int_0^{300} \pi y^2\,dx = \int_0^{300} \pi(0.16666x + 50)^2\,dx$$

$$= \pi\int_0^{300}(0.02778\,x^2 + 16.6666\,x + 2500)\,dx$$

$$= \pi\left[\frac{0.02778x^3}{3} + \frac{16.6666x^2}{2} + 2500\,x\right]_0^{300}$$

$$= \pi\left[\frac{0.02778 \times 300^3}{3} + \frac{16.6666 \times 300^2}{2} + 2500 \times 300\right]$$

$= \mathbf{5\ 497\ 840.55\ mm^3 = 5.498\ litres}$

OA = 50 mm   BD = 100 mm
OD = 300 mm

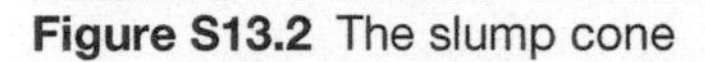

Figure S13.2 The slump cone

10. a) Force on the wall at a depth of 4 m $= \int_0^H K_a \gamma y\,dy = \int_0^4 0.33 \times 18.0\,y\,dy$

$$= 5.94\int_0^4 y\,dy = 5.94\left[\frac{y^2}{2}\right]_0^4$$

$$= 5.94\left[\frac{4^2}{2} - \frac{0^2}{2}\right] = 47.52\text{ kN}$$

b) Force on the wall at a depth of 6 m $= \int_0^6 0.33 \times 18.0\,y\,dy$

$$= 5.94\int_0^6 y\,dy = 5.94\left[\frac{y^2}{2}\right]_0^6$$

$$= 5.94\left[\frac{6^2}{2} - \frac{0^2}{2}\right] = \mathbf{106.92\text{ kN}}$$

11. The shear force diagram (SFD) and the bending moment diagram (BMD) are shown in Figure S13.3.

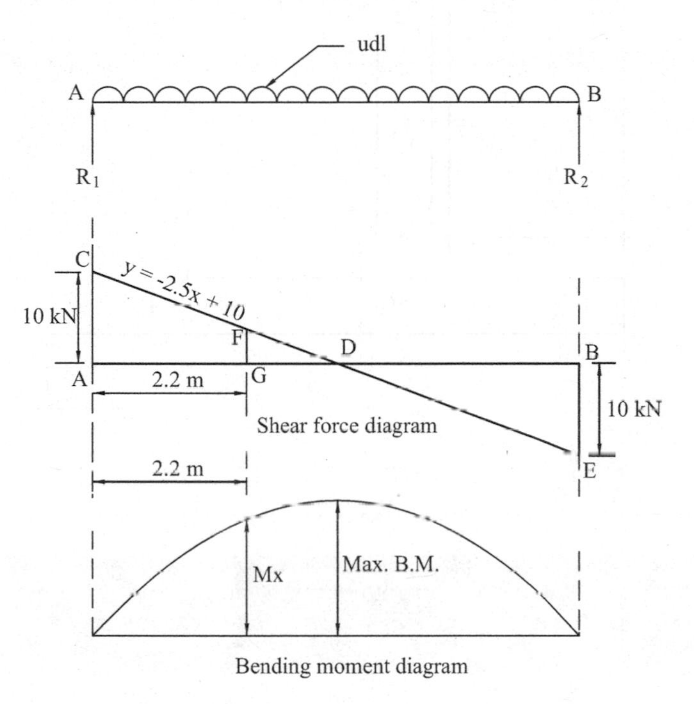

Figure S13.3

As the bending moment is equal to the area of the SFD, therefore:

Bending moment at 2.2 m from A = Area ACFG of the SFD

$$= \int_0^{2.2} (-2.5\,x + 10)\,dx$$

$$= \left[\frac{-2.5x^2}{2} + 10x\right]_0^{2.2}$$

$$= \left[\frac{-2.5 \times 2.2^2}{2} + 10 \times 2.2\right] - \left[\frac{0}{2} + 0\right]$$

$$= -6.05 + 22 = \mathbf{15.95\text{ kNm}}$$

Solutions – Exercise 14.1

1. a)

Let Ox and Oy be the reference axes.

The shape has been divided into 2 parts, as shown in Figure S14.1; the centroids of both parts (G1 and G2) are also shown in this diagram. Let G be the centroid of the whole section.

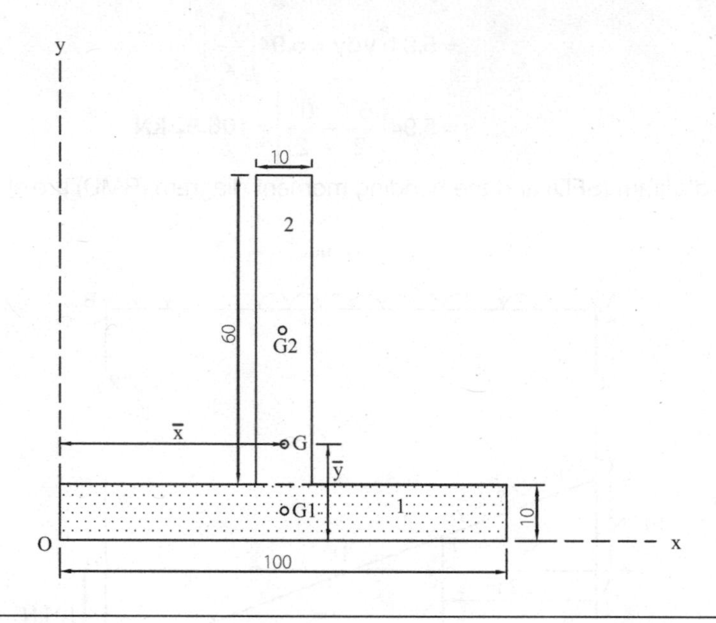

Figure S14.1

	Area	Distance to centroid		Moment of area	
		From y-axis	From x-axis	About Oy	About Ox
	A	x	y	Ax	Ay
1	100 × 10 =1000	50	5	1000 × 50 = 50 000	1000 × 5 = 5000
2	10 × 60 = 600	50	40	30 000	24 000
	ΣA = 1600			ΣAx = 80 000	ΣAy = 29 000

$$\bar{x} = \frac{\Sigma Ax}{\Sigma A} = \frac{80000}{1600} = \mathbf{50.0\ mm}$$

$$\bar{y} = \frac{\Sigma Ay}{\Sigma A} = \frac{29000}{1600} = \mathbf{18.13\ mm}$$

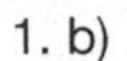

1. b)

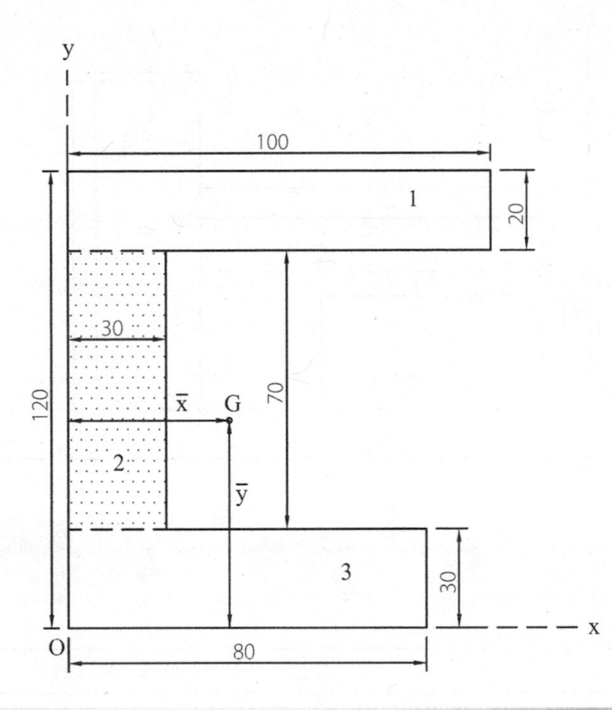

Figure S14.2

Let Ox and Oy be the reference axes.

The shape has been divided into 3 parts, as shown in Figure S14.2. The centroids of all parts are also shown.

	Area	Distance to centroid		Moment of area	
		From y-axis	From x-axis	About Oy	About Ox
	A	x	y	Ax	Ay
1	100 × 20 = 2000	50	110	2000 × 50 = 100 000	2000 × 110 = 220 000
2	70 × 30 = 2100	15	65	2100 × 15 = 31 500	2100 × 65 = 136 500
3	80 × 30 = 2400	40	15	2400 × 40 = 96 000	2400 × 15 = 36 000
	ΣA = 6500			ΣAx = 227 500	ΣAy = 392 500

$$\bar{x} = \frac{\Sigma Ax}{\Sigma A} = \frac{227\,500}{6500} = \mathbf{35.0\ mm}$$

$$\bar{y} = \frac{\Sigma Ay}{\Sigma A} = \frac{392\,500}{6500} = \mathbf{60.38\ mm}$$

1. c)

The shape has been divided into 3 parts, as shown in Figure 14.3. The centroids of all parts are also shown.

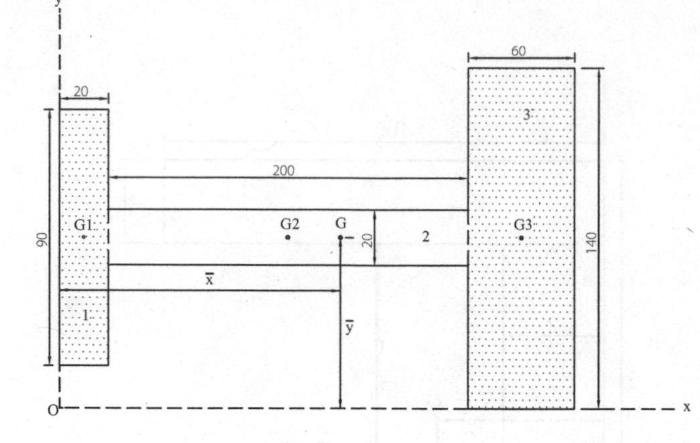

Figure S14.3

	Area	Distance to centroid		Moment of area	
		From y-axis	From x-axis	About Oy	About Ox
	A	x	y	Ax	Ay
1	20 × 90 = 1800	10	70	1800 × 10 = 18 000	1800 × 70 = 126 000
2	200 × 20 = 4000	120	70	4000 × 120 = 480 000	4000 × 70 = 280 000
3	60 × 140 = 8400	250	70	8400 × 250 = 2 100 000	8400 × 70 = 588 000
	ΣA = 14 200			ΣAx = 2 598 000	ΣAy = 994 000

$$\bar{x} = \frac{\Sigma Ax}{\Sigma A} = \frac{2598000}{14200} = \mathbf{182.96\ mm}$$

$$\bar{y} = \frac{\Sigma Ay}{\Sigma A} = \frac{994000}{14200} = \mathbf{70.0\ mm}$$

1. d)

The shape has been divided into 2 parts, as shown in Figure S14.4. The centroids of both parts are also shown.

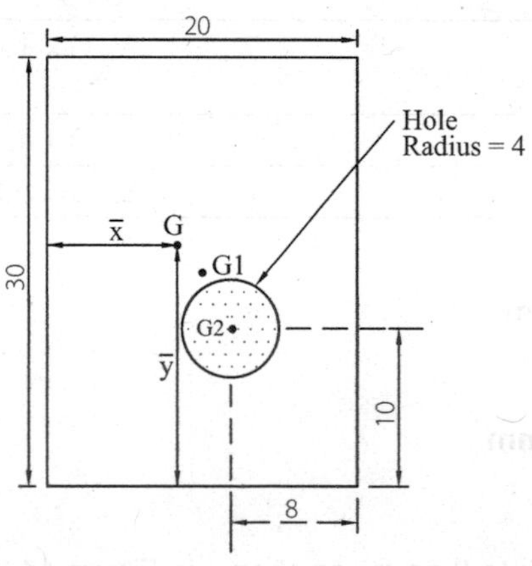

Figure S14.4

	Area	Distance to centroid		Moment of area	
		From y-axis	From x-axis	About Oy	About Ox
	A	x	y	Ax	Ay
1	30 × 20 = 600	10	15	6000	9000
2	– π × 42 = – 50.27	12	10	– 603.24	– 502.7
	ΣA = 549.73			ΣAx = 5396.76	ΣAy = 8497.30

$$\bar{x} = \frac{\Sigma Ax}{\Sigma A} = \frac{5396.76}{549.73} = \mathbf{9.82\ mm}$$

$$\bar{y} = \frac{\Sigma Ay}{\Sigma A} = \frac{8497.30}{549.73} = \mathbf{15.46\ mm}$$

2.

Let Ox and Oy be the reference axes.

The shape has been divided into 2 parts, as shown in Figure S14.5; the centroids of both parts (G1 and G2) are also shown in this diagram. Let G be the centroid of the whole section.

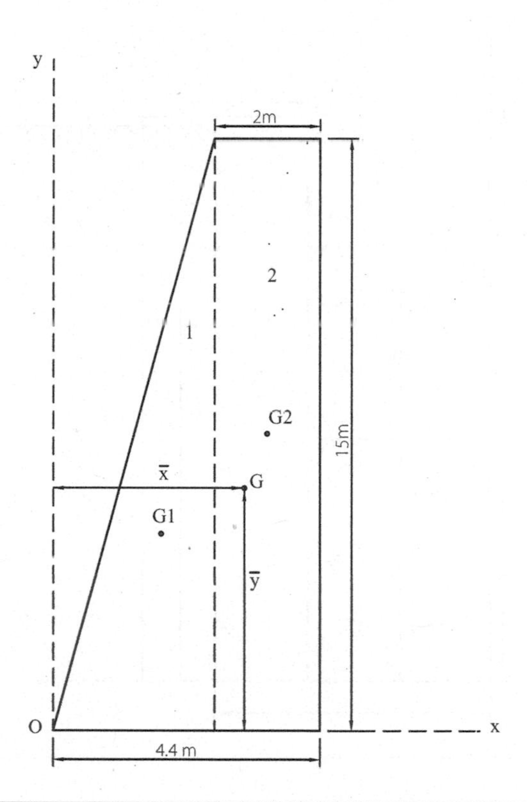

Figure S14.5

	Area	Distance to centroid		Moment of area	
		From y-axis	From x-axis	About Oy	About Ox
	A	x	y	Ax	Ay
1	2.4 × 15 ÷ 2 = 18	1.6	5	18 × 1.6 = 28.8	18 × 5 = 90
2	15 × 2 = 30	3.4	7.5	102	225
	ΣA = 48			ΣAx = 130.8	ΣAy = 315

$$\bar{x} = \frac{\Sigma Ax}{\Sigma A} = \frac{130.8}{48} = \mathbf{2.725\ m}$$

$$\bar{y} = \frac{\Sigma Ay}{\Sigma A} = \frac{315}{48} = \mathbf{6.563\ m}$$

3. a)

The shape has been divided into 3 parts, as shown in Figure S14.6. The centroids of all parts are also shown.

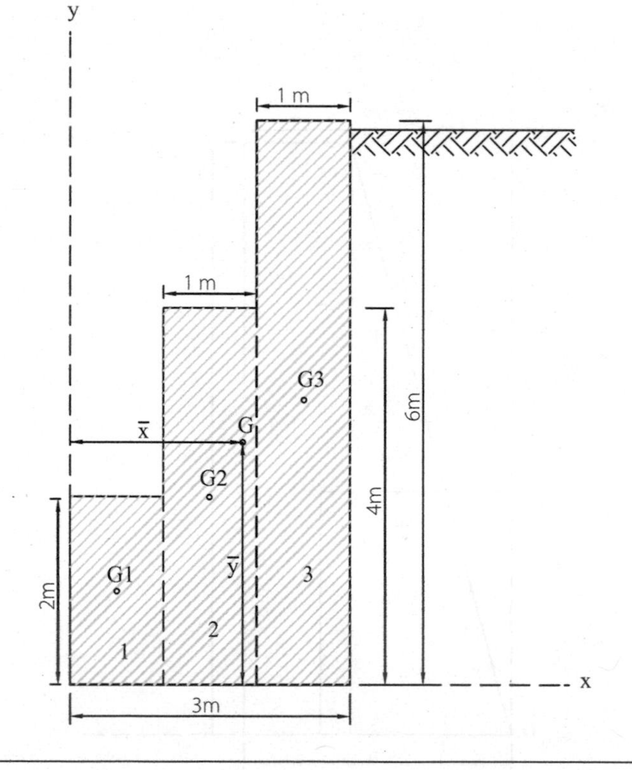

Figure S14.6

	Area	Distance to centroid		Moment of area	
		From y-axis	From x-axis	About Oy	About Ox
	A (m²)	x	y	Ax	Ay
1	2 × 1 = 2	0.5	1	2 × 0.5 = 1	2 × 1 = 2
2	4 × 1 = 4	1.5	2	4 × 1.5 = 6	4 × 2 = 8
3	6 × 1 = 6	2.5	3	6 × 2.5 = 15	6 × 3 = 18
	ΣA = 12			ΣAx = 22	ΣAy = 28

$$\bar{x} = \frac{\Sigma Ax}{\Sigma A} = \frac{22}{12} = \mathbf{1.833\ m}$$

$$\bar{y} = \frac{\Sigma Ay}{\Sigma A} = \frac{28}{12} = \mathbf{2.333\ m}$$

3. b)

The shape has been divided into three parts, as shown in Figure S14.7. The centroids of all parts are also shown.

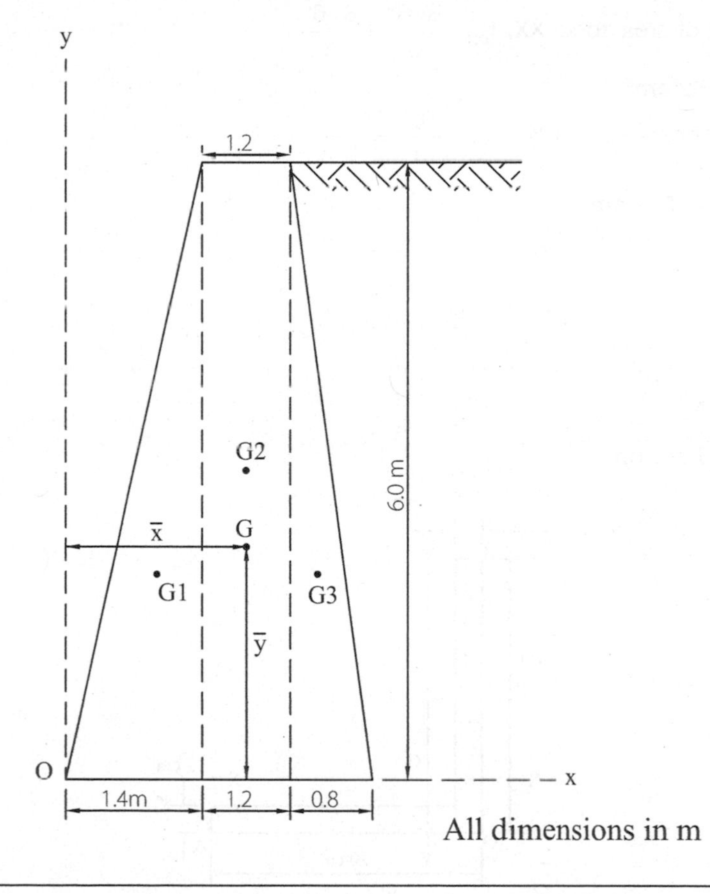

Figure S14.7

	Area	Distance to centroid		Moment of area	
		From y-axis	From x-axis	About Oy	About Ox
	A (m^2)	x	y	Ax	Ay
1	4.2	0.933	2	4.2 × 0.933 = 3.919	4.2 × 2 = 8.4
2	7.2	2.0	3	7.2 × 2.0 = 14.4	7.2 × 3 = 21.6
3	2.4	2.867	2	2.4 × 2.867 = 6.881	2.4 × 2 = 4.8
	ΣA = 13.8			ΣAx = 25.2	ΣAy = 34.8

$$\bar{x} = \frac{\Sigma Ax}{\Sigma A} = \frac{25.2}{13.8} = \mathbf{1.826\ m}$$

$$\bar{y} = \frac{\Sigma Ay}{\Sigma A} = \frac{34.8}{13.8} = \mathbf{2.522\ m}$$

4. Second moment of area about AA, $I_{AA} = \frac{bd^3}{3}$

$$= \frac{8 \times 12^3}{3} = \mathbf{4608\ cm^4}$$

Second moment of area about XX, $I_{XX} = \frac{8 \times 6^3}{3} + \frac{8 \times 6^3}{3}$

$$= 576 + 576 = \mathbf{1152\ cm^4}$$

Second moment of area about YY, $I_{YY} = \frac{db^3}{3} + \frac{db^3}{3}$

$$= \frac{12 \times 4^3}{3} + \frac{12 \times 4^3}{3} = \mathbf{512\ cm^4}$$

5. $I_{AA} = I_G + Ah^2$

$$= \frac{bd^3}{12} + bd \times h^2$$

$$= \frac{8 \times 12^3}{12} + 8 \times 12 \times 8^2$$

$$= 1152 + 6144 = \mathbf{7296\ cm^4}$$

6.

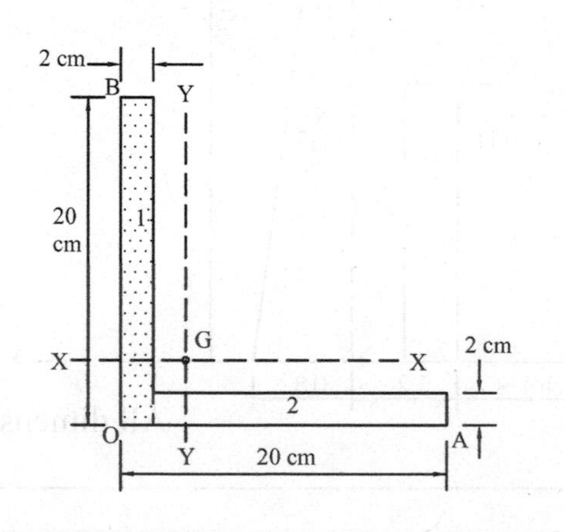

Figure S14.8

The section is divided into 2 parts, part 1 and part 2. The position of the centroid is determined by taking moments about axes OA and OB (Figure S14.8)

		Distance of centroid		Moment of area	
	Area	From axis OB	From axis OA	About OB	About OA
	A (cm²)	x (cm)	y (cm)	Ax	Ay
1	40	1	10	40	400
2	36	11	1	396	36
	ΣA = 76			ΣAx = 436	Σay = 436

$$\bar{x} = \frac{\Sigma Ax}{\Sigma A} = \frac{436}{76} = \mathbf{5.74\ cm}$$

As the section is symmetrical, centroid G is 5.74 cm from both OA and OB.

As axis OA coincides with one side of both part 1 and part 2:

$$I_{OA} = \frac{b_1 d_1^3}{3} + \frac{b_2 d_2^3}{3}$$

$$= \frac{2 \times 20^3}{3} + \frac{18 \times 2^3}{3} = 5381.33\ cm^4$$

$I_{OA} = I_{XX} + Ah^2$

$5381.33 = I_{XX} + 76 \times 5.74^2$

$I_{XX} = 5381.33 - 2504.02 = \mathbf{2877.3\ cm^4}$

Due to symmetry, $I_{XX} = I_{YY} = \mathbf{2877.3\ cm^4}$

7.

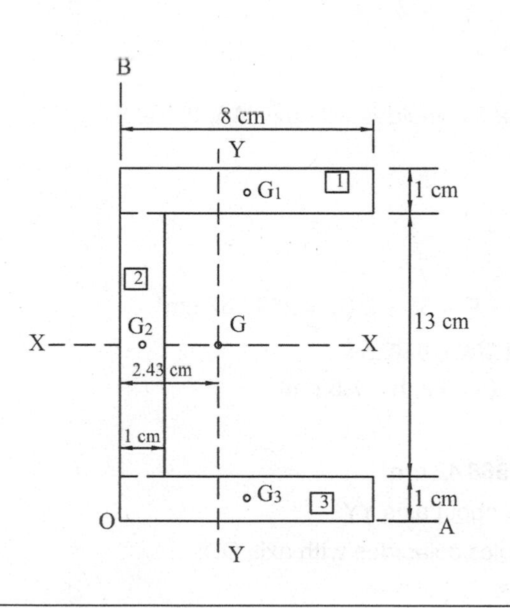

Figure S14.9

The section is divided into 3 rectangles, 1, 2 and 3, and their centroids denoted by G1, G2 and G3 (Figure S14.9). The centroid of the whole section will lie on the axis of symmetry XX but its exact position needs to be determined as the section is not perfectly symmetrical.

	Area	Distance of centroid From axis OB	Moment of area About OB
	A (cm²)	x (cm)	Ax
1	8	4	32
2	13	0.5	6.5
3	8	4	32
	ΣA = 29		ΣAy = 70.5

$$\bar{x} = \frac{\Sigma Ax}{\Sigma A} = \frac{70.5}{29} = \mathbf{2.43\ cm}$$

The centroid is positioned at mid-height, 2.43 cm from axis OB.

Second moment of area about axis OA

Area 1: The second moment of area of Area 1 about OA is given by:

$$I_{OA} = I_{G1} + A_1h_1^2$$

$$= \frac{8 \times 1^3}{12} + 8 \times 14.5^2 \quad (A_1 = 8;\ h_1 = 14.5)$$

$$= 1682.67\ cm^4$$

Area 2: The second moment of area of Area 2 about OA is given by:

$$I_{OA} = I_{G2} + A_2h_2^2$$

$$= \frac{1 \times 13^3}{12} + 13 \times 7.5^2 \quad (A_2 = 13;\ h_2 = 7.5)$$

$$= 914.33\ cm^4$$

Area 3: The base of Area 3 coincides with axis AA, therefore:

$$I_{OA} = \frac{bd^3}{3}$$

$$= \frac{8 \times 1^3}{3} = 2.67\ cm^4$$

Therefore, $I_{OA} = 1682.67 + 914.33 + 2.67 = 2599.67\ cm^4$

Second moment of area about axis XX

$$I_{OA} = I_{XX} + Ah^2 \quad (A = 29;\ h = 7.5\ cm)$$

$$I_{XX} = I_{OA} - A \times h^2$$

$$= 2599.67 - 29 \times 7.5^2 = \mathbf{968.42\ cm^4}$$

Second moment of area about axis YY

As 1 side of all 3 rectangles coincides with axis OB:

$$I_{OB} = \frac{b_1d_1^{\ 3}}{3} + \frac{b_2d_2^{\ 3}}{3} + \frac{b_3d_3^{\ 3}}{3}$$

$$= \frac{1 \times 8^3}{3} + \frac{13 \times 1^3}{3} + \frac{1 \times 8^3}{3} = 345.67\ \text{cm}^4$$

$I_{OB} = I_{YY} + Ah^2$ (A = 29; h = 2.43 cm)

$I_{YY} = I_{OB} - A \times h^2$

$= 345.67 - 29 \times 2.43^2 = \mathbf{174.43\ cm^4}$

8.

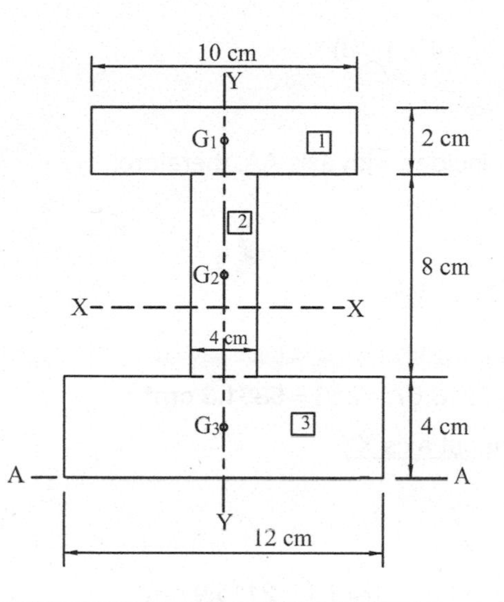

Figure S14.10

The section is divided into 3 rectangles, 1, 2 and 3, and their centroids denoted by G1, G2 and G3 (see Figure S14.10). The centroid of the whole section will lie on the axis of symmetry YY but its exact position needs to be determined as the section is not perfectly symmetrical.

	Area	Distance of centroid	Moment of area
		From axis AA	About AA
	A (cm²)	y (cm)	Ay
1	20	13	260
2	32	8	256
3	48	2	96
	ΣA = 100		ΣAy = 612

$$\bar{y} = \frac{\Sigma Ay}{\Sigma A} = \frac{612}{100} = \mathbf{6.12\ cm}$$

The centroid is positioned on axis YY at a distance of 6.12 cm from axis AA.

Second moment of area about axis AA

Area 1: The second moment of area of Area 1 about AA is given by:

$I_{AA} = I_{G1} + A_1 h_1^2$

$= \frac{10 \times 2^3}{12} + 20 \times 13^2 \quad (A_1 = 20;\ h_1 = 4 + 8 + 1 = 13)$

$= 3386.67\ cm^4$

Area 2: The second moment of area of Area 2 about AA is given by:

$I_{AA} = I_{G2} + A_2 h_2^{\ 2}$

$= \frac{4 \times 8^3}{12} + 32\ \times 8^2 (A_2 = 32;\ h_2 = 4 + 4 = 8)$

$= 2218.67\ cm^4$

Area 3: The base of Area 3 coincides with axis AA, therefore:

$I_{AA} = \frac{bd^3}{3}$

$= \frac{12 \times 4^3}{3} = 256\ cm^4$

Therefore, $I_{AA} = 3386.67 + 2218.67 + 256 =$ **5861.3 cm⁴**

Second moment of area about axis XX

$I_{AA} = I_{XX} + Ah^2$

$I_{XX} = I_{AA} - A \times h^2$

$= 5861.3 - 100 \times 6.12^2 = 5861.3 - 3745.4 = \mathbf{2115.9\ cm^4}$

Second moment of area about axis YY

Axis YY passes through the centroids of all three parts

$I_{YY} = \frac{b_1 d_1^{\ 3}}{12} + \frac{b_2 d_2^{\ 3}}{12} + \frac{b_3 d_3^{\ 3}}{12}$

$= \frac{2 \times 10^3}{12} + \frac{8 \times 4^3}{12} + \frac{4 \times 12^3}{12} = 166.67 + 42.67 + 576 = \mathbf{785.3\ cm^4}$

Solutions – Exercise 15.1

1. a) $\begin{pmatrix} 2 & -3 \\ 6 & 4 \end{pmatrix} + \begin{pmatrix} -2 & 0 \\ -3 & 5 \end{pmatrix} = \begin{pmatrix} 2+(-2) & -3+0 \\ 6+(-3) & 4+5 \end{pmatrix} = \begin{pmatrix} 0 & -3 \\ 3 & 9 \end{pmatrix}$

b) $\begin{pmatrix} 3 & 2 & -4 \\ 3 & 4 & 1 \\ 2 & 3 & -3 \end{pmatrix} + \begin{pmatrix} 1 & 6 & -5 \\ -2 & 1 & 0 \\ 5 & 3 & 3 \end{pmatrix} = \begin{pmatrix} 3+1 & 2+6 & -4+(-5) \\ 3+(-2) & 4+1 & 1+0 \\ 2+5 & 3+3 & -3+3 \end{pmatrix}$

$= \begin{pmatrix} 4 & 8 & -9 \\ 1 & 5 & 1 \\ 7 & 6 & 0 \end{pmatrix}$

2. a) $\begin{pmatrix} 2 & -3 \\ 6 & 4 \end{pmatrix} - \begin{pmatrix} -2 & 0 \\ -3 & 5 \end{pmatrix} = \begin{pmatrix} 2-(-2) & -3-0 \\ 6-(-3) & 4-5 \end{pmatrix} = \begin{pmatrix} 4 & -3 \\ 9 & -1 \end{pmatrix}$

b) $\begin{pmatrix} 3 & 2 & -4 \\ 3 & 4 & 1 \\ 2 & 3 & -3 \end{pmatrix} - \begin{pmatrix} 1 & 6 & -5 \\ -2 & 1 & 0 \\ 5 & 3 & 3 \end{pmatrix} = \begin{pmatrix} 3-1 & 2-6 & -4-(-5) \\ 3-(-2) & 4-1 & 1-0 \\ 2-5 & 3-3 & -3-3 \end{pmatrix}$

$= \begin{pmatrix} 2 & -4 & 1 \\ 5 & 3 & 1 \\ -3 & 0 & -6 \end{pmatrix}$

3. $B = \begin{pmatrix} -2 & 0 \\ 6 & -3 \end{pmatrix}$; $2B = 2\begin{pmatrix} -2 & 0 \\ 6 & -3 \end{pmatrix} = \begin{pmatrix} -4 & 0 \\ 12 & -6 \end{pmatrix}$

$A - 2B = \begin{pmatrix} 2 & -2 \\ -6 & 3 \end{pmatrix} - \begin{pmatrix} -4 & 0 \\ 12 & -6 \end{pmatrix} = \begin{pmatrix} 2-(-4) & -2-0 \\ -6-(12) & 3-(-6) \end{pmatrix} = \begin{pmatrix} 6 & -2 \\ -18 & 9 \end{pmatrix}$

4. $A = \begin{pmatrix} 2 & 5 \\ 4 & 1 \end{pmatrix}$, therefore $3A = \begin{pmatrix} 6 & 15 \\ 12 & 3 \end{pmatrix}$

$B^2 = \begin{pmatrix} 3 & 1 \\ 2 & 0 \end{pmatrix} \times \begin{pmatrix} 3 & 1 \\ 2 & 0 \end{pmatrix} = \begin{pmatrix} (3\times3)+(1\times2) & (3\times1)+(1\times0) \\ (2\times3)+(0\times2) & (2\times1)+(0\times0) \end{pmatrix}$

$= \begin{pmatrix} 11 & 3 \\ 6 & 2 \end{pmatrix}$

$3A + B^2 = \begin{pmatrix} 6 & 15 \\ 12 & 3 \end{pmatrix} + \begin{pmatrix} 11 & 3 \\ 6 & 2 \end{pmatrix} = \begin{pmatrix} 17 & 18 \\ 18 & 5 \end{pmatrix}$

5. $\mathbf{X} - \begin{pmatrix} 5 & 3 \\ 7 & 0 \end{pmatrix} = \begin{pmatrix} 2 & 4 \\ 1 & -8 \end{pmatrix}$

Add $\begin{pmatrix} 5 & 3 \\ 7 & 0 \end{pmatrix}$ to both sides, $\mathbf{X} - \begin{pmatrix} 5 & 3 \\ 7 & 0 \end{pmatrix} + \begin{pmatrix} 5 & 3 \\ 7 & 0 \end{pmatrix} = \begin{pmatrix} 2 & 4 \\ 1 & -8 \end{pmatrix} + \begin{pmatrix} 5 & 3 \\ 7 & 0 \end{pmatrix}$

$\mathbf{X} = \begin{pmatrix} 2 & 4 \\ 1 & -8 \end{pmatrix} + \begin{pmatrix} 5 & 3 \\ 7 & 0 \end{pmatrix} = \begin{pmatrix} 7 & 7 \\ 8 & -8 \end{pmatrix}$

6. $\begin{pmatrix} 3 & 2 \\ 6 & 1 \end{pmatrix} \times \begin{pmatrix} 5 & 3 \\ 6 & 0 \end{pmatrix} = \begin{pmatrix} (3\times5)+(2\times6) & (3\times3)+(2\times0) \\ (6\times5)+(1\times6) & (6\times3)+(1\times0) \end{pmatrix}$

$= \begin{pmatrix} 27 & 9 \\ 36 & 18 \end{pmatrix}$

7. $\begin{pmatrix} 3 & 4 \\ 7 & 2 \end{pmatrix} \times \begin{pmatrix} 5 \\ 6 \end{pmatrix} = \begin{pmatrix} (3\times5) + (4\times6) \\ (7\times5) + (2\times6) \end{pmatrix} = \begin{pmatrix} 39 \\ 47 \end{pmatrix}$

8. a) $\begin{vmatrix} -5 & -6 \\ 4 & 3 \end{vmatrix} = -5 \times 3 - 4 \times (-6) = \mathbf{9}$

b) $\begin{vmatrix} -2 & 3 & 0 \\ -3 & 1 & 1 \\ -2 & 3 & 4 \end{vmatrix} = -2\begin{vmatrix} 1 & 1 \\ 3 & 4 \end{vmatrix} - 3\begin{vmatrix} -3 & 1 \\ -2 & 4 \end{vmatrix} + 0\begin{vmatrix} -3 & 1 \\ -2 & 3 \end{vmatrix}$

$= -2(4 - 3) - (3)(-12 + 2) + 0(-9 + 2) = \mathbf{28}$

9. $|A| = 3 \times 2 - 1 \times 2 = 4$

$$A^{-1} = \frac{1}{4}\begin{pmatrix} 2 & -2 \\ -1 & 3 \end{pmatrix} = \begin{pmatrix} \frac{2}{4} & \frac{-2}{4} \\ \frac{-1}{4} & \frac{3}{4} \end{pmatrix} = \begin{pmatrix} 0.5 & -0.5 \\ -0.25 & 0.75 \end{pmatrix}$$

Verification: $\mathbf{AA}^{-1} = \begin{pmatrix} 3 & 2 \\ 1 & 2 \end{pmatrix}\begin{pmatrix} 0.5 & -0.5 \\ -0.25 & 0.75 \end{pmatrix}$

$$= \begin{pmatrix} 3 \times 0.5 + 2 \times (-0.25) & 3 \times (-0.5) + 2 \times 0.75 \\ 1 \times 0.5 + 2 \times (-0.25) & 1 \times (-0.5) + 2 \times 0.75 \end{pmatrix}$$

$$= \begin{pmatrix} 1.5 - 0.5 & -1.5 + 1.5 \\ 0.5 - 0.5 & -0.5 + 1.5 \end{pmatrix} = \begin{pmatrix} 1 & 0 \\ 0 & 1 \end{pmatrix} = \mathbf{I}$$

10. $2x + 3y = 4$ and

$x - y = -3$

$\det \mathbf{D} = \begin{vmatrix} 2 & 3 \\ 1 & -1 \end{vmatrix} = -2 - 3 = -5$

$\det \mathbf{D}_x = \begin{vmatrix} 4 & 3 \\ -3 & -1 \end{vmatrix} = -4 - (-9) = 5$

$\det \mathbf{D}_y = \begin{vmatrix} 2 & 4 \\ 1 & -3 \end{vmatrix} = -6 - 4 = -10$

$\mathbf{x} = \frac{D_x}{D} = \frac{5}{-5} = \mathbf{-1}$ $\mathbf{y} = \frac{D_y}{D} = \frac{-10}{-5} = \mathbf{2}$

11. $2x + 2y = -2$ and

$3x - y = 9$

$\det \mathbf{D} = \begin{vmatrix} 2 & 2 \\ 3 & -1 \end{vmatrix} = -2 - 6 = -8$

$\det \mathbf{D}_x = \begin{vmatrix} -2 & 2 \\ 9 & -1 \end{vmatrix} = 2 - 18 = -16$

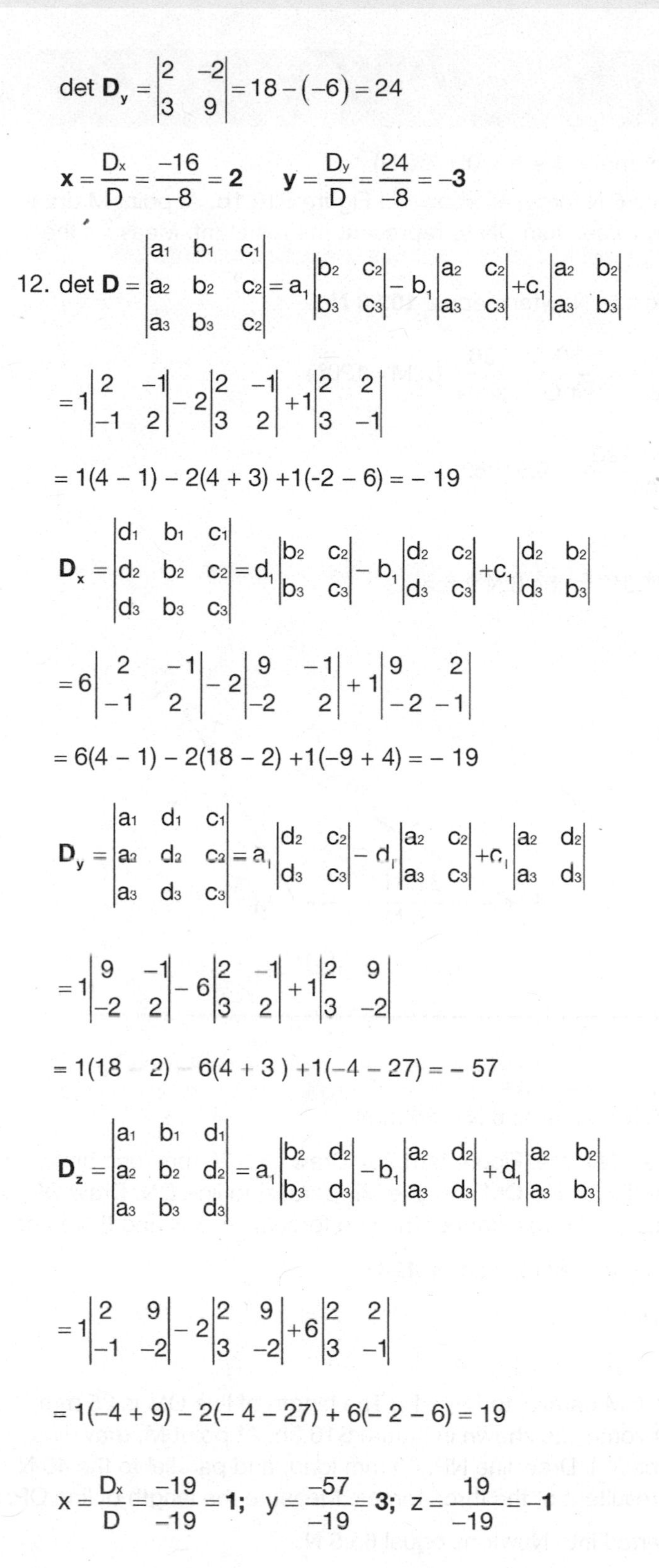

$$\det \mathbf{D_y} = \begin{vmatrix} 2 & -2 \\ 3 & 9 \end{vmatrix} = 18 - (-6) = 24$$

$$\mathbf{x} = \frac{D_x}{D} = \frac{-16}{-8} = \mathbf{2} \qquad \mathbf{y} = \frac{D_y}{D} = \frac{24}{-8} = \mathbf{-3}$$

12. $$\det \mathbf{D} = \begin{vmatrix} a_1 & b_1 & c_1 \\ a_2 & b_2 & c_2 \\ a_3 & b_3 & c_2 \end{vmatrix} = a_1 \begin{vmatrix} b_2 & c_2 \\ b_3 & c_3 \end{vmatrix} - b_1 \begin{vmatrix} a_2 & c_2 \\ a_3 & c_3 \end{vmatrix} + c_1 \begin{vmatrix} a_2 & b_2 \\ a_3 & b_3 \end{vmatrix}$$

$$= 1\begin{vmatrix} 2 & -1 \\ -1 & 2 \end{vmatrix} - 2\begin{vmatrix} 2 & -1 \\ 3 & 2 \end{vmatrix} + 1\begin{vmatrix} 2 & 2 \\ 3 & -1 \end{vmatrix}$$

$$= 1(4 - 1) - 2(4 + 3) + 1(-2 - 6) = -19$$

$$\mathbf{D_x} = \begin{vmatrix} d_1 & b_1 & c_1 \\ d_2 & b_2 & c_2 \\ d_3 & b_3 & c_3 \end{vmatrix} = d_1 \begin{vmatrix} b_2 & c_2 \\ b_3 & c_3 \end{vmatrix} - b_1 \begin{vmatrix} d_2 & c_2 \\ d_3 & c_3 \end{vmatrix} + c_1 \begin{vmatrix} d_2 & b_2 \\ d_3 & b_3 \end{vmatrix}$$

$$= 6\begin{vmatrix} 2 & -1 \\ -1 & 2 \end{vmatrix} - 2\begin{vmatrix} 9 & -1 \\ -2 & 2 \end{vmatrix} + 1\begin{vmatrix} 9 & 2 \\ -2 & -1 \end{vmatrix}$$

$$= 6(4 - 1) - 2(18 - 2) + 1(-9 + 4) = -19$$

$$\mathbf{D_y} = \begin{vmatrix} a_1 & d_1 & c_1 \\ a_2 & d_2 & c_2 \\ a_3 & d_3 & c_3 \end{vmatrix} = a_1 \begin{vmatrix} d_2 & c_2 \\ d_3 & c_3 \end{vmatrix} - d_1 \begin{vmatrix} a_2 & c_2 \\ a_3 & c_3 \end{vmatrix} + c_1 \begin{vmatrix} a_2 & d_2 \\ a_3 & d_3 \end{vmatrix}$$

$$= 1\begin{vmatrix} 9 & -1 \\ -2 & 2 \end{vmatrix} - 6\begin{vmatrix} 2 & -1 \\ 3 & 2 \end{vmatrix} + 1\begin{vmatrix} 2 & 9 \\ 3 & -2 \end{vmatrix}$$

$$= 1(18 - 2) - 6(4 + 3) + 1(-4 - 27) = -57$$

$$\mathbf{D_z} = \begin{vmatrix} a_1 & b_1 & d_1 \\ a_2 & b_2 & d_2 \\ a_3 & b_3 & d_3 \end{vmatrix} = a_1 \begin{vmatrix} b_2 & d_2 \\ b_3 & d_3 \end{vmatrix} - b_1 \begin{vmatrix} a_2 & d_2 \\ a_3 & d_3 \end{vmatrix} + d_1 \begin{vmatrix} a_2 & b_2 \\ a_3 & b_3 \end{vmatrix}$$

$$= 1\begin{vmatrix} 2 & 9 \\ -1 & -2 \end{vmatrix} - 2\begin{vmatrix} 2 & 9 \\ 3 & -2 \end{vmatrix} + 6\begin{vmatrix} 2 & 2 \\ 3 & -1 \end{vmatrix}$$

$$= 1(-4 + 9) - 2(-4 - 27) + 6(-2 - 6) = 19$$

$$x = \frac{D_x}{D} = \frac{-19}{-19} = \mathbf{1}; \quad y = \frac{-57}{-19} = \mathbf{3}; \quad z = \frac{19}{-19} = \mathbf{-1}$$

Solutions – Exercise 16.1

1. Select a suitable scale, i.e. 1 N = 6 mm. (1 mm = 1 ÷ 6 = 0.1666 N)

   Draw, line OM = 36 mm, and parallel to the 6 N force, as shown in Figure S16.1b. At point M draw, line MN = 39 mm and parallel to the 6.5 N force. Join ON to represent the resultant. Measure the length of line ON:

   Resultant ON = 65 mm. 65 mm converted into Newtons equal **10.83 N.**

   Use the sine rule to determine $\angle O$: $\dfrac{65}{\sin 120^\circ} = \dfrac{39}{\sin O} = \dfrac{36}{\sin N}$ $(\angle M = 120^\circ)$

   $$\frac{65}{\sin 120^\circ} = \frac{39}{\sin O}\text{; therefore, } \sin O = \frac{39 \times \sin 120^\circ}{65} = 0.51962$$

   $\angle O = \sin^{-1} 0.51962 =$ **31.31°**

   The resultant (**10.83 N**) acts at angle of **31.31°** to the 6 N force.

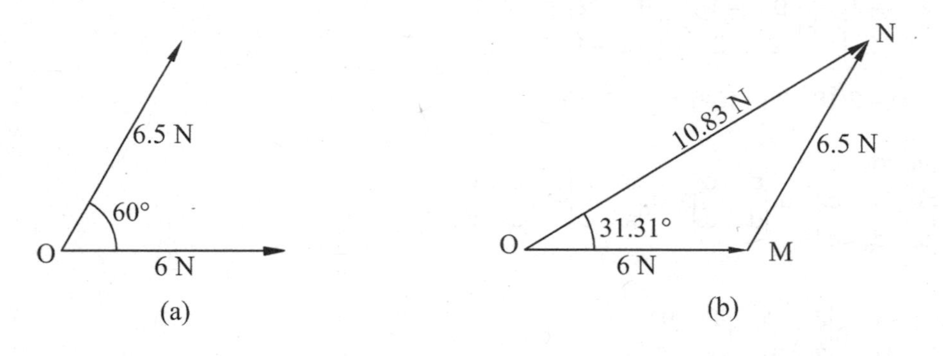

Figure S16.1

2. Select a suitable scale, i.e. 1 N = 6 mm. (7N = 42 mm; 6 N = 36 mm)

   Draw line OM = 42 mm and parallel to the 7 N force (Figure S16.2b). Draw ON, 36 mm long and parallel to the 6 N force. Draw line NP parallel to line OM and line MP parallel to line ON. Draw OP, the diagonal of parallelogram OMPN, which is the resultant of the two forces, i.e. 7 N and 6 N force.

   Resultant OP = 72 mm. 72 mm converted into Newtons equal **12 N.**

   The resultant is inclined at **21°** to the horizontal.

3. Select a suitable scale, i.e. 1 N = 1 mm.

   To represent the 25 N force (F_1) draw line OM parallel to force F_1. The length of line OM is 25 mm, and it makes an angle of 36° with the horizontal, as shown in Figure S16.3b. At point M, draw line MN = 30 mm and parallel to the 30 N force (F_2). Draw line NP, 40 mm long, and parallel to the 40 N force (F_3). Draw line OP to represent the resultant of the three forces. Measure the length of line OP:

   Resultant OP = 65.5 mm. 65.5 mm converted into Newtons equal **65.5 N.**

   The resultant is inclined at **65° to the 25N** force.

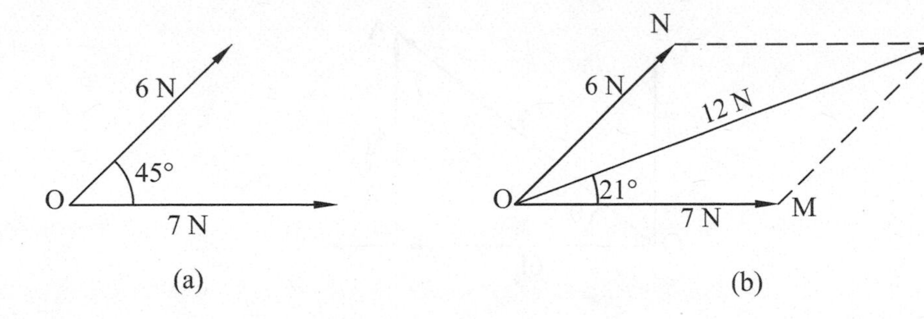

Figure S16.2

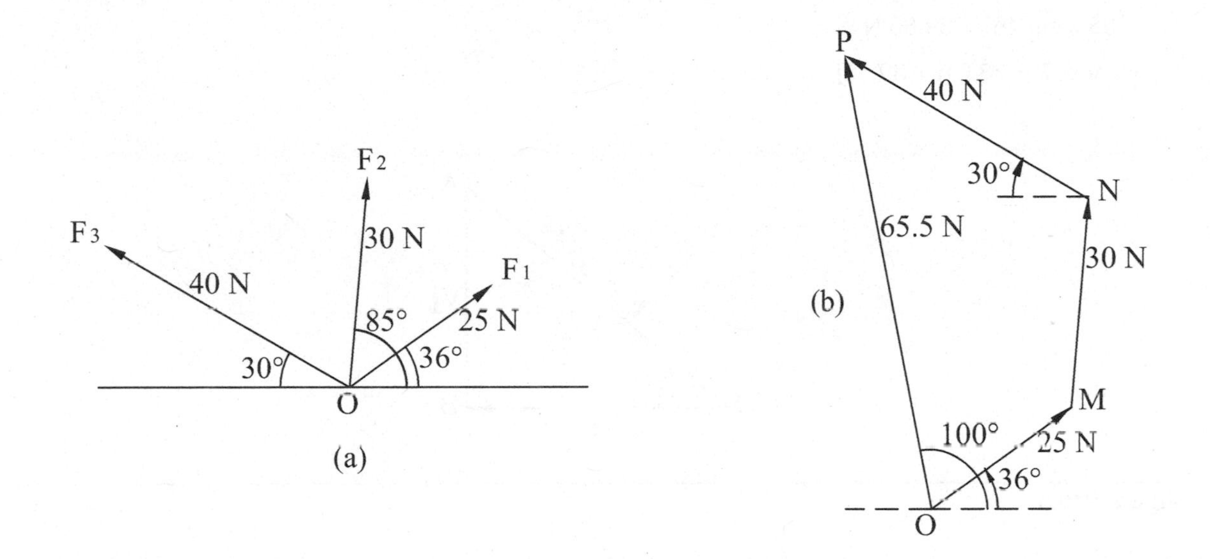

Figure S16.3

4. Figure S16.4 shows the vector

   $|\mathbf{v}| = \sqrt{6^2 + 4^2} = \sqrt{52} = \mathbf{7.21}$

   $\tan\theta = \dfrac{4}{6}$, therefore $\theta = \tan^{-1} = \left(\dfrac{4}{6}\right) = \mathbf{33.69°}$

   Magnitude = **7.21** units, acting at **33.69°** to the horizontal

5. Let the horizontal component of the force be x**i** and the vertical component be y**j**

   F = x**i** + y**j**

   Triangle AOB (Figure S16.5) is a right-angled triangle, therefore:

   $|OB| = x = |F| \cos 46°$

   $= 55 \times \cos 46° = 38.21$ N

   $|BA| = y = |F| \sin 46°$

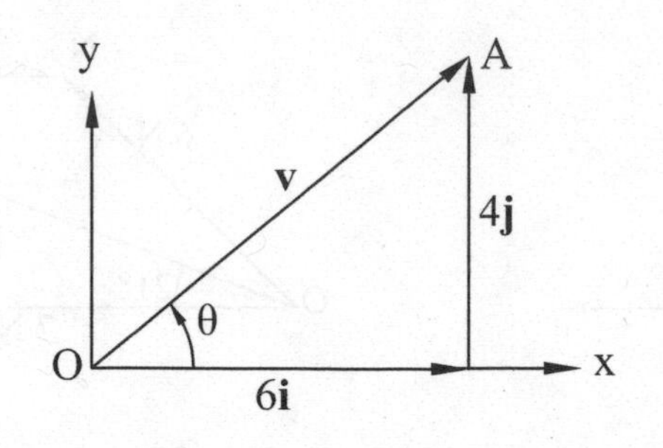

Figure S16.4

$= 55 \times \sin 46° = 39.56$ N

Hence, **F** = 38.21**i** + 39.56**j**

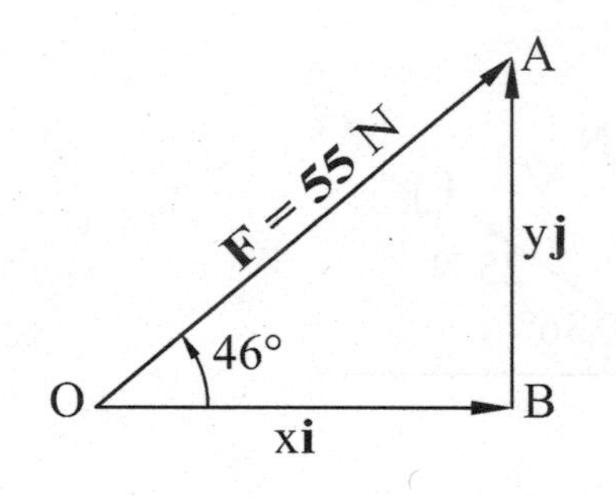

Figure S16.5

6.

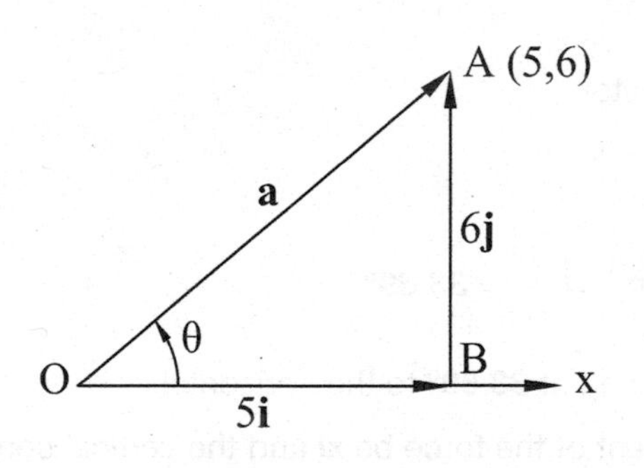

Figure S16.6

Refer to Figure S16.6, $|\mathbf{a}| = \sqrt{5^2 + 6^2} = \sqrt{61} = \mathbf{7.81}$

In ΔAOB, $\tan\theta = \dfrac{6}{5}$, therefore $\theta = \tan^{-1} = \left(\dfrac{6}{5}\right) = \mathbf{50.19°}$

The magnitude of resultant vector **a** is **7.81** units that acts at an angle of **50.19°** to the horizontal.

7. a) $\begin{pmatrix} 0 \\ 6 \end{pmatrix}$ b) $\begin{pmatrix} 6 \\ 4 \end{pmatrix}$ c) $\begin{pmatrix} 6 \\ 0 \end{pmatrix}$ d) $\begin{pmatrix} 6 \\ -2 \end{pmatrix}$

8. Refer to Figure S16.7, $|\mathbf{v}| = \sqrt{5^2 + (-3)^2} = \sqrt{34} = \mathbf{5.83}$

   In ΔOAB, $\tan\theta = \frac{-3}{5}$, therefore $\theta = \tan^{-1} = \left(\frac{-3}{5}\right) = -\ \mathbf{30.96°}$

The magnitude of vector **v** is **5.83** units that acts at an angle of **–30.96°** to the horizontal.

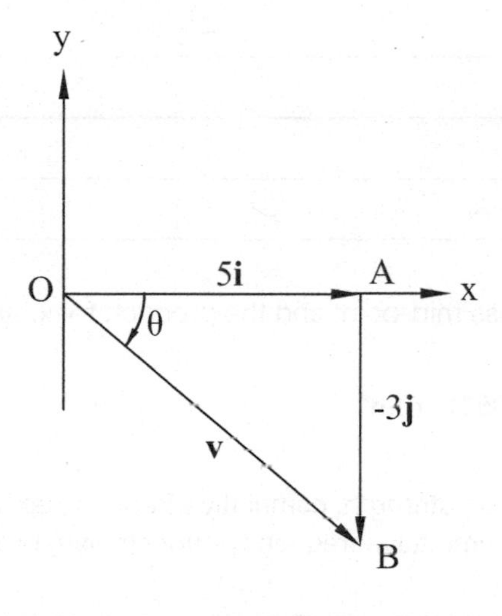

Figure S16.7

Solution – Exercise 17.1

1. a) Mean, $\bar{x} = \frac{\Sigma x}{n}$

   $= \frac{142.2}{12} = \mathbf{11.85}$

   b) The data has been arranged in an ascending order to find the median (middle number):

   10.0, 10.4, 10.8, 11.0, 11.4, 12.0, 12.2, 12.2, 12.3, 13.0, 13.0, 13.9

   There are two middle numbers, 12.0 and 12.2, therefore:

   $\text{Median} = \frac{12.0+12.2}{2} - \mathbf{12.1}$

   c) There are 2 modes, **12.2** and **13.0,** as both of them occur twice

d) Range = maximum value – minimum value

= 13.9 – 10.0 = **3.9**

2. The minimum and the maximum strengths are 34 N/mm² and 51 N/mm² respectively.

a) The 6 classes and their frequencies are shown in the following table:

Class interval (crushing strength)	Frequency (f)	Class mid-point (x)	f × x
34-36	4	35	4 × 35 = 140
37-39	7	38	7 × 38 = 266
40–42	10	41	10 × 41 = 410
43–45	9	44	9 × 44 = 396
46–48	6	47	6 × 47 = 282
49–51	4	50	4 × 50 = 200
	Σ f = 40		Σ fx = 1694

b) To find the mean, the class mid-point and the product, f × x, are included the table.

$$\text{Mean}, \bar{x} = \frac{\Sigma fx}{\Sigma f} = \frac{1694}{40} = \mathbf{42.35\ N / mm^2}$$

c) For determining the median strength, cumulative frequencies are calculated as shown in the following table, and the cumulative frequency curve (ogive) produced as shown in Figure S17.1.

Class interval (Crushing strength)	Frequency (f)	Crushing strength less than	Cumulative frequency
		33.5	0
34-36	4	36.5	0 + 4 = 4
37-39	7	39.5	4 + 7 = 11
40–42	10	42.5	11 + 10 = 21
43–45	9	45.5	21 + 9 = 30
46–48	6	48.5	30 + 6 = 36
49–51	4	51.5	36 + 4 = 40

From the Ogive, Q_2 = Median = **42.3 N/mm²**

d) The cumulative frequency corresponding to 44 N/mm² is 25.6, therefore

Number of bricks having strength less than 44 N/mm² = 25.6 or **26.**

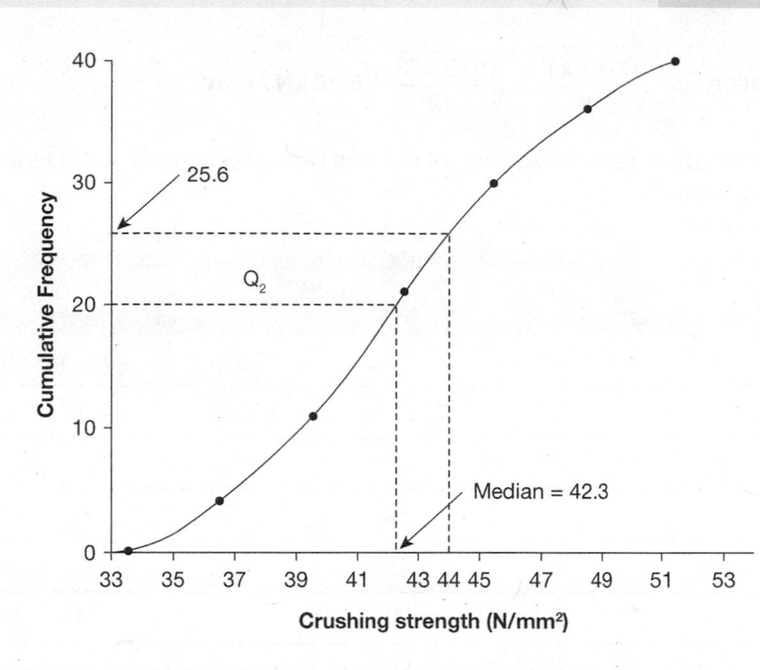

Figure S17.1

Standard deviation, $\sigma = \sqrt{\frac{\Sigma f(x-\bar{x})^2}{\Sigma f}} = \sqrt{\frac{755.11}{40}} = \mathbf{4.34\ N/mm^2}$

3. Mean strength = 204.75 kN/mm²

   The deviations from the mean and their squares are shown in the following table

Sample No.	Young's modulus (x)	Mean ($\bar{x}$)		$(x-\bar{x})^2$
1	200	204.75	–4.75	22.56
2	198	204.75	–6.75	45.56
3	210	204.75	5.25	27.56
4	215	204.75	10.25	105.06
5	202	204.75	–2.75	7.56
6	195	204.75	–9.75	95.06
7	200	204.75	–4.75	22.56
8	205	204.75	0.25	0.063
9	203	204.75	–1.75	3.06
10	212	204.75	7.25	52.56
11	214	204.75	9.25	85.56
12	203	204.75	–1.75	3.06
				Σ = 470.223

Standard deviation, $\sigma = \sqrt{\frac{\Sigma(x-\bar{x})^2}{n}} = \sqrt{\frac{470.223}{12}} = \mathbf{6.26\ kN / mm^2}$

4. Some of the information from the solution of Q.2 has been reproduced in the table below. Mean of the data = 42.35 N/mm^2

Moisture content (class interval	Frequency (f)	Class mid-point (x)	$f \times x$	Mean $(\bar{x})$	$(x-\bar{x})$	$(x-\bar{x})^2$	$f \times (x-\bar{x})^2$
34–36	4	35	140	42.35	–7.35	54.02	216.09
37–39	7	38	266	42.35	–4.35	18.92	132.46
40–42	10	41	410	42.35	–1.35	1.82	18.23
43–45	9	44	396	42.35	1.65	2.72	24.50
46–48	6	47	282	42.35	4.65	21.62	129.74
49–51	4	50	200	42.35	7.65	58.52	234.09
	$\Sigma f = 40$		$\Sigma fx =$ 1694				$\Sigma =$ 755.11

Standard deviation, $\sigma = \sqrt{\frac{\Sigma f(x-\bar{x})^2}{\Sigma f}} = \sqrt{\frac{755.11}{40}} = \mathbf{4.34\ N / mm^2}$

Range = 51 – 34 = **17 N/mm^2**

Interquartile range = $Q_3 - Q_1$ = 45.4 – 39.0 = **6.4 N/mm^2**

(Q_3 and Q_1 have been obtained from Figure S17.1)

5. a) P(1 fault) = 0.10 and P(2 faults) = 0.05

   Probability of 1or 2 faults = 0.10 + 0.05 = **0.15**

   b) Probability of more than 2 faults = 1 – [P(0 fault) + P(1 fault) + P(2 faults)]

   = 1 – [0.82 + 0.10 + 0.05] = **0.03**

   c) Probability of at least 1 fault = [P(1 fault) + P(2 faults) + P(>2 faults)]

   = [0.10 + 0.05 + 0.03] = **0.18**

6. The defective component can be picked from any sample, as shown:

   (D = defective; ND = not defective)

Company A	Company B	Company C	Outcome
D (0.1)	ND (0.8)	ND (0.75)	0.1×0.8×0.75 = 0.06
ND (0.9)	D (0.2)	ND (0.75)	0.9×0.2×0.75 = 0.135
ND (0.9)	ND (0.8)	D (0.25)	0.9×0.8×0.25 = 0.18

   Probability of picking 1 defective component = 0.06 + 0.135 + 0.18 = **0.375**

7. The tree diagram is shown in Figure S17.2

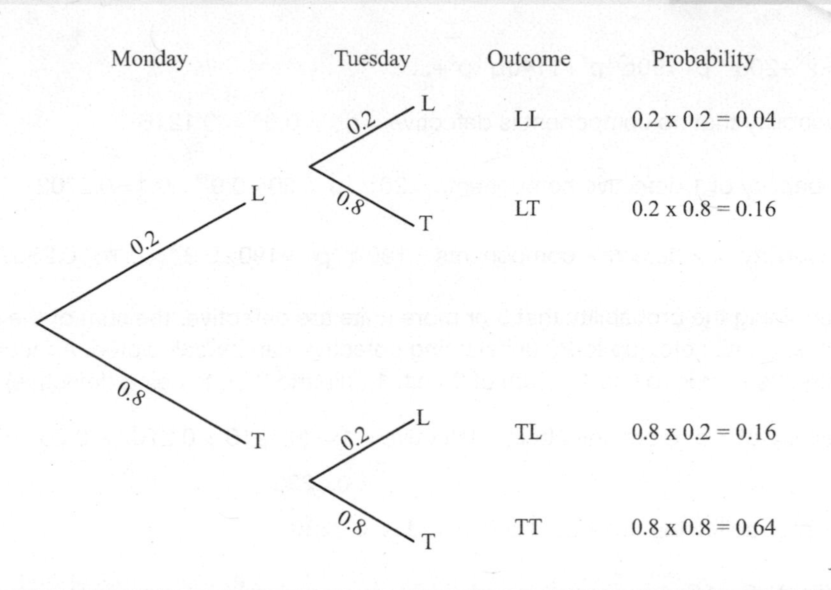

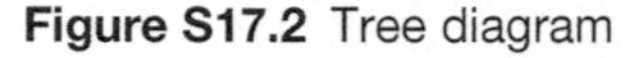
Figure S17.2 Tree diagram

Let L = late, T = on time

a) From the tree diagram: the probability that the delivery is late on both Monday and Tuesday is LL or **0.04**

b) Probability that the delivery is late on just one of the two days is:

$(L \times T) + (T \times L) = 0.16 + 0.16 =$ **0.32**

c) Probability that the delivery is on time on Monday and Tuesday is:

TT or **0.64**

8. The probability of a component being defective, p, is 3 in 75 or 4 in 100. This is same as 4% or 0.04.

The probability of a component being non-defective, q, is 1 – 0.04, or 0.96

The probability of 0, 1, 2, 3, ... defective components is given by the successive terms of $(q + p)^5$ (the power 5 is same as the sample size of 5):

$$(q+p)^5 = q^5 + 5q^{5-1}p + \frac{5(5-1)}{2!}q^{5-2}p^2 + \frac{5(5-1)(5-2)}{3!}q^{5-3}p^3 + ...$$

$$= q^5 + 5q^4p + 10q^3p^2 + 10q^2p^3 + ...$$

a) The probability that no component is defective $= q^5 = 0.96^5 =$ **0.8154**

b) The probability of 2 defective components $= 10q^3p^2 = 10 \times 0.96^3 \times 0.04^2 =$ **0.0142**

9. The probability of a unit being defective, p, is 10% or 0.1

The probability of a unit being non-defective, q, is 1 – 0.1, or 0.9

The probability of 0, 1, 2, 3, ... defective components is given by the successive terms of $(q + p)^{20}$:

$$(q+p)^{20} = q^{20} + 20q^{20-1}p + \frac{20(20-1)}{2!}q^{20-2}p^2\ \frac{20(20-1)(20-2)}{3!}q^{20-3}p^3 + ...$$

$=q^{20}+20q^{19}p+190q^{18}p^{2}+1140q^{17}p^{3}+......$

– The probability that no component is defective = $q^{20} = 0.9^{20} = 0.1216$

– The probability of 1 defective components $= 20q^{19}p = 20\times 0.9^{19}\times 0.1 = 0.2702$

– The probability of 2 defective components $= 190q^{18}p^{2} = 190\times 0.9^{18}\times 0.1^{2} = 0.2852$

– For determining the probability that 3 or more units are defective, the sum of the probabilities of 3 units, 4 units, 5 units etc. (up to 20 units) being defective can be calculated. A much easier method to calculate this sum is to find 1– (sum of 0 unit, 1 unit and 2 units being defective). Therefore:

The probability of 3 or more units being defective = 1 – (0.1216 + 0.2702 + 0.2852)

= **0.3230**

Hence the probability that the batch is rejected is: **0.3230.**

10. $\bar{x} = np = 75\times 0.02 = 1.5$

a) Probability of 0 defective $= P(0) = e^{-\bar{x}} = e^{-1.5} =$ **0.2231**

b) Probability of 1 defective $= P(1) = e^{-\bar{x}}\dfrac{\bar{x}}{1!} = e^{-1.5}\dfrac{1.5}{1!} =$ **0.3347**

c) Probability of 2 defective $= P(2) = e^{-\bar{x}}\dfrac{(\bar{x})^{2}}{2!} = e^{-\bar{x}}\dfrac{(\bar{x})^{2}}{2\times 1} = e^{-1.5}\dfrac{1.5^{2}}{2\times 1} =$ **0.251**

11. Mean diameter, $\bar{x} = 22$ mm. Its z value is 0 on the standardised normal curve

$(z = \dfrac{x-\bar{x}}{\sigma} = \dfrac{22-22}{0.2} = 0;$ here x is also $= 22)$

z – value of 21.6 mm $= \dfrac{x-\bar{x}}{\sigma} = \dfrac{21.6-22}{0.2} = -\dfrac{0.4}{0.2} = -2$

a) The area between z = 0 and z = – 2, from Table 17.9 is 0.4772

The area under the curve (Figure S17.3a) up to the z-value of – 2 represents the number of components having diameter < 21.6 mm. This is equal to:

0.5 – 0.4772 = 0.0228. Therefore the components having diameter less than 21.6 mm is: **2.28 or 2%** (2% = 0.0228 × 100)

b) z – value of 21.7 mm $= \dfrac{x-\bar{x}}{\sigma} = \dfrac{21.7-22}{0.2} = -1.5$

z – value of 22.2 mm $= \dfrac{x-\bar{x}}{\sigma} = \dfrac{22.2-22}{0.2} = 1.0$

Using Table 17.9, z-value of – 1.5 corresponds to an area of 0.4332 between the mean value and the vertical line at – 1.5. The negative value shows that it lies to the left of z = 0, i.e. the mean value.

22.2 mm has a z-value of 1.0 standard deviation. From Table 17.9 the area under the curve is 0.3413; the positive value shows that it lies to the right of the mean value, as shown in Figure S17.3b.

The total area under the curve is: 0.4332 + 0.3413 = 0.7745.

Thus the probability that the diameter of the components lies between 21.7 mm and 22.2 mm is **0.7745.**

Therefore the proportion of components having diameter between 21.7 mm and 22.2 mm is: **77.45 or 77%** (77% = 0.7745 × 100).

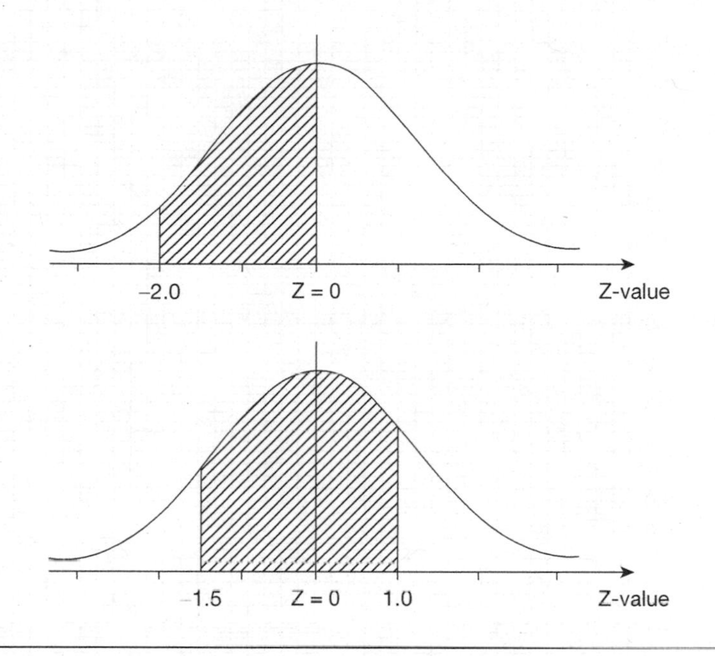

Figure S17.3

12. The cumulative frequency and percentage cumulative frequency are shown in the table below:

Class interval (Compressive strength)	Frequency	Cumulative frequency	Percentage cumulative frequency
32–34	4	4	4×100 ÷ 85 = 4.7
35–37	12	16	18.8
38–40	16	32	37.6
41–43	20	52	61.2
44–46	17	69	81.2
47–49	13	82	96.5
50–52	3	85	100

The percentage cumulative frequencies are plotted against the upper class boundaries and a best fit straight line drawn as illustrated in Figure S17.4. Since the points lie close to the straight line, the results may be considered to be normally distributed.

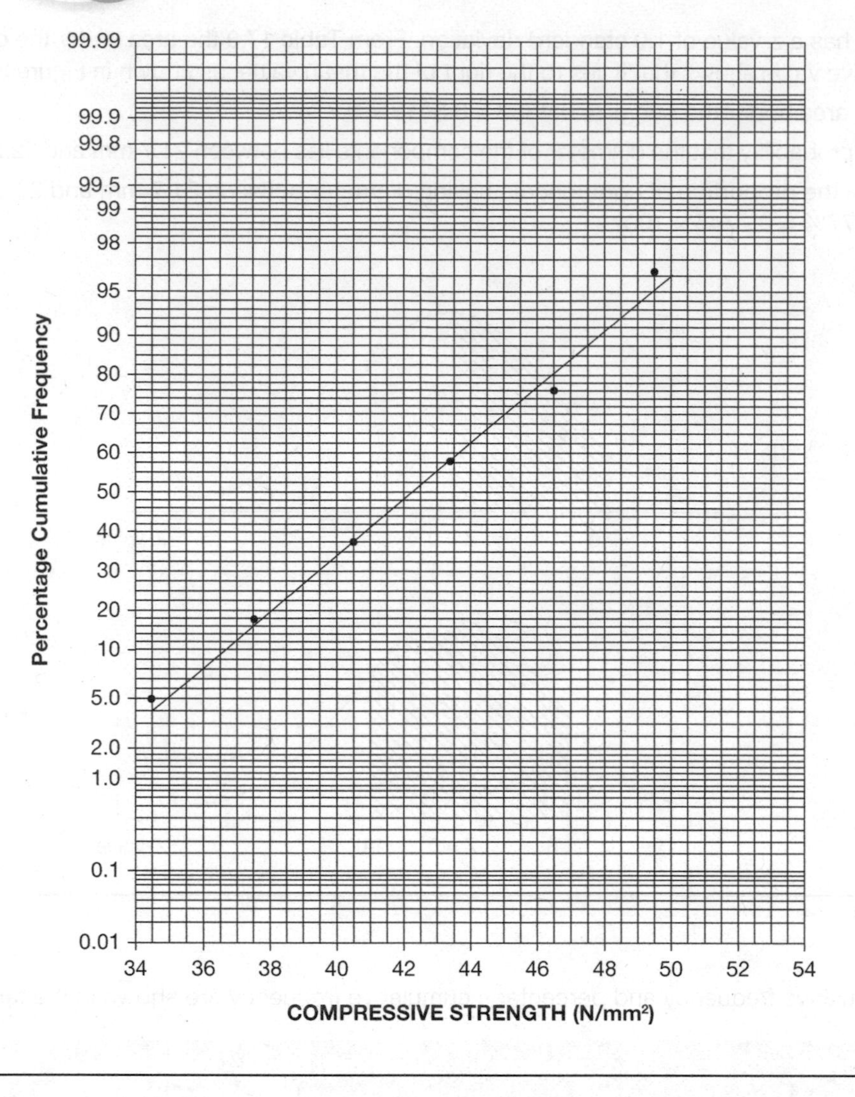

Figure S17.4

From the graph, Mean $(\bar{x}) = \mathbf{41.9\ N/mm^2}$ (mean at 50% PCF)

x_s (at 84.13% PCF) = 46.4 N/mm^2

Therefore, standard deviation $= x_s - \bar{x} = 46.4 - 41.9 = \mathbf{4.5\ N/mm^2}$

Solution – Exercise 18.1

For questions 1 to 8 enter the data and the formulae as shown in the relevant figure.

1. Figure S18.1 shows the formulae to be used.

Task 1 [Compatibility Mode] - Excel
File Home Insert Page Layout Formulas Data Review View Tell me what you want to do... Surinder Virdi Share
Paste Arial 10 Wrap Text Merge & Center General Conditional Formatting Format as Table Cell Styles Insert Delete Format AutoSum Fill Clear Sort & Filter Find & Select
Clipboard Font Alignment Number Styles Cells Editing
J33

	A	B	C	D	E	F	G
2	Q. 1						
4		Length (m)	Width (m)	Area (m²)	Number of packs	Wastage @ 10%	Number of packs required
6	DINING	3.5	3.5	=B6*C6	=D6/2.106	=10/100*E6	=E6+F6
8	LIVING	6.1	4.2	=B8*C8	=D8/2.106	=10/100*E8	=E8+F8
10	HALL	4	2	=B10*C10	=D10/2.106	=10/100*E10	=E10+F10
12	TOTAL						=SUM(G6:G10)

Sheet1 Sheet2 Sheet3
Ready
Search Windows ENG 12:51 28/09/2018

Figure S18.1

2. Figure S18.2 shows the formulae to be used.

Task 2 [Compatibility Mode] - Excel
File Home Insert Page Layout Formulas Data Review View Tell me what you want to do... Surinder Virdi Share
Paste Arial 10 Wrap Text Merge & Center General Conditional Formatting Format as Table Cell Styles Insert Delete Format AutoSum Fill Sort & Find &
Clipboard Font Alignment Number Styles Cells Editing
G34

	A	B	C	D	E
2	Q. 2				
4	Component	U-value (W/m² °C)	Area (m²)	Temperature difference (°C)	Heat Loss (Watts) = U x A x T
5	Walls	0.25	54	20	=B5*C5*D5
6	Floor	0.25	20	20	=B6*C6*D6
7	Roof	0.25	20	20	=B7*C7*D7
8	Patio door	2	5	20	=B8*C8*D8
9	Door	0.46	1.7	20	=B9*C9*D9
11	Total				=SUM(E5:E9)

Sheet1 Sheet3 Sheet2
Ready
Search Windows ENG 12:52 28/09/2018

Figure S18.2

3. Figure S18.3 shows the formulae to be used.

Q. 3
Maximum Tensile Force (kN)
Mean
Range
9.9 12.1 10 12 10.2 11.9 10.3 11.6 10.5 11.5 11.3 10.9 11.3 11.6 11
=AVERAGE(A5:O5)
=MAX(A5:O5)-MIN(A5:O5)

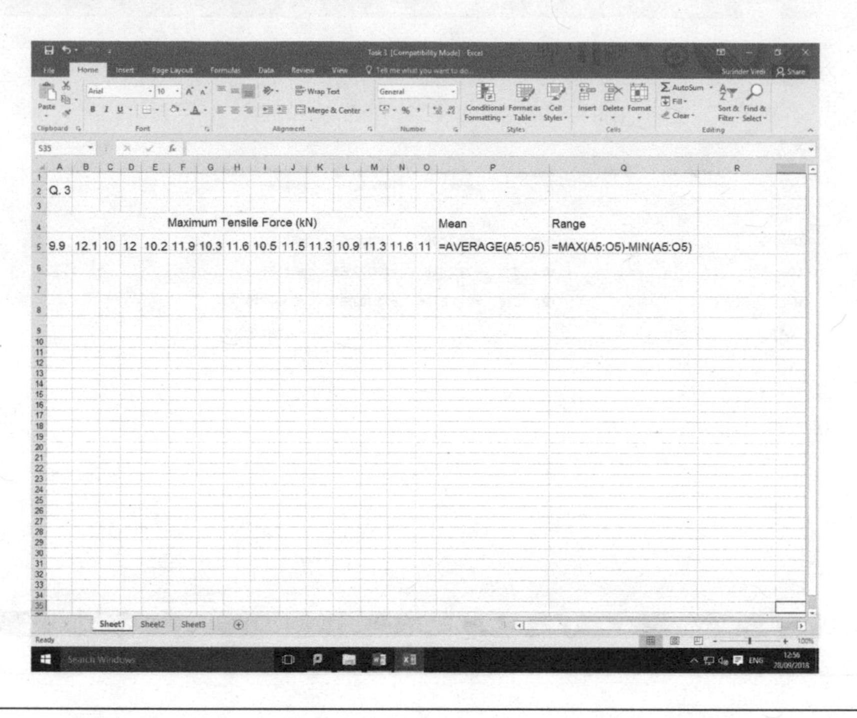

Figure S18.3

4. Figure S18.4 shows the formulae to be used.

Q. 4
class interval
frequency (f)
class mid-point (x)
fx
Mean (Σfx/Σf)
60 - 64 2 62 =B6*C6
65 - 69 6 67 =B7*C7
70 - 74 11 72 =B8*C8
75 - 79 5 77 =B9*C9
80 - 84 1 82 =B10*C10
Σf =
Σfx =
=SUM(B6:B10)
=SUM(D6:D10)
=D13/B13

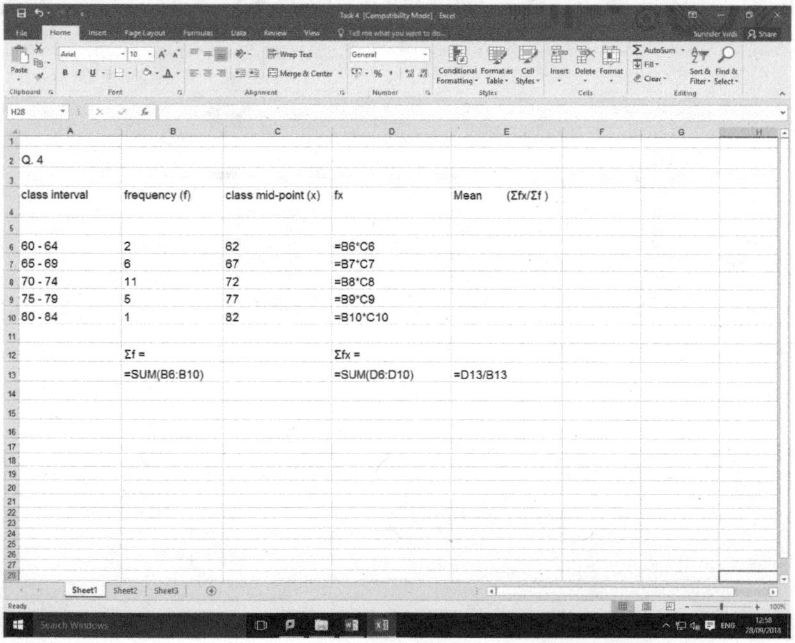

Figure S18.4

5. Figure S18.5 shows the formulae to be used.

Q. 5
Crushing strength of concrete cubes (N/mm2)
25 27 31 32 30 26 29 24 25 27 31 32 30 27 29 =STDEV(A5:O5)

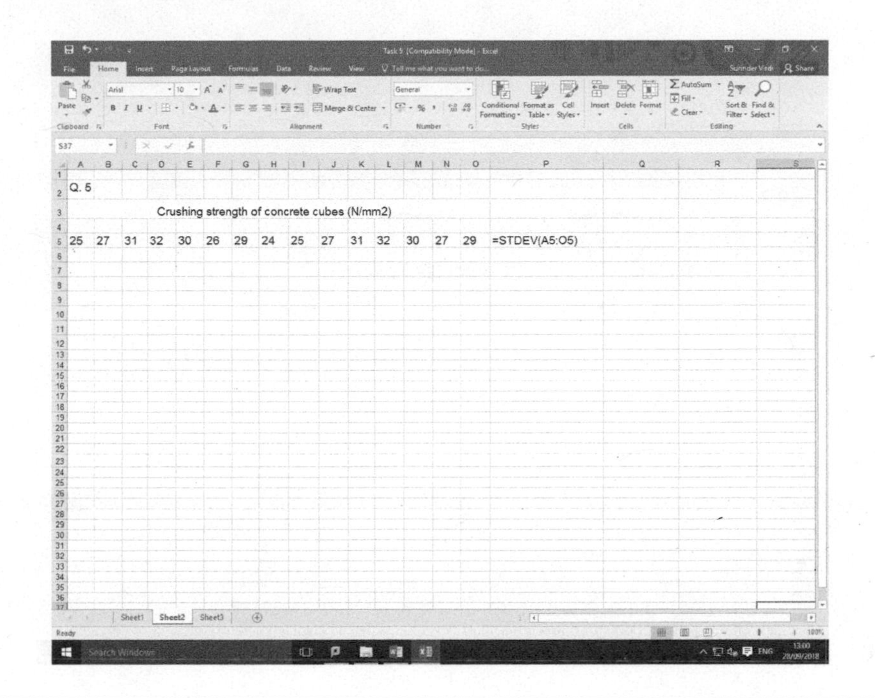

Figure S18.5

6. Figure S18.6 shows the formula to be used.

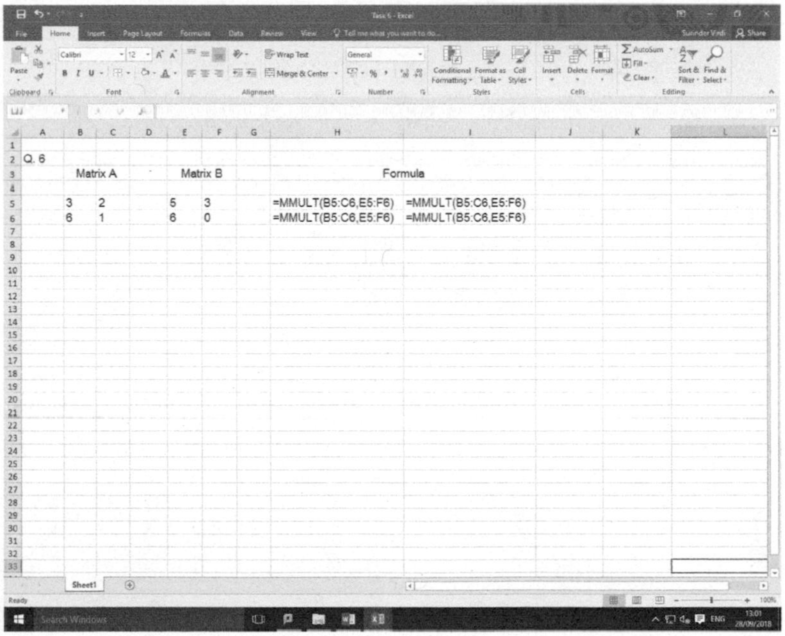

Q. 6
Matrix A
Matrix B
Formula
3 2
6 1
5 3
6 0
=MMULT(B5:C6,E5:F6) =MMULT(B5:C6,E5:F6)
=MMULT(B5:C6,E5:F6) =MMULT(B5:C6,E5:F6)

Figure S18.6

7. Figure S18.7 shows the formula to be used.

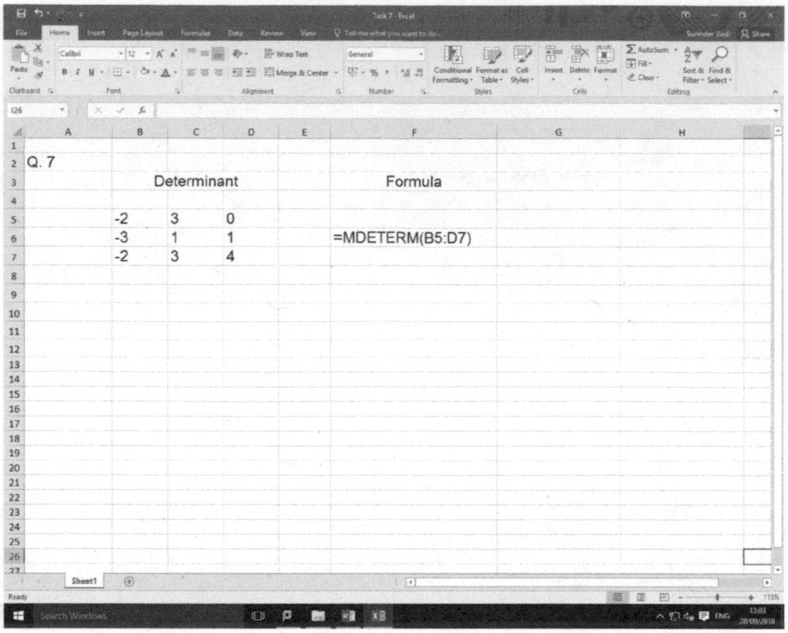

Q. 7

	Determinant			Formula
	-2	3	0	
	-3	1	1	=MDETERM(B5:D7)
	-2	3	4	

Figure S18.7

8. Figure S18.8 shows the formulae to be used.

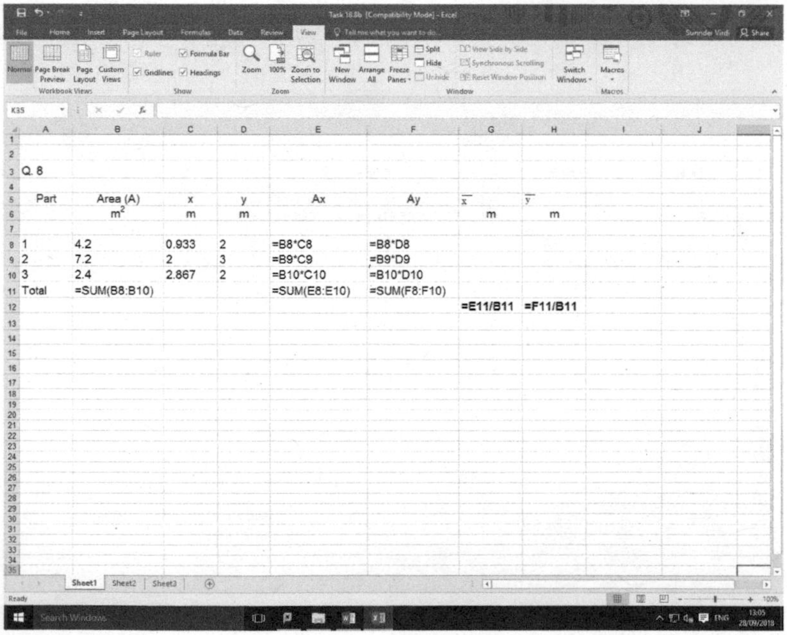

Q. 8

Part	Area (A)	x	y	Ax	Ay	$\bar{x}$	$\bar{y}$
	m^2	m	m			m	m
1	4.2	0.933	2	=B8*C8	=B8*D8		
2	7.2	2	3	=B9*C9	=B9*D9		
3	2.4	2.867	2	=B10*C10	=B10*D10		
Total	=SUM(B8:B10)			=SUM(E8:E10)	=SUM(F8:F10)		
						=E11/B11	**=F11/B11**

Figure S18.8

Index

absolute error 4
addition of sine waves 127
addition of matrices 218
addition of vectors 230
algebra 9
  brackets 11
  division 9
  multiplication 9
angles 67
antilogarithm 134
application of differentiation 167
  maximum and minimum 171
  second derivatives 169
  structural mechanics 167
  velocity and acceleration 170
approximation 34
area between a curve and a straight line 189
area under a curve 187
area 75
  circle 78
  irregular shapes 80
    mid-ordinate rule 80
    Simpson's rule 81
    trapezoidal rule 81
  regular shapes 76
    quadrilaterals 77
    circle 78
    beams 77
arithmetic progression 17
averages 243

bending moments 195, 289
binomial distribution 260
binomial series for negative indices 33
binomial theorem 31; approximation 34; practical problems 35
brackets 11

centroids by integration 202
centroids of simple shapes 199
circle 71
  major sector 72
  minor sector 72
  major segment 72
  minor segment 72
cofactors 220
compound interest 138
computer techniques 271
  area 273
  centroids 278
  determinants 281
  matrices 278
  median 277
  mean 275
  range 275
  standard deviation 277
cone 84
continuous data 242
correlation 62
cosine rule 104
cubic equations 61
cuboid 83
cumulative frequency curves 250
curve fitting 62
cylinder 83

decay of sound energy 141
definite integrals 180
determinants 220
  properties of 222
diagonal matrix 218
differential coefficients 161
differentiation 149
  chain rule 157
  exponential function 157
  first principles 150
  function of a function 155
  logarithms 157
  of a product 159
  of a quotient 160
  trigonometric functions 154
dimensional analysis 14
discrete data 241
dispersion of data 251
distribution curves 256

earth pressure on retaining walls 192
equations 12
estimation 4
evaluation of formulae 30
exponential function 137
  compound interest 138

decay of sound energy 141
growth and decay 140
newton's law of cooling 139

factorisation 12
frames 108
frequency distribution 247

geometric progression 19
sums of series 19
geometry 67
angles 67
triangles 68
circle 71
graph of sin x 124
graph of cos x 124
graphical solutions 53
cubic equations 61
linear equations 55
quadratic equations 60
simultaneous equations 56
straight line 58
grouped data 242

hot water cylinder 84
hyperbolic functions 142

indices 5
division 5
multiplication 5
negative powers 6
power of a power 6
zero index 6
integration 177
applications of 187
area under a curve 187
by parts 183
by substitution 181
change of limits 182
definite integrals 180
indefinite integrals 178
inverse of square matrix 221

law of straight line 58
laws of growth and decay 140
linear equation 55
linear regression 63
logarithmic function 133

mass-haul diagram 87
matrices 217
addition and subtraction 218
application of 223
transpose 220
maximum and minimum 171
mean 243
mean deviation 251
median 243
mid-ordinate rule 80
mode 243

normal distribution 262
normal distribution test 266

order of operation 1

parallel axis theorem 209
Pascal's triangle 31
permeability of soils 258
Poisson distribution 260
probability 258
independent events 259
mutually exclusive events 258
pyramid 84
Pythagoras' theorem 71

quadratic equations 45
application of 49
completing the square 48
factorising 46
quadratic formula 48

radius of gyration 234
range 251
raw data 242
resolution of vectors 234
roof terminology 101
rounding 2
significant figures 2
to nearest whole number 2
truncation 2

second derivatives 169
second moment of area 206
similar triangles 69
simple equations 12
Simpson's rule 81, 86
simultaneous equations 41
application of 44
elimination method 41
substitution method 42
sine rule 103
ambiguous case 107
sine wave 124
amplitude 124
phase angle 124
period 124
sinusoidal waveform 124

spreadsheet 271
square matrix 217
standard deviation 252
standard form 3
statistics 241
statistical diagrams 246
  bar chart 246
  frequency polygon 249
  histogram 249
  pie chart 247

theorem of Pappus 88
thermal movement of building
  components 140
transposition of formulae 25
trapezoidal rule 81, 86
tree diagrams 260
triangles 68
  acute-angled 69
  equilateral 69
  isosceles 69
  obtuse angled 69
  right-angled 69
  scalene 69
trigonometric equations 121
trigonometric graphs 124
trigonometric identities 117
trigonometrical ratios 98
  of compound angles 119
trigonometry 97

unit vectors 233

vectors 229
  addition of 230
  polygon of 231
  resultant of 234
  subtraction of 232
variance 251
volume 83
  irregular objects 86
    embankment 86
    Simpson's rule 86
    trapezoidal rule 86
  regular objects 83
    hot water cylinder 84
    retaining wall 85
  volume of revolution 190
  cone 190
  slump cone 197